AN INTRODUCTION TO SEMIGROUP THEORY

L.M.S. MONOGRAPHS

Editors: P. M. Cohn *and* G. E. H. Reuter

1. Surgery on Compact Manifolds *by* C. T. C. Wall, F.R.S.
2. Free Rings and Their Relations *by* P. M. Cohn.
3. Abelian Categories with Applications to Rings and Modules *by* N. Popescu.
4. Sieve Methods *by* H. Halberstam and H.-E. Richert.
5. Maximal Orders *by* I. Reiner.
6. On Numbers and Games *by* J. H. Conway.
7. An Introduction to Semigroup Theory *by* J. M. Howie.
8. Matroid Theory *by* D. J. A. Welsh.
9. Subharmonic Functions Volume 1 *by* W. K. Hayman and P. B. Kennedy

Published for the London Mathematical Society
by Academic Press Inc. (London) Ltd.

AN INTRODUCTION TO SEMIGROUP THEORY

J. M. HOWIE
University of St. Andrews,
Scotland

1976

ACADEMIC PRESS

LONDON NEW YORK SAN FRANCISCO

A Subsidiary of Harcourt Brace Jovanovich, Publishers

ACADEMIC PRESS INC. (LONDON) LTD.
24/28 Oval Road
London NW1

United States Edition published by
ACADEMIC PRESS INC.
111 Fifth Avenue
New York, New York 10003

Library of Congress Catalog Card Number: 0-12-356950-8
ISBN: 75-46333

PRINTED IN GREAT BRITAIN BY
PAGE BROS (NORWICH) LTD, NORWICH

Preface

Semigroup theory is an elementary subject, in the sense that it could very easily be taught to undergraduate students at a fairly early stage in their course. It would indeed be perfectly possible (as for group theory) to write a book on the subject that would be accessible to a reasonably gifted school pupil. I have, however, judged that at the time of writing there is no place at all for such a book. In practice, semigroup theory is encountered only by fairly advanced students of mathematics, and there seems no reason to suppose that this situation will change substantially in the near future. Accordingly, while very little actual preliminary knowledge of abstract algebra is required of the reader of this book, I have felt able to assume that he has a modest degree of mathematical sophistication, corresponding perhaps to the level of a senior undergraduate or a junior postgraduate student. The book has in fact grown out of various courses of lectures at postgraduate and final year undergraduate level given in the State University of New York at Buffalo and in the University of St. Andrews.

The metamorphosis from a course of lectures into a book has involved me in much more labour that I would have believed possible, and the amount of material covered has grown in the process. In the earlier chapters there are one or two sections that might safely be omitted at a first reading; these are marked with a star.

In some respects the theory of semigroups is similar to group theory and ring theory, and the earliest major contributions to the theory (by Suškevič (1928), Rees (1940), Clifford (1941) and Dubreil (1941)) were strongly motivated by comparisons with groups and rings. In more recent years, however, the subject has developed characteristic aims and methods that are harder to link to these other parts of abstract algebra, largely because of the necessity of studying congruences. In a group a congruence is determined if we know a single congruence class, in particular if we know the normal subgroup which is the class containing the identity. Similarly, in a ring a congruence is determined if we know the ideal which is the congruence class containing the zero. In semigroups there is no such fortunate occurrence, and we are therefore faced with the necessity of studying congruences as such. More than anything else, it is this necessity that gives semigroup theory its characteristic flavour. Semigroups are in fact the first and simplest type of algebra to which the methods of universal algebra must be applied, and any mathematician interested in universal algebra will find semigroup theory a rewarding study.

The publication in 1961 of Volume I of Clifford and Preston's *The Algebraic Theory of Semigroups* stimulated a considerable growth of interest in the subject. When Volume II appeared in 1967 all semigroup theorists had in their possession an extremely valuable reference work, containing a wealth of information, and presented in an admirably readable style. Their book will be for many years to come an indispensable source of material for semigroup theorists, and the present volume in no way sets out to replace it.

Why, then, did I consider that a new book on semigroup theory was necessary? Partly it was because *The Algebraic Theory of Semigroups* has become the victim of its own success. By stimulating so effectively the growth of the subject it has hastened its own obsolescence, and there is now a large corpus of material published since 1965 that every serious student of semigroup theory needs to know. Also, it seemed to me that there was a need for a more compact, less comprehensive introduction to the subject that would perhaps be less forbidding to the beginner and that would by its selection of material present a more closely defined point of view as to which parts of the subject appeared most significant. I hope that the present volume satisfies these criteria.

Since obviously I have had to be highly selective in my choice of material, it may be appropriate to make a few remarks about the principles that guided my selection. To some extent there was the highly subjective principle of including topics that at the time of writing happened to interest me. But I have been guided also by the more objective principle of studying those types of semigroup for which there is a well-developed and coherent theory. Semigroup theory is an enormously diffuse subject and has advanced on a very broad front. Inevitably, progress along this front has not been uniform, and the greatest interest is to be found in those parts of the subject where significant and coherent advances in understanding have been achieved.

There can be little doubt that the most coherent part of semigroup theory at the present time is the part concerned with the structure of regular semigroups of various kinds, and it is to the study of such semigroups that the greater part of this book is devoted. In particular, the longest chapter (Chapter V) is devoted to inverse semigroups, which have been for a number of years (and continue to be) perhaps the most rewarding object of study in semigroup theory.

I have allowed myself a final chapter on semigroup amalgams, which have been a special interest of mine for many years. The recent work of T. E. Hall (1975) on inverse semigroup amalgams does provide a striking link both in content and in method with the rest of the book, but a chapter on amalgams seemed to be justified also on other grounds. It demonstrates the importance for semigroup theory of general ideas, such as coproduct and pushout, belonging to the theory of categories, and also illustrates a combinatorial style of argument that is not found elsewhere in the book, but which is an important tool in uncovering facts about semigroups.

I have adopted the standard mathematical practice of making formal italicized statements of the main results. The word 'lemma' has the usual meaning of a result that contributes towards the proof of some deeper or more striking result. As for the other italicized statements, I have attempted to guide the reader by distinguishing between theorems and propositions, the former term being reserved for results that are either deeper or more important in the development of the theory. In deciding whether to call results theorems or propositions, however, I have not lost any sleep, and the reader should neither deduce too much from the title given to a result nor be unduly disturbed by any inconsistencies he may discover.

The system of reference within the book is, I hope, fairly obvious. A result numbered $x.y$ in Section x of (say) Chapter II is referred to within Chapter II as Theorem (or Proposition or Lemma or formula) $x.y$, but outside Chapter II it is given the number II.$x.y$. The symbol ■ marks the end of a proof or the end of an italicized statement for which no proof (or no further proof) is required.

I have not used logical symbolism ($\exists$, $\forall$, $\Rightarrow$, $\Leftrightarrow$, etc.) in any very systematic way, but have felt free to use it whenever it appears to contribute to clarity. The symbolism is by now so standard in mathematical writing that it scarely seems necessary to explain it in a book of this type.

It is a pleasure to record thanks to Mr J. E. Jones and Dr. Mario Petrich who read the entire manuscript, to a team of secretaries led by Miss K. P. Dunne who typed it, to Dr. T. S. Blyth and Professor G. B. Preston who read the proofs, and (last but not least) to my long-suffering wife and family, who learned most graciously to accept 'THE BOOK' as an excuse for neglect over a period that lengthened into years.

July 1975 J.M.H.
University of St. Andrews,
Scotland.

Contents

Preface v

Chapter I. Introductory Ideas

1. Basic definitions . . . 1
2. Monogenic semigroups . . . 7
3. Ordered sets, semilattices and lattices . . . 11
4. Binary relations; equivalences . . . 14
5. Congruences . . . 21
6. Free semigroups . . . 29
7. Ideals and Rees congruences . . . 30
8*. Lattices of equivalences and congruences . . . 31
Exercises . . . 35

Chapter II. Green's Equivalences

Introduction . . . 38
1. The equivalences $\mathscr{L}$, $\mathscr{R}$, $\mathscr{H}$, $\mathscr{J}$ and $\mathscr{D}$ . . . 38
2. The structure of $\mathscr{D}$-classes . . . 42
3. Regular $\mathscr{D}$-classes . . . 44
4. Regular semigroups . . . 48
Exercises . . . 53

Chapter III. 0-Simple Semigroups

Introduction . . . 57
1. Simple and 0-simple semigroups; principal factors . . . 57
2. Rees's Theorem . . . 60
3. Primitive idempotents . . . 68
4*. Congruences on completely 0-simple semigroups . . . 72
5*. The lattice of congruences on a completely 0-simple semigroup . . . 81
6*. Finite congruence-free semigroups . . . 83
Exercises . . . 85

Chapter IV. Unions of Groups

Introduction . . . 89
1. Unions of groups . . . 90
2. Semilattices of groups . . . 93
3. Bands . . . 95
4*. Free bands . . . 103
5*. Varieties of bands . . . 109
Exercises . . . 125

Chapter V. Inverse Semigroups

Introduction 128
1. Preliminaries 129
2. The natural order relation on an inverse semigroup 137
3. Congruences on inverse semigroups 139
4. Fundamental inverse semigroups 142
5. Anti-uniform semilattices 146
6. Bisimple inverse semigroups 150
7. Simple inverse semigroups 159
8*. Representations of inverse semigroups 168
Exercises 180

Chapter VI. Orthodox Semigroups

Introduction 186
1. Basic properties of orthodox semigroups 186
2. The analogue of the Munn semigroup 193
3. Uniform and anti-uniform bands 203
4. The structure of orthodox semigroups 205
Exercises 209

Chapter VII. Semigroup Amalgams

Introduction 212
1. Free products 212
2. Dominions and zigzags 220
3. The embedding of amalgams 229
4. Inverse semigroup amalgams 243
Exercises 253

References 257

List of Special Symbols 265

Index 268

Chapter I

Introductory Ideas

In this chapter certain basic definitions and results are presented. Reference will be made to these throughout the book, though it should be noted that the starred Section 8 is referred to only in Section III.5.

1. BASIC DEFINITIONS

By a *groupoid* (S, μ) we shall mean a non-empty set S on which a binary operation μ is defined. That is to say, we have a mapping $\mu: S \times S \to S$. We shall say that (S, μ) is a *semigroup* if μ is *associative*, i.e. if

$$(\forall x, y, z \in S) \qquad ((x, y)\mu, z)\mu = (x, (y, z)\mu)\mu. \tag{1.1}$$

(Here and elsewhere we shall write function symbols on the *right*.) We shall not in fact use this rather cumbersome notation. Following the usual practice in algebra we shall write $(x, y)\mu$ simply as xy and usually refer to the semigroup operation as *multiplication*. The formula (1.1) then becomes

$$(xy)z = x(yz),$$

the familiar associative law of elementary algebra, which allows us to write expressions such as $xyz, x_1x_2 \ldots x_n$ with unambiguous meaning. The notation $x^n (n \in \mathbf{N})$ will be used for the product of n elements each equal to x. The cardinal number $|S|$ (see, e.g., Halmos (1960) for this and other items of basic set theory) will be called the *order* of the semigroup S.

We shall write a multiplicative semigroup as $(S, .)$ or often simply as S. If $(S, .)$ has the additional property that

$$(\forall x, y \in S) \qquad xy = yx,$$

we shall say that it is a *commutative* semigroup.

If there exists an element 1 of S such that

$$(\forall x \in S) \qquad x1 = 1x = x,$$

we say that 1 is an *identity* (*element*) of S and that S is a *semigroup with identity*, or *monoid.* A semigroup S has at most one such element, since if $1' \in S$ also has the property

$$(\forall x \in S) \qquad x1' = 1'x = x,$$

then

$$\begin{aligned} 1' &= 11' \quad \text{(since 1 is an identity)} \\ &= 1 \quad\ \ \text{(since } 1' \text{ is an identity).} \end{aligned}$$

If S has no identity element it is very easy indeed to adjoin an extra element 1 to the set S. Then if we define

$$(\forall s \in S) \qquad 1s = s1 = s,$$

and

$$11 = 1,$$

$S \cup \{1\}$ becomes a semigroup with identity element 1. We shall consistently use the notation S^1 with the following meaning:

$$S^1 = \begin{cases} S \text{ if } S \text{ has an identity element} \\ S \cup \{1\} \text{ otherwise.} \end{cases}$$

S^1 is called *the semigroup obtained from S by adjoining an identity if necessary.*

If a semigroup S with at least two elements contains an element 0 such that

$$(\forall x \in S) \qquad x0 = 0x = 0,$$

we say that 0 is a *zero* (*element*) of S and that S is a *semigroup with zero.* Once again there can be at most one such element, since if

$$(\forall x \in S) \qquad x0' = 0'x = 0',$$

then

$$\begin{aligned} 0' &= 00' \quad \text{(since } 0' \text{ is a zero)} \\ &= 0 \quad\ \ \text{(since 0 is a zero).} \end{aligned}$$

The proviso that S should have at least two elements means merely that we shall not want to refer to the single element of the *trivial* semigroup $\{e\}$ (in which $e^2 = e$) as a zero.

If S has no zero element, then again it is easy to adjoin an extra element 0 to the set S. Then we define

$$(\forall s \in S) \qquad 0s = s0 = 0,$$

and

$$00 = 0,$$

making $S \cup \{0\}$ into a semigroup with zero element 0. Continuing the analogy with the case of the identity element, we write

$$S^0 = \begin{cases} S \text{ if } S \text{ has a zero element} \\ S \cup \{0\} \text{ otherwise} \end{cases}$$

and refer to S^0 as *the semigroup obtained from S by adjoining a zero if necessary*.

Despite the great ease with which we may adjoin an identity and a zero to a semigroup, we cannot altogether reduce the theory of semigroups to the theory of semigroups with identity and zero, for in adjoining the extra elements we may sacrifice some crucial property of the semigroup. To take a very trivial example, if we adjoin a zero to a semigroup which is a *group*, then the resulting semigroup is *not* a group.

Among semigroups with zero we find the very trivial *null* semigroups, in which the product of any two elements is zero. Only slightly less trivially, if S is any non-empty set whatever, the multiplication defined on S by the rule that

$$xy = x \qquad (x, y \in S)$$

is easily seen to be associative. The semigroup $(S, .)$ formed in this way is called a *left zero semigroup. Right zero semigroups* are defined analogously.

Both left and right zero semigroups are special cases of a type of semigroup which we now describe. If I and Λ are arbitrary non-empty sets, an associative multiplication can be defined on the Cartesian product $I \times \Lambda$ by the rule that

$$(i, \lambda)(j, \mu) = (i, \mu) \qquad (i, j \in I, \lambda, \mu \in \Lambda).$$

We call $(I \times \Lambda, .)$ a *rectangular band.* If $|\Lambda| = 1$ it is a left zero semigroup, while if $|I| = 1$ it is a right zero semigroup.

If A and B are subsets of a semigroup S, we write AB for $\{ab : a \in A, b \in B\}$. It is easy to see that

$$(\forall A, B, C \subseteq S) \qquad (AB)C = A(BC);$$

hence once again notations such as ABC, $A_1 A_2 \ldots A_n$ are meaningful. We write A^2, A^3, etc., respectively for AA, AAA, etc. When dealing with singleton sets we shall use the notational simplifications that are customary in algebra, thus writing (for example) Ab rather than $A\{b\}$.

If a is an element of a semigroup S without identity then Sa will not in general contain a. We write

$$\left.\begin{array}{lll} S^1a & \text{for} & Sa \cup \{a\} \\ aS^1 & \text{for} & aS \cup \{a\} \\ \text{and} & & \\ S^1aS^1 & \text{for} & SaS \cup Sa \cup aS \cup \{a\}. \end{array}\right\} \qquad (1.2)$$

This short-hand will be extremely useful at times. Notice that S^1a, aS^1 and S^1aS^1 are all subsets of S, i.e. they do not contain 1.

If a semigroup S has the property

$$(\forall a \in S) \qquad aS = S \quad \text{and} \quad Sa = S \tag{1.3}$$

we call it a *group*. This is not the commonest definition of a group, but it is an easy exercise to show that it is equivalent to the more usual definition of a group as a semigroup S for which

$$\left.\begin{array}{ll} (\exists e \in S)(\forall a \in S) & ea = a \\ \text{and} & \\ (\forall a \in S)(\exists a^{-1} \in S) & a^{-1}a = e. \end{array}\right\} \tag{1.4}$$

The definition (1.3) is given first because it is the one that seems to occur most often in semigroup theory. It is easily seen to be equivalent to

$$(\forall a, b \in S)(\exists x, y \in S) \qquad ax = b \quad \text{and} \quad ya = b, \tag{1.5}$$

and we shall sometimes want to refer to it in that form.

If G is a group then $G^0 = G \cup \{0\}$ is a semigroup. It is not of course a group: we call it a 0-*group*. Closely related to (1.3) we have

Proposition 1.6. *A semigroup S with zero is a* 0*-group if and only if*

$$(\forall a \in S\backslash\{0\}) \qquad aS = Sa = S.$$

Proof. If $S = G^0$, a 0-group, then $aG = Ga = G$ for all a in $G = S\backslash\{0\}$. Since $aS = aG \cup \{0\}$ and $Sa = Ga \cup \{0\}$, it follows that $aS = Sa = S$ for all a in $G = S\backslash\{0\}$.

Conversely, if S has the given property, let us consider the subset $G = S\backslash\{0\}$ of S. Since S has by implication at least two elements, we have that $G \neq \varnothing$. To show that G is a group we must first show that it is closed with respect to multiplication, i.e. that $(\forall a, b \in G)\ ab \in G$. To see that this is so, suppose by way of contradiction that there exist a, b in G such that $ab = 0$. Then

$$S^2 = S\,.\,S = (Sa)(bS) = S(ab)S = S0S = \{0\},$$

and so

$$S = aS \subseteq S^2 = \{0\},$$

a contradiction. Hence G is closed with respect to multiplication. Also, for all a, b in G there exist, by the given property, x, y in S such that $ax = b$, $ya = b$. The elements x and y cannot be zero and so are in G. Thus the semigroup G satisfies (1.5) and so is a group. ■

If $(S, .)$ is a semigroup, then a non-empty subset T of S is called a *subsemigroup* of S if it is closed with respect to multiplication, i.e. if

$$(\forall x, y \in T) \qquad xy \in T. \tag{1.7}$$

Alternatively, writing the condition in terms of subset multiplication, we have that T is a subsemigroup if $T^2 \subseteq T$. The associative property that holds throughout S certainly holds throughout T; hence $(T, .)$ is a semigroup. Among special subsemigroups of S worthy of mention are S itself, $\{0\}$ and $\{1\}$ where appropriate, and also more generally $\{e\}$, where e is any element of S that is *idempotent*, i.e. for which $e^2 = e$.

There will often be special interest in considering those subsemigroups of S that are *groups*. A subsemigroup of this type will be called a *subgroup*. It is easy to show that a non-empty subset T of a semigroup S is a subgroup of S if and only if

$$(\forall a \in T) \qquad aT = Ta = T. \tag{1.8}$$

A non-empty subset A of a semigroup S is called a *left ideal* if $SA \subseteq A$, a *right ideal* if $AS \subseteq A$, and a (two-sided) *ideal* if it is both a left and a right ideal. Evidently every ideal (whether one- or two-sided) is a subsemigroup, but not every subsemigroup is an ideal. Among the ideals of S are S itself and (if S has a zero element) $\{0\}$. An ideal I of S such that $\{0\} \subset I \subset S$ (strictly) is called *proper*.

If ϕ is a mapping from a semigroup $(S, .)$ into a semigroup $(T, .)$ we say that ϕ is a *homomorphism* if

$$(\forall x, y \in S) \qquad (xy)\phi = (x\phi)(y\phi).$$

We refer to S as the *domain* of ϕ, to T as the *codomain* of ϕ, and to the subset

$$S\phi = \{s\phi : s \in S\}$$

of T as the *range* of ϕ. If ϕ is one–one we shall call it a *monomorphism*, and if it is both one–one and onto we shall call it an *isomorphism*. If ϕ is a homomorphism from S into S we call it an *endomorphism* of S. An isomorphism from S onto S will be called an *automorphism* of S. If there exists an isomorphism ϕ from S onto T we say that S and T are *isomorphic* and write $S \simeq T$.

If S and T are semigroups, the Cartesian product $S \times T$ becomes a semigroup if we define

$$(s, t)(s', t') = (ss', tt').$$

We shall refer to this semigroup as the *direct product* of S and T.

The proof of the following very elementary result is left as an exercise:

Proposition 1.9. *A semigroup is isomorphic to a rectangular band if and only if it is isomorphic to the direct product of a left zero semigroup and a right zero semigroup.* ■

We shall have occasion in Chapter IV to make use of a more general notion of direct product. If $\{S_i : i \in I\}$ is a family of semigroups, then P, the *direct product* of the family, is defined as the set of all mappings $p: I \to \bigcup\{S_i : i \in I\}$ such that $ip \in S_i$ for each i in I. If multiplication in P is defined by the "component-wise" rule

$$i(pq) = (ip)(iq) \qquad (i \in I),$$

then P becomes a semigroup.

If $I = \{1, 2\}$, then P essentially coincides with the direct product $S_1 \times S_2$ as previously defined, since the mapping $p \mapsto (1p, 2p)$ from P onto $S_1 \times S_2$ is an isomorphism. More generally, if $I = \{1, 2, \ldots, n\}$, a finite set, then in the same way we may think of P as consisting of all n-tuples $(x_1, x_2, \ldots, x_n)$ for which $x_i \in S_i$ $(i = 1, 2, \ldots, n)$, with multiplication given by

$$(x_1, x_2, \ldots, x_n)(y_1, y_2, \ldots, y_n) = (x_1y_1, x_2y_2, \ldots, x_ny_n).$$

Just as groups arise most naturally as groups of permutations of a set, so semigroups arise from more general mappings from a set into itself. The analogue of the *symmetric group* $(\mathcal{G}(X), \circ)$ of all permutations of a set X is the *full transformation semigroup* $(\mathcal{T}(X), \circ)$ consisting of all mappings from X into X. The semigroup operation is *composition of mappings*: if $\alpha, \beta \in \mathcal{T}(X)$ then $\alpha \circ \beta \in \mathcal{T}(X)$ is defined by

$$x(\alpha \circ \beta) = (x\alpha)\beta \qquad (x \in X).$$

It is an elementary and well-known fact that the operation of composition of mappings is associative. Thus $(\mathcal{T}(X), \circ)$ is a semigroup. Among the elements of $\mathcal{T}(X)$ are all the permutations of X, i.e. all the mappings $\gamma: X \to X$ that are *bijections* (i.e. both one–one and onto). Thus $\mathcal{G}(X)$ is a subset of $\mathcal{T}(X)$—and indeed it is easy to show that it is a *subgroup*. Except in the trivial case where $|X| = 1$, the semigroup $\mathcal{T}(X)$ contains elements that are not in $\mathcal{G}(X)$; indeed in the finite case where $|X| = n$ it is easy to see that $|\mathcal{G}(X)| = n!$ while $|\mathcal{T}(X)| = n^n$.

We shall usually find it convenient in this context to drop the notation $\circ$ for composition of mappings and to write about $\mathcal{T}(X)$ in the same multiplicative notation as for other semigroups.

If S is, for some X, a subsemigroup of $\mathcal{T}(X)$, we call S a *semigroup of mappings.* A homomorphism ϕ from a semigroup S into some $\mathcal{T}(X)$ will be called a *representation of S (by mappings).* If ϕ is a monomorphism we call the representation *faithful*; in such a case $S\phi$ is a semigroup of mappings isomorphic to S.

The following theorem, closely analogous to Cayley's Theorem for groups (see M. Hall (1959) for this and other items of elementary group theory), shows that every semigroup is isomorphic to a semigroup of mappings:

Theorem 1.10. *If S is a semigroup and $X = S^1$ then there is a monomorphism $\rho: S \to \mathcal{T}(X)$.*

Proof. For each a in S we define a mapping $\rho_a\colon S^1 \to S^1$ by

$$x\rho_a = xa \qquad (x \in S^1).$$

Thus $\rho_a \in \mathscr{T}(X)$ and so there is a mapping $\rho\colon S \to \mathscr{T}(X)$ given by

$$a\rho = \rho_a \qquad (a \in S).$$

The mapping ρ is one–one, since, for all a, b in S,

$$a\rho = b\rho \Rightarrow \rho_a = \rho_b \Rightarrow xa = xb \text{ for all } x \text{ in } S^1$$
$$\Rightarrow 1a = 1b \Rightarrow a = b.$$

(Notice that the argument might break down at this stage if S rather than S^1 were used.) Also, ρ is a homomorphism, since

$$(\forall x \in S^1)\ x(\rho_a\rho_b) = (x\rho_a)\rho_b = (xa)b = x(ab) = x\rho_{ab},$$

and hence $(a\rho)(b\rho) = (ab)\rho$. ∎

The representation ρ introduced here will be called the *extended right regular representation* of S. The word "extended" betokens that S^1 is used for the set X rather than (as in group theory) the more obvious S. This is necessary to ensure the faithfulness of the representation. (See Exercise 5.)

2. MONOGENIC SEMIGROUPS

If $\{U_i\colon i \in I\}$ is a non-empty family of subsemigroups of a semigroup S, then it is easy to see that $\bigcap\{U_i\colon i \in I\}$ is either empty or is itself a subsemigroup of S. If A is an arbitrary non-empty subset of S, then the family of subsemigroups of S containing A is non-empty, since S itself is one such semigroup; hence the intersection of the family is a subsemigroup of S containing A. We denote it by $\langle A\rangle$. It is characterized within the set of subsemigroups of S by the properties:

(2.1) $A \subseteq \langle A\rangle$;

(2.2) if U is a subsemigroup of S containing A, then $\langle A\rangle \subseteq U$.

The subsemigroup $\langle A\rangle$ consists of all elements of S that can be expressed as finite products of elements in A. If $\langle A\rangle = S$ we shall say that A is a *set of generators* for S or a *generating set* of S.

Particular interest attaches to the case where A is finite. If $A = \{a_1, a_2, \ldots, a_n\}$ we shall write $\langle A\rangle$ as $\langle a_1, a_2, \ldots, a_n\rangle$. Especially interesting is the case where $A = \{a\}$, when

$$\langle a\rangle = \{a, a^2, a^3, \ldots\}.$$

We refer to $\langle a \rangle$ as the *monogenic* subsemigroup of S generated by the element a. The *order* of a is defined as the order of the subsemigroup $\langle a \rangle$. If a semigroup S has the property that $S = \langle a \rangle$ for some a in S, we say that S is *monogenic*.

Clifford and Preston (1961) follow the group-theoretic terminology and use the term "cyclic" where I have used "monogenic". Certainly "cyclic" is a less fussy word, but the reader may judge from what now follows whether monogenic semigroups are "round" enough to merit the description "cyclic".

Let a be an element of a semigroup S and consider the monogenic subsemigroup

$$\langle a \rangle = \{a, a^2, a^3, \ldots\}$$

of S generated by a. If there are no repetitions in the list $a, a^2, a^3, \ldots$, i.e. if

$$a^m = a^n \Rightarrow m = n,$$

then evidently $(\langle a \rangle, .)$ is isomorphic to the semigroup $(\mathbf{N}, +)$ of natural numbers with respect to addition. In such a case we say that $\langle a \rangle$ is an *infinite* monogenic semigroup and that the element a has *infinite order*.

If repetitions do occur among the powers of a, then the set

$$\{x \in \mathbf{N} : (\exists y \in \mathbf{N})\, a^x = a^y, x \neq y\}$$

is non-empty and so has a least element. Let us denote this least element by m and call it the *index* of a. Then the set

$$\{x \in \mathbf{N} : a^{m+x} = a^m\}$$

is non-empty and so it too has a least element r which we call the *period* of a. We shall also refer to m and r as the index and period respectively of the monogenic semigroup $\langle a \rangle$. We thus have

$$a^m = a^{m+r}. \tag{2.3}$$

We now investigate the structure of $\langle a \rangle$, given that a has index m and period r. First notice that (2.3) implies that

$$a^m = a^{m+r} = a^m a^r = a^{m+r} a^r = a^{m+2r},$$

and more generally that

$$(\forall q \in \mathbf{N})\, a^m = a^{m+qr}.$$

The powers $a, a^2, \ldots, a^m, a^{m+1}, \ldots, a^{m+r-1}$ are all distinct. For any $s \geqslant m$ we can by the division algorithm write $s = m + qr + u$, where $q \geqslant 0$ and $0 \leqslant u \leqslant r - 1$. It then follows that

$$a^s = a^{m+qr} a^u = a^m a^u = a^{m+u};$$

thus $\langle a \rangle = \{a, a^2, \ldots, a^{m+r-1}\}$ and $|\langle a \rangle| = m + r - 1$. We say that a has *finite order* in this case; the order is given by the rule

$$\text{order of } a = (\text{index of } a) + (\text{period of } a) - 1.$$

The subset $K_a = \{a^m, a^{m+1}, \ldots, a^{m+r-1}\}$ of $\langle a \rangle$ is a subsemigroup of $\langle a \rangle$. We call it the *kernel* of $\langle a \rangle$, and we shall see (Exercise III.2) that this use of "kernel" does not conflict with the more general use of "kernel" in Chapter III. In fact K_a is a *subgroup* of $\langle a \rangle$; for if a^{m+u}, a^{m+v} are any two elements of K_a, then we can find an element a^{m+x} in K_a for which

$$a^{m+u}\, a^{m+x} = a^{m+v}$$

simply by choosing x so that

$$x \equiv v - u - m \pmod{r} \quad \text{and} \quad 0 \leqslant x \leqslant r - 1.$$

Since K_a is commutative, it follows by (1.5) that it is a group.

Indeed K_a is a cyclic group. To see this, notice that the integers

$$m, m+1, \ldots, m+r-1$$

form a complete set of incongruent residues modulo r. (For this and other elementary number-theoretic ideas see, e.g., Hardy and Wright (1938).) Hence there exists g such that

$$0 \leqslant g \leqslant r - 1 \quad \text{and} \quad m + g \equiv 1 \pmod{r}. \tag{2.4}$$

Thus $k(m + g) \equiv k \pmod{r}$ for any k in $\mathbf{N}$ and so the powers $(a^{m+g})^k$ of a^{m+g}, for $k = 1, \ldots, r$, exhaust K_a. That is, K_a is a cyclic group of order r generated by the element a^{m+g}.

If we choose z so that

$$0 \leqslant z \leqslant r - 1 \quad \text{and} \quad m + z \equiv 0 \pmod{r}, \tag{2.5}$$

then a^{m+z} is idempotent and so is the identity of K_a.

We may illustrate these ideas by considering the element

$$\alpha = \begin{pmatrix} 1 & 2 & 3 & 4 & 5 & 6 & 7 \\ 2 & 3 & 4 & 5 & 6 & 7 & 5 \end{pmatrix}$$

of $\mathscr{T}(\{1, 2, \ldots, 7\})$. (The notation for α is an obvious generalization of the standard notation for permutations; the import is that $1\alpha = 2$, $2\alpha = 3, \ldots$, $6\alpha = 7$ and $7\alpha = 5$.) We have

$$\alpha^2 = \begin{pmatrix} 1 & 2 & 3 & 4 & 5 & 6 & 7 \\ 3 & 4 & 5 & 6 & 7 & 5 & 6 \end{pmatrix}, \qquad \alpha^3 = \begin{pmatrix} 1 & 2 & 3 & 4 & 5 & 6 & 7 \\ 4 & 5 & 6 & 7 & 5 & 6 & 7 \end{pmatrix},$$

$$\alpha^4 = \begin{pmatrix} 1 & 2 & 3 & 4 & 5 & 6 & 7 \\ 5 & 6 & 7 & 5 & 6 & 7 & 5 \end{pmatrix}, \qquad \alpha^5 = \begin{pmatrix} 1 & 2 & 3 & 4 & 5 & 6 & 7 \\ 6 & 7 & 5 & 6 & 7 & 5 & 6 \end{pmatrix},$$

$$\alpha^6 = \begin{pmatrix} 1 & 2 & 3 & 4 & 5 & 6 & 7 \\ 7 & 5 & 6 & 7 & 5 & 6 & 7 \end{pmatrix}, \qquad \alpha^7 = \begin{pmatrix} 1 & 2 & 3 & 4 & 5 & 6 & 7 \\ 5 & 6 & 7 & 5 & 6 & 7 & 5 \end{pmatrix},$$

and so α has index 4 and period 3. The kernel K_α is equal to $\{\alpha^4, \alpha^5, \alpha^6\}$ and has Cayley table

	α^4	α^5	α^6
α^4	α^5	α^6	α^4
α^5	α^6	α^4	α^5
α^6	α^4	α^5	α^6.

Thus α^6 is the identity element of K_α (in accord with formula (2.5), since $6 \equiv 0 \pmod 3$). Also, in accord with formula (2.4), since $4 \equiv 1 \pmod 3$, the element α^4 is a generator of the cyclic group K_α:

$$(\alpha^4)^2 = \alpha^5, \qquad (\alpha^4)^3 = \alpha^6.$$

We can visualize $\langle\alpha\rangle$ as

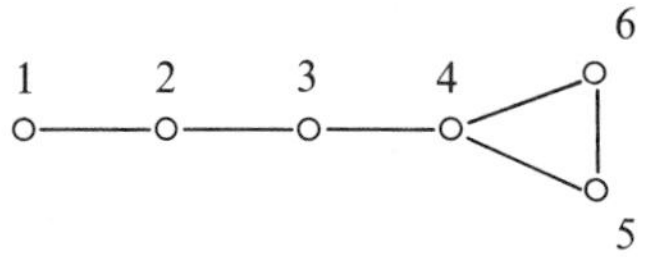

We summarize our results in a theorem:

Theorem 2.6. *Let a be an element of a semigroup S. Then either: (i) all powers of a are distinct and the monogenic subsemigroup $\langle a\rangle$ of S is isomorphic to the semigroup* $(\mathbf{N}, +)$ *of natural numbers under addition; or (ii) there exist positive integers m (the index of a) and r (the period of a) with the following properties:*

(A) $a^m = a^{m+r}$;

(B) $(\forall u, v \in \mathbf{N})\ a^{m+u} = a^{m+v}$ *if and only if* $m + u \equiv m + v \pmod r$;

(C) $\langle a\rangle = \{a, a^2, \ldots, a^{m+r-1}\}$;

(D) $K_a = \{a^m, a^{m+1}, \ldots, a^{m+r-1}\}$ *is a cyclic subgroup of* $\langle a\rangle$. ■

Nothing that we have said so far makes it clear that for every pair (m, r) of positive integers there does in fact exist a semigroup S containing an element a of index m and period r. This, however, is the case: it is a routine matter to verify that the element

$$a = \begin{pmatrix} 1 & 2 & 3\ldots & m & m+1\ldots m+r-1 & m+r \\ 2 & 3 & 4\ldots m+1 & m+2\ldots & m+r & m+1 \end{pmatrix}$$

of the semigroup $\mathscr{T}(\{1, 2, \ldots, m + r\})$ has index m and period r. It is easy too to show that if a and b are elements of finite order in the same or in different semigroups, then $\langle a \rangle \simeq \langle b \rangle$ if and only if a and b have the same index and period. The conclusion is that for any (m, r) in $\mathbf{N} \times \mathbf{N}$ there is one and (up to isomorphism) only one monogenic semigroup with index m and period r. We shall sometimes therefore talk of *the monogenic semigroup* $M(m, r)$ with index m and period r. Notice that $M(1, r)$ is the cyclic group of order r.

A semigroup is called *periodic* if all its elements are of finite order. A finite semigroup is necessarily periodic.

Proposition 2.7. *In a periodic semigroup every element has a power which is idempotent. Hence in a periodic semigroup (in particular, in a finite semigroup) there is at least one idempotent.*

Proof. If $a \in S$, a periodic semigroup, then $\langle a \rangle$ is finite and so some power a^n of $\langle a \rangle$ serves as the identity of the group K_a. The result follows. ■

If the hypothesis of periodicity is dropped then we can no longer guarantee the existence of an idempotent. (See Exercise 3.)

3. ORDERED SETS, SEMILATTICES AND LATTICES

A binary relation ω on a set X (i.e. a subset of $X \times X$) will be called a *(partial) order* if

(O1) $(x, x) \in \omega$ for all x in X—i.e. ω is *reflexive*;
(O2) $(\forall x, y \in X)\,(x, y) \in \omega$ and $(y, x) \in \omega \Rightarrow x = y$—i.e. ω is *antisymmetric*;
(O3) $(\forall x, y, z \in X)\,(x, y) \in \omega$ and $(y, z) \in \omega \Rightarrow (x, z) \in \omega$—i.e. ω is *transitive*.

Traditionally one writes $x \leqslant y$ rather than $(x, y) \in \omega$. We shall follow this convention, and also write $x \geqslant y$, $x < y$, $x > y$ to mean (respectively) $(y, x) \in \omega$, $(x, y) \in \omega$ and $x \neq y$, $(y, x) \in \omega$ and $x \neq y$.

A partial order having the extra property

(O4) $(\forall x, y \in X)\, x \leqslant y \quad \text{or} \quad y \leqslant x$

will be called a *total order*. We shall refer to $(X, \leqslant)$, or just to X, as a *(partially) ordered set* or (where appropriate) as a *totally ordered set*.

If Y is a non-empty subset of a partially ordered set $(X, \leqslant)$, then an element a of Y is called *minimal* if there is no element of Y strictly less than a, i.e. if

$$(\forall y \in Y) \qquad y \leqslant a \Rightarrow y = a.$$

An element b of Y is called *minimum* if

$$(\forall y \in Y) \qquad b \leqslant y.$$

Evidently a minimum element is minimal, but in a partially ordered set it is perfectly possible to have minimal elements that are not minimum. The following elementary facts are easily verified:

Proposition 3.1. *Let Y be a non-empty subset of a partially ordered set X. Then*

(i) *Y has at most one minimum element;*
(ii) *if Y is totally ordered, then the terms "minimal" and "minimum" are equivalent.* ■

We shall say that $(X, \leqslant)$ *satisfies the minimal condition* if every non-empty subset of X has a minimal element. A totally ordered set X satisfying the minimal condition is said to be *well-ordered.*

We leave it to the reader to provide the analogous definitions of *maximal, maximum* and the *maximal condition.*

If Y is a non-empty subset of a partially ordered set X, we say that an element c in X is a *lower bound* for Y if $c \leqslant y$ for every y in Y. If the set of lower bounds of Y is non-empty and has a maximum element d, we refer to d as the *greatest lower bound*, or *meet*, of Y. The element d is unique if it exists; we write

$$d = \bigwedge \{y : y \in Y\}.$$

If $Y = \{a, b\}$, we write $d = a \wedge b$. If $(X, \leqslant)$ is such that $a \wedge b$ exists for every a, b in X we say that $(X, \leqslant)$ is a *lower semilattice.* If we have the stronger property that $\bigwedge\{y\colon y \in Y\}$ exists for every non-empty subset Y of X, then we say that $(X, \leqslant)$ is a *complete lower semilattice.* In a lower semilattice $(X, \leqslant)$ we have that, for all a, b in X,

$$a \leqslant b \quad \text{if and only if} \quad a \wedge b = a. \tag{3.2}$$

Analogous definitions exist for the *least upper bound* or *join*

$$\bigvee \{y : y \in Y\}$$

of a non-empty subset Y of X, for the *least upper bound* or *join* $a \vee b$ of $\{a, b\}$, for an *upper semilattice* and for a *complete upper semilattice.* If $(X, \leqslant)$ is both a (complete) lower semilattice and a (complete) upper semilattice we call it a (*complete*) *lattice.* A lattice L with partial order $\leqslant$ in which the greatest lower bound of a and b is $a \wedge b$ and the least upper bound of a and b is $a \vee b$ is written $L = (L, \leqslant, \wedge, \vee)$. By a *sublattice* of L we shall mean a non-empty subset M of L with the property that

$$a, b \in M \;\Rightarrow\; a \wedge b, \quad a \vee b \in M.$$

If $(E, \leqslant)$ is a lower semilattice, we have a binary operation $\wedge$ defined on E. If $a, b, c \in E$ then

$$(a \wedge b) \wedge c \leqslant a \wedge b \leqslant a, \qquad (a \wedge b) \wedge c \leqslant a \wedge b \leqslant b,$$

and

$$(a \wedge b) \wedge c \leqslant c.$$

Thus $(a \wedge b) \wedge c$ is a lower bound of $\{a, b, c\}$. If d is a lower bound of $\{a, b, c\}$ then

$$d \leqslant a, \qquad d \leqslant b \quad \text{and} \quad d \leqslant c.$$

Hence $d \leqslant a \wedge b$, $d \leqslant c$, and so $d \leqslant (a \wedge b) \wedge c$. Thus $(a \wedge b) \wedge c$ is the unique greatest lower bound of $\{a, b, c\}$. Equally, we may show that $a \wedge (b \wedge c)$ is the unique greatest lower bound of $\{a, b, c\}$. Hence

$$(a \wedge b) \wedge c = a \wedge (b \wedge c)$$

and so $(E, \wedge)$ is a semigroup. Since it is obvious that $a \wedge a = a$ for every a in E and that $a \wedge b = b \wedge a$ for every a, b in E, we now invoke formula (3.2) and observe that we have proved half of the following proposition:

Proposition 3.3. *Let $(E, \leqslant)$ be a lower semilattice. Then $(E, \wedge)$ is a commutative semigroup of idempotents and*

$$(\forall a, b \in E) \quad a \leqslant b \quad \textit{if and only if} \quad a \wedge b = a.$$

Let $(E, .)$ be a commutative semigroup of idempotents. Then the relation $\leqslant$ on E defined by

$$a \leqslant b \quad \textit{if and only if} \quad ab = a$$

is a partial order on E with respect to which E is a lower semilattice. In $(E, \leqslant)$, the meet of a and b is their product ab.

Proof. Since $a^2 = a$ we have $a \leqslant a$. If $a \leqslant b$ and $b \leqslant a$ then $ab = a$ and $ba = b$. Hence

$$a = ab = ba = b.$$

If $a \leqslant b$ and $b \leqslant c$ then $ab = a$, $bc = b$. Hence

$$ac = abc = ab = a$$

and so $a \leqslant c$. Thus $\leqslant$ is a partial order. Since

$$a(ab) = a^2b = ab \quad \text{and} \quad b(ab) = ab^2 = ab,$$

we have $ab \leqslant a$, $ab \leqslant b$. If $c \leqslant a$ and $c \leqslant b$ then $c(ab) = cb = c$, giving $c \leqslant ab$. Thus $(E, \leqslant)$ is a lower semilattice, and the meet of a and b is ab. ■

The effect of this proposition is that the notions of "lower semilattice" and "commutative semigroup of idempotents" are equivalent and interchangeable. We shall use the term *semilattice* with either meaning, making free and frequent transfers between the semigroup and the ordered set points of view.

When describing an ordered set $(S, \leqslant)$, particularly if S is finite, we shall sometimes use so-called *Hasse diagrams.* In such a diagram, elements of the set are represented by small black circles, and two elements a, b in S for which $a < b$ and for which there is no x in S such that $a < x < b$ are depicted thus:

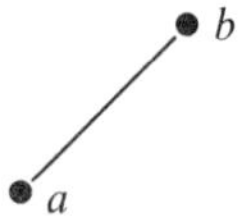

That is, b appears above a and a line connects the two. We thus build up diagrams such as

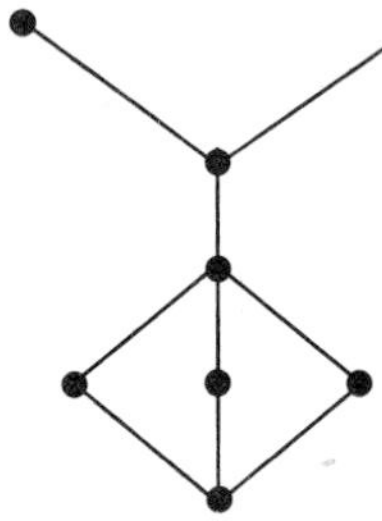

which we can label if necessary.

4. BINARY RELATIONS; EQUIVALENCES

If X is a non-empty set, then a subset ρ of $X \times X$ is called a (*binary*) *relation* on X. The empty subset $\varnothing$ of $X \times X$ will be included among the binary relations. Other special binary relations worthy of mention are the *universal* relation $X \times X$ and the *equality* relation

$$1_X = \{(x, x) : x \in X\}. \tag{4.1}$$

A binary operation on the set $\mathscr{B}(X)$ of all binary relations on X is defined as follows: if $\rho, \sigma \in \mathscr{B}(X)$, then

$$\rho \circ \sigma = \{(x, y) \in X \times X : (\exists z \in X)\,(x, z) \in \rho \quad \text{and} \quad (z, y) \in \sigma\}. \tag{4.2}$$

It is easy to see that for all ρ, σ, τ in $\mathscr{B}(X)$,

$$\rho \subseteq \sigma \Rightarrow \rho \circ \tau \subseteq \sigma \circ \tau \text{ and } \tau \circ \rho \subseteq \tau \circ \sigma. \tag{4.3}$$

Also,

Proposition 4.4. $(\mathscr{B}(X), \circ)$ *is a semigroup.*

Proof. We show that the operation $\circ$ is associative.

If $\rho, \sigma, \tau \in \mathscr{B}(X)$, then, for all x, y in X,

$$(x, y) \in (\rho \circ \sigma) \circ \tau \Leftrightarrow (\exists z \in X)\,(x, z) \in \rho \circ \sigma, (z, y) \in \tau,$$

i.e. $\Leftrightarrow (\exists z \in X)\,(\exists u \in X)\,(x, u) \in \rho, (u, z) \in \sigma, (z, y) \in \tau,$
i.e. $\Leftrightarrow (\exists u \in X)\,(x, u) \in \rho, (u, y) \in \sigma \circ \tau,$
i.e. $\Leftrightarrow (x, y) \in \rho \circ (\sigma \circ \tau).$

Thus $(\rho \circ \sigma) \circ \tau = \rho \circ (\sigma \circ \tau)$ as required. ■

While we shall not normally revert to simple multiplicative notation when discussing the semigroup $(\mathscr{B}(X), \circ)$, we shall allow ourselves to write ρ^2, ρ^3, etc., instead of $\rho \circ \rho$, $\rho \circ \rho \circ \rho$, etc.

If $\rho \in \mathscr{B}(X)$ we define the *domain* $\text{dom}(\rho)$ of ρ by

$$\text{dom}(\rho) = \{x \in X : (\exists y \in X)\,(x, y) \in \rho\} \tag{4.5}$$

and the *range* $\text{ran}(\rho)$ of ρ by

$$\text{ran}(\rho) = \{y \in X : (\exists x \in X)\,(x, y) \in \rho\}. \tag{4.6}$$

It is immediate that, for all ρ, σ in $\mathscr{B}(X)$,

$$\rho \subseteq \sigma \Rightarrow \text{dom}(\rho) \subseteq \text{dom}(\sigma),\ \text{ran}(\rho) \subseteq \text{ran}(\sigma). \tag{4.7}$$

If $x \in X$ we define

$$x\rho = \{y \in X : (x, y) \in \rho\}; \tag{4.8}$$

thus $x\rho \neq \varnothing$ if and only if $x \in \text{dom}(\rho)$. If A is a subset of X, we define

$$A\rho = \bigcup\{a\rho : a \in A\}. \tag{4.9}$$

For each ρ in $\mathscr{B}(X)$ we define ρ^{-1}, the *converse* of ρ, by

$$\rho^{-1} = \{(x, y) \in X \times X : (y, x) \in \rho\}. \tag{4.10}$$

Then $\rho^{-1} \in \mathscr{B}(X)$, and it is easy to verify that for all $\rho, \sigma, \rho_1, \ldots, \rho_n$ in $\mathscr{B}$(X),

$$(\rho^{-1})^{-1} = \rho, \tag{4.11}$$

$$(\rho_1 \circ \rho_2 \circ \ldots \rho_n)^{-1} = \rho_n^{-1} \circ \ldots \circ \rho_2^{-1} \circ \rho_1^{-1}, \tag{4.12}$$

$$\rho \subseteq \sigma \Rightarrow \rho^{-1} \subseteq \sigma^{-1}. \tag{4.13}$$

Notice also that

$$\text{dom}(\rho^{-1}) = \text{ran}(\rho), \qquad \text{ran}(\rho^{-1}) = \text{dom}(\rho). \tag{4.14}$$

By analogy with the remark following (4.8) we have that

$$x\rho^{-1} \neq \varnothing \quad \text{if and only if} \quad x \in \operatorname{ran}(\rho).$$

An element ϕ of $\mathscr{B}(X)$ is called a *partial mapping* of X if $|x\phi| = 1$ for every x in $\operatorname{dom}(\phi)$, i.e. if, for all x, y_1, y_2 in X,

$$[(x, y_1) \in \phi \quad \text{and} \quad (x, y_2) \in \phi] \Rightarrow y_1 = y_2. \tag{4.15}$$

It will not conflict at all with (4.8) if we decide in such a case to let $x\phi$ denote the unique element y such that $(x, y) \in \phi$ (rather than the set consisting of that element). Notice that the condition (4.15) is fulfilled (vacuously) by the empty relation $\varnothing$, which is therefore included among the partial mappings.

If ϕ, ψ are partial mappings of X such that $\phi \subseteq \psi$, we sometimes say that ϕ is a *restriction* of ψ or that ψ is an *extension* of ϕ. If $\operatorname{dom}(\phi) = A \subset \operatorname{dom}(\psi)$, we often denote ϕ by $\psi|A$ (ψ *restricted* to A).

Proposition 4.16. *The subset $\mathscr{PT}(X)$ of $\mathscr{B}(X)$ consisting of all partial mappings of X is a subsemigroup of $(\mathscr{B}(X), \circ)$.*

Proof. Let $\phi, \psi \in \mathscr{PT}(X)$. If (x, y_1) and (x, y_2) are in $\phi \circ \psi$ then there exist z_1, z_2 in X such that

$$(x, z_1) \in \phi, \qquad (z_1, y_1) \in \psi, \qquad (x, z_2) \in \phi, \qquad (z_2, y_2) \in \psi.$$

The condition (4.15) on ϕ implies that $z_1 = z_2$, and then the same condition on ψ implies that $y_1 = y_2$. Thus $\phi \circ \psi$ is a partial mapping as required. ■

It is important to note that the converse ϕ^{-1} of a partial mapping ϕ need not be a partial mapping. For example, if $X = \{1, 2\}$, then $\phi = \{(1, 1), (2, 1)\} \in \mathscr{PT}(X)$, but $\phi^{-1} \notin \mathscr{PT}(X)$.

In view of Proposition 4.16 we can talk of $(\mathscr{PT}(X), \circ)$ as the *semigroup of partial mappings of* X. The composition law $\circ$ in this semigroup is in fact a fairly natural composition law for partial mappings:

Proposition 4.17. *If $\phi, \psi \in \mathscr{PT}(X)$, then*

$$\operatorname{dom}(\phi \circ \psi) = [\operatorname{ran}(\phi) \cap \operatorname{dom}(\psi)]\,\phi^{-1},$$

$$\operatorname{ran}(\phi \circ \psi) = [\operatorname{ran}(\phi) \cap \operatorname{dom}(\psi)]\,\psi,$$

and

$$(\forall x \in \operatorname{dom}(\phi \circ \psi)) \quad x(\phi \circ \psi) = (x\phi)\,\psi.$$

Proof. Before proving this, we illustrate it in a diagram as follows:

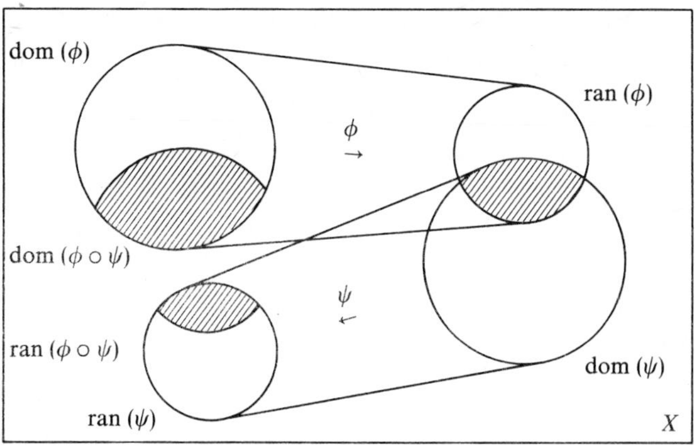

If $x \in \mathrm{dom}(\phi \circ \psi)$ then there exist y and z in X such that $(x, z) \in \phi$, $(z, y) \in \psi$. Now $z \in \mathrm{ran}(\phi) \cap \mathrm{dom}(\psi)$ and $(z, x) \in \phi^{-1}$; hence

$$x \in z\phi^{-1} \subseteq [\mathrm{ran}(\phi) \cap \mathrm{dom}(\psi)]\, \phi^{-1}.$$

Conversely, if $x \in [\mathrm{ran}(\phi) \cap \mathrm{dom}(\psi)]\, \phi^{+1}$ then there exists z in $\mathrm{ran}(\phi) \cap \mathrm{dom}(\psi)$ such that $x \in z\phi^{-1}$, i.e. such that $(x, z) \in \phi$. Since $z \in \mathrm{dom}(\psi)$, there exists y in X such that $(z, y) \in \psi$. Thus $(x, y) \in \phi \circ \psi$ and so $x \in \mathrm{dom}(\phi \circ \psi)$. Hence

$$\mathrm{dom}(\phi \circ \psi) = [\mathrm{ran}(\phi) \cap \mathrm{dom}(\psi)]\, \phi^{-1}$$

as stated. The proof that $\mathrm{ran}(\phi \circ \psi) = [\mathrm{ran}(\phi) \cap \mathrm{dom}(\psi)]\, \psi$ is similar. (Notice that no special properties of partial mappings have been used so far, and that the characterization established for $\mathrm{dom}(\phi \circ \psi)$ and $\mathrm{ran}(\phi \circ \psi)$ apply equally well to general relations in $\mathscr{B}(X)$.)

Finally, $(x, z) \in \phi \circ \psi$ if and only if there exists y in X such that $(x, y) \in \phi$, $(y, z) \in \psi$. Since ϕ, ψ and $\phi \circ \psi$ are partial mappings, we have $y = x\phi$, $z = y\psi$ and $z = x(\phi \circ \psi)$. Hence

$$x(\phi \circ \psi) = (x\phi)\, \psi. \blacksquare$$

An element ϕ of $\mathscr{PT}(X)$ is called a *mapping* or *function* if $\mathrm{dom}(\phi) = X$. If ϕ and ψ are mappings then it is easy to see that $\phi \circ \psi$ is also a mapping and that the operation $\circ$ defined by (4.2) coincides with ordinary composition of mappings. In fact,

Proposition 4.18. *The set $\mathscr{T}(X)$ of all mappings from X into itself is a subsemigroup of $(\mathscr{B}(X), \circ)$.* $\blacksquare$

Once again it is important to note that in general

$$\phi \in \mathscr{T}(X) \not\Rightarrow \phi^{-1} \in \mathscr{T}(X).$$

If $\phi \in \mathscr{T}(X)$ and $x \in X$ then $x\phi$ always denotes a single element of X, but $x\phi^{-1}$ denotes a subset of X which will be empty if $x \notin \operatorname{ran}(\phi)$, and may contain more than one element of X if $x \in \operatorname{ran}(\phi)$. The following proposition is easy to prove:

Proposition 4.19. *(i) If $\phi \in \mathscr{PT}(X)$ then $\phi^{-1} \in \mathscr{PT}(X)$ if and only if ϕ is one-one.*

(ii) If $\phi \in \mathscr{T}(X)$, then $\phi^{-1} \in \mathscr{T}(X)$ if and only if ϕ is bijective (i.e. both one-one and onto). ■

Partial mappings and mappings are not the only special relations that are of interest. We say that a relation ρ on X is *reflexive* if $1_X \subseteq \rho$, i.e. if $(x, x) \in \rho$ for every x in X, that ρ is *symmetric* if $\rho^{-1} = \rho$, i.e. if

$$(\forall x, y \in X) \qquad (x, y) \in \rho \Rightarrow (y, x) \in \rho,$$

and that ρ is *transitive* if $\rho \circ \rho \subseteq \rho$, i.e. if

$$(\forall x, y, z \in X) \qquad [(x, y) \in \rho \text{ and } (y, z) \in \rho] \Rightarrow (x, z) \in \rho.$$

A relation is called an *equivalence relation* if it is reflexive, symmetric and transitive. If ρ is an equivalence relation then

$$\operatorname{dom}(\rho) \supseteq \operatorname{dom}(1_X) = X, \qquad \operatorname{ran}(\rho) \supseteq \operatorname{ran}(1_X) = X$$

and so $\operatorname{dom}(\rho) = \operatorname{ran}(\rho) = X$.

A family $\pi = \{A_i : i \in I\}$ of subsets of X is said to form a *partition* of X if

(a) each A_i is non-empty;
(b) for all i, j in I, either $A_i = A_j$ or $A_i \cap A_j = \varnothing$;
(c) $\bigcup\{A_i : i \in I\} = X$.

The two apparently different notions of "equivalence" and "partition" are in fact interchangeable. The following proposition is elementary, and its proof is omitted.

Proposition 4.20. *Let ρ be an equivalence relation on a set X. Then the family*

$$\Phi(\rho) = \{x\rho : x \in X\}$$

of subsets of X is a partition of X.

Conversely, if $\pi = \{A_i : i \in I\}$ is a partition of X, then the relation

$$\Psi(\pi) = \{(x, y) \in X \times X : (\exists i \in I)\, x, y \in A_i\}$$

is an equivalence relation on X.

If ρ is an equivalence relation on X, then

$$\Psi(\Phi(\rho)) = \rho.$$

If π is a partition of X, then

$$\Phi(\Psi(\pi)) = \pi. \blacksquare$$

If ρ is an equivalence relation on X, we shall sometimes write $x\rho y$ or $x \equiv y$ (mod ρ) as alternatives to $(x, y) \in \rho$. The sets $x\rho$ that form the associated partition of the equivalence relation are called *ρ-classes* or *equivalence classes.*

The set of ρ-classes, i.e. the set whose elements are the sets $x\rho$, is called the *quotient set* of X by ρ and is denoted by X/ρ. We shall have occasion in the next section to examine the natural mapping $\rho^\natural$ from X onto X/ρ defined by

$$x\rho^\natural = x\rho \quad (x \in X). \tag{4.21}$$

An important connection between mappings and equivalences is given by

Proposition 4.22. *If $\phi: X \to Y$ is a mapping, then $\phi \circ \phi^{-1}$ is an equivalence.*

Proof. The easiest way to see this is to note first that

$$\begin{aligned} \phi \circ \phi^{-1} &= \{(x, y) \in X \times X : (\exists z \in X)\,(x, z) \in \phi, (y, z) \in \phi\} \\ &= \{(x, y) \in X \times X : x\phi = y\phi\}. \end{aligned}$$

It is then obvious that $\phi \circ \phi^{-1}$ is reflexive, symmetric and transitive. $\blacksquare$

We call the equivalence $\phi \circ \phi^{-1}$ the *kernel of ϕ* and write $\phi \circ \phi^{-1} = \ker \phi$. Notice that $\ker \rho^\natural = \rho$.

If $\{\rho_i : i \in I\}$ is a non-empty family of equivalence relations on a set X, then it is an entirely routine matter to verify that $\bigcap\{\rho_i : i \in I\}$ is an equivalence relation on X. If $\mathbf{R}$ is any relation whatever on X (even the empty relation will do) then the family of equivalence relations containing $\mathbf{R}$ is non-empty, since $X \times X$ is one such equivalence; hence the intersection of all the equivalences on X containing $\mathbf{R}$ is an equivalence, the unique minimum equivalence on X containing $\mathbf{R}$. We call it the equivalence on X *generated by* $\mathbf{R}$ and denote it by $\mathbf{R}^e$.

It is frequently useful to be able to describe $\mathbf{R}^e$ for a given $\mathbf{R}$, and the foregoing general description is not particularly helpful. It is necessary therefore to develop an alternative description, and for this some preliminaries are required.

First, if $\mathbf{S}$ is any relation on X, we define $\mathbf{S}^\infty$, the *transitive closure* of $\mathbf{S}$, by

$$\mathbf{S}^\infty = \bigcup_{n=1}^{\infty} \mathbf{S}^n. \tag{4.23}$$

The term "transitive closure" is justified by the following lemma:

Lemma 4.24. *If* $\mathbf{S}$ *is a relation on a set* X, *then* $\mathbf{S}^\infty$ *is the smallest transitive relation on* X *containing* $\mathbf{S}$.

Proof. First, $\mathbf{S}^\infty$ is transitive, for if (x, y) and (y, z) belong to $\mathbf{S}^\infty$ then there exist positive integers m, n such that

$$(x, y) \in \mathbf{S}^m, \qquad (y, z) \in \mathbf{S}^n.$$

Hence $(x, z) \in \mathbf{S}^m \circ \mathbf{S}^n = \mathbf{S}^{m+n} \subseteq \mathbf{S}^\infty$. Also clearly $\mathbf{S}^\infty$ contains $\mathbf{S}^1 = \mathbf{S}$.

If $\mathbf{T}$ is a transitive relation on X containing $\mathbf{S}$, then

$$\mathbf{S}^2 = \mathbf{S} \circ \mathbf{S} \subseteq \mathbf{T} \circ \mathbf{T} \subseteq \mathbf{T},$$

and more generally $\mathbf{S}^n \subseteq \mathbf{T}$ for $n = 1, 2, \ldots$. Thus $\mathbf{S}^\infty \subseteq \mathbf{T}$. ∎

We now have

Proposition 4.25. *If* $\mathbf{R}$ *is a relation on a set* X, *then*

$$\mathbf{R}^e = [\mathbf{R} \cup \mathbf{R}^{-1} \cup 1_X]^\infty.$$

Proof. As in the proof of the lemma, we have that the relation $\mathbf{E} = [\mathbf{R} \cup \mathbf{R}^{-1} \cup 1_X]^\infty$ is transitive and contains $\mathbf{R}$. Since

$$1_X \subseteq \mathbf{R} \cup \mathbf{R}^{-1} \cup 1_X \subseteq \mathbf{E}$$

we have that $\mathbf{E}$ is also reflexive. Certainly the relation $\mathbf{S} = \mathbf{R} \cup \mathbf{R}^{-1} \cup 1_X$ is symmetric. Hence, for every n in $\mathbf{N}$ we have

$$\mathbf{S}^n = (\mathbf{S}^{-1})^n = (\mathbf{S}^n)^{-1}$$

(the second equality being a specialization of formula (4.12)), and so $\mathbf{S}^n$ is symmetric. Thus

$$\begin{aligned}(x, y) \in \mathbf{E} &\Rightarrow (\exists n \in \mathbf{N})\,(x, y) \in \mathbf{S}^n \\ &\Rightarrow (\exists n \in \mathbf{N})\,(y, x) \in \mathbf{S}^n \\ &\Rightarrow (y, x) \in \mathbf{E},\end{aligned}$$

and so $\mathbf{E}$ is symmetric. That is, $\mathbf{E}$ is an equivalence relation on X containing $\mathbf{R}$.

Conversely, if σ is an equivalence relation on X containing $\mathbf{R}$ then $1_X \subseteq \sigma$ and $\mathbf{R}^{-1} \subseteq \sigma^{-1} = \sigma$. Hence $\mathbf{S} = \mathbf{R} \cup \mathbf{R}^{-1} \cup 1_X \subseteq \sigma$. Moreover,

$$\mathbf{S} \circ \mathbf{S} \subseteq \sigma \circ \sigma \subseteq \sigma,$$

and more generally $\mathbf{S}^n \subseteq \sigma$ for every n in $\mathbf{N}$. Thus $\mathbf{E} \subseteq \sigma$. We have thus shown that $\mathbf{E}$ $(= [\mathbf{R} \cup \mathbf{R}^{-1} \cup 1_X]^\infty)$ is the smallest equivalence relation on X containing $\mathbf{R}$, i.e. that $\mathbf{E}$ coincides with $\mathbf{R}^e$. ∎

In more down-to-earth terms, Proposition 4.25 may be rewritten thus:

Proposition 4.26. *If* $\mathbf{R}$ *is a relation on a set* X *and* $\mathbf{R}^e$ *is the smallest equivalence relation on* X *containing* $\mathbf{R}$, *then* $(x, y) \in \mathbf{R}^e$ *if and only if either* $x = y$ *or for some* n *in* $\mathbf{N}$ *there is a sequence*

$$x = z_1 \to z_2 \ldots \to z_n = y$$

in which for each $i \in \{1, \ldots, n-1\}$ *either* $(z_i, z_{i+1}) \in \mathbf{R}$ *or* $(z_{i+1}, z_i) \in \mathbf{R}$. ■

5. CONGRUENCES

Let $(S, .)$ be a semigroup. A relation $\mathbf{R}$ on the set S is called *left compatible* (with the operation on S) if

$$(\forall s, t, a \in S) \qquad (s, t) \in \mathbf{R} \Rightarrow (as, at) \in \mathbf{R},$$

and *right compatible* if

$$(\forall s, t, a \in S) \qquad (s, t) \in \mathbf{R} \Rightarrow (as, ta) \in \mathbf{R}.$$

It is called *compatible* if

$$(\forall s, t, s', t' \in S) \qquad (s, s') \in \mathbf{R} \text{ and } (t, t') \in \mathbf{R} \ \Rightarrow \ (st, s't') \in \mathbf{R}.$$

A left [right] compatible equivalence relation is called a *left* [*right*] *congruence.* A compatible equivalence relation is called a *congruence.*

Proposition 5.1. *A relation* ρ *on a semigroup* S *is a congruence if and only if it is both a left and a right congruence.*

Proof. Suppose first that ρ is a congruence. If $(s, t) \in \rho$ and $a \in S$ then $(a, a) \in \rho$ by reflexivity and so $(as, at) \in \rho$ and $(sa, ta) \in \rho$ by compatibility. Thus ρ is both a left and a right congruence.

Conversely, if ρ is both a left and a right congruence and if (s, s'), $(t, t') \in \rho$, it follows that $(st, s't) \in \rho$ by right compatibility and that $(s't, s't') \in \rho$ by left compatibility. Hence $(st, s't') \in \rho$ by transitivity and so ρ is a congruence. ■

Exercise 10 below explores in more detail the relationship between left and right compatibility on the one hand and compatibility on the other.

If ρ is a congruence on a semigroup S then we can define a binary operation on the quotient set S/ρ in a natural way as follows:

$$(a\rho)(b\rho) = (ab)\rho. \tag{5.2}$$

This is well-defined, since (for all a, a', b, b' in S)

$$a\rho = a'\rho \quad \text{and} \quad b\rho = b'\rho \Rightarrow (a, a') \in \rho \quad \text{and} \quad (b, b') \in \rho$$
$$\Rightarrow (ab, a'b') \in \rho \Rightarrow (ab)\rho = (a'b')\rho.$$

The operation is easily seen to be associative, and so $(S/\rho, .)$ is a semigroup. The natural mapping $\rho^\natural$ from S onto S/ρ given by (4.21) is a homomorphism.
We have proved part of the following theorem:

Theorem 5.3. *If ρ is a congruence on a semigroup S, then S/ρ is a semigroup with respect to the operation defined by* (5.2) *and the mapping $\rho^\natural : S \to S/\rho$ defined by*

$$x\rho^\natural = x\rho \qquad (x \in S)$$

is a homomorphism.
If $\phi: S \to T$ is a homomorphism, where S and T are semigroups, then the relation

$$\ker \phi = \phi \circ \phi^{-1} = \{(a, b) \in S \times S : a\phi = b\phi\}$$

is a congruence on S and there is a monomorphism $\alpha: S/\ker \phi \to T$ such that $\operatorname{ran}(\alpha) = \operatorname{ran}(\phi)$ *and the diagram*

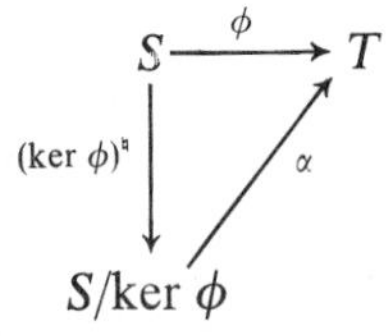

commutes.

Proof. To establish the second half of this result, observe first that $\ker \phi$ is an equivalence by Proposition 4.22. If a, b, c, d in S are such that (a, b), $(c, d) \in \ker \phi$, then $a\phi = b\phi$, $c\phi = d\phi$. It follows that

$$(ac)\phi = (a\phi)(c\phi) = (b\phi)(d\phi) = (bd)\phi,$$

and hence $(ac, bd) \in \ker \phi$. Thus $\ker \phi$ is a congruence.
Now we define $\alpha: S/\ker \phi \to T$ by

$$[a(\ker \phi)]\,\alpha = a\phi \qquad (a \in S).$$

Then α is both well-defined and one–one since, for all a, b in S,

$$a(\ker \phi) = b(\ker \phi) \Leftrightarrow (a, b) \in \ker \phi \Leftrightarrow a\phi = b\phi.$$

It is homomorphic, since, for all a, b in S,

$$[a(\ker \phi) \,.\, b(\ker \phi)]\alpha = [(ab)(\ker \phi)]\alpha = (ab)\phi$$
$$= (a\phi)(b\phi) = [a(\ker \phi)]\alpha \,.\, [b(\ker \phi)]\alpha.$$

Clearly $\operatorname{ran}(\alpha) = \operatorname{ran}(\phi)$, and from the definition of α it is obvious that, for all a in S,

$$a(\ker \phi)^{\natural}\alpha = a\phi. \blacksquare$$

The next theorem is concerned with a more general situation and will frequently be useful:

Theorem 5.4. *Let ρ be a congruence on a semigroup S. If $\phi: S \to T$ is a homomorphism such that $\rho \subseteq \ker \phi$ then there is a unique homomorphism $\beta: S/\rho \to T$ such that $\operatorname{ran}(\beta) = \operatorname{ran}(\phi)$ and the diagram*

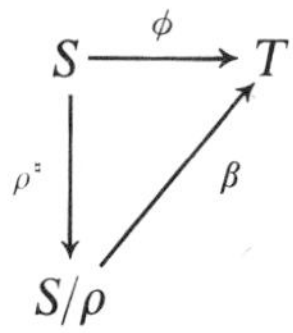

commutes.

Proof. Define $\beta: S/\rho \to T$ by

$$(s\rho)\beta = s\phi \qquad (s\rho \in S/\rho). \tag{5.5}$$

Then β is well-defined, since for all s, s' in S,

$$s\rho = s'\rho \Rightarrow (s, s') \in \rho \Rightarrow (s, s') \in \ker \phi \Rightarrow s\phi = s'\phi.$$

It is now a routine matter to verify that β is a homomorphism, that $\operatorname{ran}(\beta) = \operatorname{ran}(\phi)$ and that $\rho^{\natural} \circ \beta = \phi$. The uniqueness of β is also obvious, since any homomorphism satisfying $\rho^{\natural} \circ \beta = \phi$ *must* be defined by the rule (5.5). $\blacksquare$

One application of this theorem is to the situation where ρ and σ are congruences on S with $\rho \subseteq \sigma$. The theorem implies that there is a homomorphism β from S/ρ onto S/σ such that the diagram

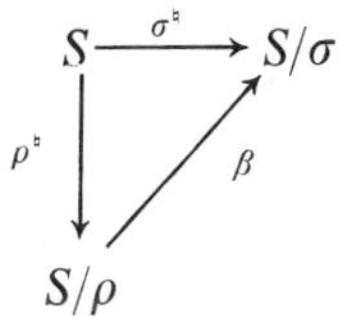

commutes. The homomorphism β is given by

$$(a\rho)\beta = a\sigma \qquad (a\rho \in S/\rho)$$

and the congruence $\ker \beta$ on S/ρ is given by

$$\ker \beta = \{(a\rho, b\rho) \in S/\rho \times S/\rho : (a, b) \in \sigma\}.$$

It is usual to write $\ker \beta$ as σ/ρ. From Theorem 5.3 it now follows that there is an isomorphism $\alpha : (S/\rho)/(\sigma/\rho) \to S/\sigma$ defined by

$$[(a\rho)(\sigma/\rho)]\alpha = a\sigma \qquad (a \in S)$$

and such that the diagram

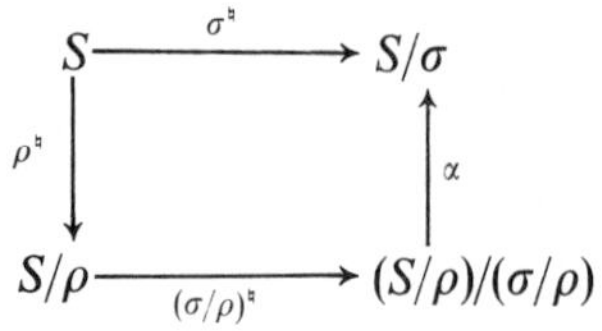

commutes. We summarize in a theorem:

Theorem 5.6. *Let ρ, σ be congruences on a semigroup S such that $\rho \subseteq \sigma$. Then*

$$\sigma/\rho = \{(x\rho, y\rho) \in S/\rho \times S/\rho : (x, y) \in \sigma\}$$

is a congruence on S/ρ, and $(S/\rho)/(\sigma/\rho) \simeq S/\sigma$. ■

Since the intersection of a non-empty family of congruences on a semigroup S is a congruence on S, we can argue exactly as in §4 and deduce that for any relation $\mathbf{R}$ on S there is a unique smallest congruence relation $\mathbf{R}^\#$ containing $\mathbf{R}$, namely the intersection of the family of *all* congruences on S containing $\mathbf{R}$. We now seek a result analogous to Proposition 4.25 which will give us a usable description of $\mathbf{R}^\#$.

First, for any relation $\mathbf{R}$ on S we define

$$\mathbf{R}^c = \{(xay, xby) : x, y \in S^1, (a, b) \in \mathbf{R}\}.$$

Then

Lemma 5.7. *Let $\mathbf{R}, \mathbf{S} \in \mathscr{B}(S)$, where S is a semigroup. Then*

(i) $\mathbf{R} \subseteq \mathbf{R}^c$;
(ii) $(\mathbf{R}^c)^{-1} = (\mathbf{R}^{-1})^c$;
(iii) $\mathbf{R} \subseteq \mathbf{S} \Rightarrow \mathbf{R}^c \subseteq \mathbf{S}^c$;
(iv) $(\mathbf{R}^c)^c = \mathbf{R}^c$;
(v) $(\mathbf{R} \cup \mathbf{S})^c = \mathbf{R}^c \cup \mathbf{S}^c$;
(vi) $\mathbf{R} = \mathbf{R}^c$ *if and only if* $\mathbf{R}$ *is left and right compatible.*

Proof. The proofs of (i), (ii) and (iii) are routine. To prove (iv), notice first that it follows from (i) and (iii) that $\mathbf{R}^c \subseteq (\mathbf{R}^c)^c$. To show the reverse inclusion, suppose that $(c, d) \in (\mathbf{R}^c)^c$. Then $c = xay$, $d = xby$ for some x, y in S^1 and some (a, b) in $\mathbf{R}^c$. Now $a = zpt$, $b = zqt$ for some z, t in S^1 and some (p, q) in $\mathbf{R}$. Hence

$$c = (xz)p(ty), \qquad d = (xz)q(ty),$$

where $xz, ty \in S^1$ and $(p, q) \in \mathbf{R}$; that is, $(c, d) \in \mathbf{R}^c$.

To prove (v) notice first that by (iii)

$$\mathbf{R}^c \subseteq (\mathbf{R} \cup \mathbf{S})^c, \qquad \mathbf{S}^c \subseteq (\mathbf{R} \cup \mathbf{S})^c;$$

hence $\mathbf{R}^c \cup \mathbf{S}^c \subseteq (\mathbf{R} \cup \mathbf{S})^c$. Conversely, if $(c, d) \in (\mathbf{R} \cup \mathbf{S})^c$ then for some x, y in S^1 we have $c = xay$, $d = xby$, where $(a, b) \in \mathbf{R} \cup \mathbf{S}$. That is, either $(a, b) \in \mathbf{R}$ or $(a, b) \in \mathbf{S}$, and so either $(c, d) \in \mathbf{R}^c$ or $(c, d) \in \mathbf{S}^c$. Thus $(c, d) \in \mathbf{R}^c \cup \mathbf{S}^c$ as required.

To prove (vi), suppose first that $\mathbf{R} = \mathbf{R}^c$. Then, for all s, t, a in S,

$$(s, t) \in \mathbf{R} \Rightarrow (as, at) \in \mathbf{R}^c = \mathbf{R}$$

and

$$(s, t) \in \mathbf{R} \Rightarrow (sa, ta) \in \mathbf{R}^c = \mathbf{R}.$$

Thus $\mathbf{R}$ is left and right compatible. Conversely, if $\mathbf{R}$ is left and right compatible and if $(c, d) \in \mathbf{R}^c$, we have $c = xay$, $d = xby$ for some x, y in S^1 and some (a, b) in $\mathbf{R}$. But then $(xay, xby) \in \mathbf{R}$ by left and right compatibility and so $\mathbf{R} = \mathbf{R}^c$ as required. ■

Next we have

Lemma 5.8. *Let* $\mathbf{R}$ *be a relation on a semigroup* S. *If* $\mathbf{R}$ *is left and right compatible, then so is* $\mathbf{R}^n$ *for any* n *in* $\mathbf{N}$.

Proof. Suppose that $\mathbf{R}$ is left and right compatible and that $(s, t) \in \mathbf{R}^n$. Then there exist $z_1, \ldots, z_{n-1}$ in S such that

$$(s, z_1), (z_1, z_2), \ldots, (z_{n-1}, t) \in \mathbf{R}.$$

Hence, for all a in S,

$$(as, az_1), (az_1, az_2), \ldots, (az_{n-1}, at) \in \mathbf{R},$$

$$(sa, z_1a), (z_1a, z_2a), \ldots, (z_{n-1}a, ta) \in \mathbf{R},$$

and so $(as, at), (sa, ta) \in \mathbf{R}^n$. ■

We can now easily obtain the following characterization of $\mathbf{R}^\#$, the congruence on S generated by $\mathbf{R}$:

Proposition 5.9. *If* $\mathbf{R}$ *is a relation on a semigroup* S, *then* $\mathbf{R}^\# = (\mathbf{R}^c)^e$.

Proof. From Proposition 4.25 we know that $(\mathbf{R}^c)^e$ is an equivalence relation containing $\mathbf{R}^c$ and so certainly containing $\mathbf{R}$. To show that $(\mathbf{R}^c)^e$ is a congruence, suppose that $(s, t) \in (\mathbf{R}^c)^e$. Then $(s, t) \in \mathbf{S}^n$ for some n in $\mathbf{N}$, where $\mathbf{S} = \mathbf{R}^c \cup (\mathbf{R}^c)^{-1} \cup 1_S$. Now, by Lemma 5.7,

$$\mathbf{S} = \mathbf{R}^c \cup (\mathbf{R}^{-1})^c \cup 1_S^c = (\mathbf{R} \cup \mathbf{R}^{-1} \cup 1_S)^c$$

Hence $\mathbf{S}$ is left and right compatible and hence, by Lemma 5.8, so also is $\mathbf{S}^n$. It follows that for every a in S we have

$$(as, at) \in \mathbf{S}^n \subseteq (\mathbf{R}^c)^e \quad \text{and} \quad (sa, ta) \in \mathbf{S}^n \subseteq (\mathbf{R}^c)^e;$$

thus $(\mathbf{R}^c)^e$ is a congruence on S containing $\mathbf{R}$.

If κ is any congruence on S containing $\mathbf{R}$, then, by Lemma 5.7 ((i) and (vi)),

$$\mathbf{R}^c \subseteq \kappa^c = \kappa.$$

Since κ is then an equivalence on S containing $\mathbf{R}^c$ it follows that $(\mathbf{R}^c)^e \subseteq \kappa$. Thus $(\mathbf{R}^c)^e$ is the smallest congruence on S containing $\mathbf{R}$ and so coincides with $\mathbf{R}^\#$. ■

We may rewrite Proposition 5.9 in more elementary terms. First, if $c, d \in S$ are such that

$$c = xpy, \qquad d = xqy$$

for some x, y in S^1, where either $(p, q) \in \mathbf{R}$ or $(q, p) \in \mathbf{R}$, we say that c is connected to d by an *elementary* **R**-*transition*. Then we have the following alternative version of Proposition 5.9:

Proposition 5.10. *Let* $\mathbf{R}$ *be a relation on a semigroup* S. *If* $a, b \in S$ *then* $(a, b) \in \mathbf{R}^\#$ *if and only if either* $a = b$ *or for some* n *in* $\mathbf{N}$ *there is a sequence*

$$a = z_1 \to z_2 \to \dots \to z_n = b$$

of elementary **R**-*transitions connecting* a *to* b. ■

The next result is rather technical and will not be required until Section VII.4. It is convenient, however, to establish it now. If $\mathbf{T}$ is a relation on a semigroup S and ρ is congruence, then $\mathbf{T}/\rho$ will denote the relation $\{(x\rho, y\rho) : (x, y) \in \mathbf{T}\}$ on the semigroup S/ρ. (Notice that if $\mathbf{T} \subseteq \rho$ then $\mathbf{T}/\rho = 1_{S/\rho}$.)

Proposition 5.11. *Let* ρ *and* σ *be congruences on a semigroup* S such that $\rho \subseteq \sigma$. *If the relation* $\mathbf{T}$ *on* S *is such that* $\sigma = \mathbf{T}^\#$, *then* $\sigma/\rho = (\mathbf{T}/\rho)^\#$.

Proof. Certainly σ/ρ is a congruence on S/ρ containing $\mathbf{T}/\rho$; hence $(\mathbf{T}/\rho)^{\#} \subseteq \sigma/\rho$. Conversely, if $(a\rho, b\rho) \in \sigma/\rho$ then $(a,b) \in \sigma$ and so there exist $p_1, \ldots, p_n, q_1, \ldots, q_n \in S^1$ and $(s_1, t_1), \ldots, (s_n, t_n) \in \mathbf{T} \cup \mathbf{T}^{-1}$ giving a sequence

$$a = p_1 s_1 q_1 \to p_1 t_1 q_1 = p_2 s_2 q_2 \to p_2 t_2 q_2 = \ldots \to p_n t_n q_n = b$$

of elementary $\mathbf{T}$-transitions connecting a to b. The "image" sequence

$$a\rho = (p_1\rho)(s_1\rho)(q_1\rho) \to (p_1\rho)(t_1\rho)(q_1\rho) = (p_2\rho)(s_2\rho)(q_2\rho) \to \ldots$$
$$\ldots \to (p_n\rho)(t_n\rho)(q_n\rho) = b\rho$$

is then a sequence of elementary $(\mathbf{T}/\rho)$-transitions connecting $a\rho$ to $b\rho$. Hence $(a\rho, b\rho) \in (\mathbf{T}/\rho)^{\#}$. ■

Having used the musical symbols ♮ and ♯, we now complete the scene by defining, for any equivalence relation $\mathbf{E}$ on a semigroup S,

$$\mathbf{E}^{\flat} = \{(a, b) \in S \times S : (\forall x, y \in S^1)\,(xay, xby) \in \mathbf{E}\}. \tag{5.12}$$

Then we have the following result:

Proposition 5.13. *If* $\mathbf{E}$ *is an equivalence on a semigroup* S *then* $\mathbf{E}^{\flat}$ *is the largest congruence on* S *contained in* $\mathbf{E}$.

Proof. First, it is clear that $\mathbf{E}^{\flat}$ is an equivalence. Also, if $(a, b) \in \mathbf{E}^{\flat}$ and $c \in S$ then $(xcay, xcby) \in \mathbf{E}$ for all choices of x and y in S^1. Hence $(ca, cb) \in \mathbf{E}^{\flat}$, and similarly $(ac, bc) \in \mathbf{E}^{\flat}$. Thus $\mathbf{E}^{\flat}$ is a congruence, and clearly $\mathbf{E}^{\flat} \subseteq \mathbf{E}$, since

$$(a, b) \in \mathbf{E}^{\flat} \Rightarrow (1a1, 1b1) \in \mathbf{E} \Rightarrow (a, b) \in \mathbf{E}.$$

If η is a congruence on S contained in $\mathbf{E}$, then for all a, b in S,

$$\begin{aligned}(a, b) \in \eta &\Rightarrow (\forall x, y \in S^1)\,(xay, xby) \in \eta \\ &\Rightarrow (\forall x, y \in S^1)\,(xay, xby) \in \mathbf{E} \\ &\Rightarrow (a, b) \in \mathbf{E}^{\flat};\end{aligned}$$

thus $\mathbf{E}^{\flat}$ is the largest congruence on S contained in $\mathbf{E}$. ■

Notice carefully that while $\mathbf{R}^{\#}$ is defined for an arbitrary relation $\mathbf{R}$, the congruence $\mathbf{E}^{\flat}$ is defined only for an *equivalence* relation $\mathbf{E}$.

If S is a semigroup then both the set $\mathscr{E}(S)$ of equivalences on S and the set $\mathscr{C}(S)$ of the congruences on S are partially ordered by inclusion. In fact both are lattices: if $\rho, \sigma \in \mathscr{E}(S)$ then $\rho \cap \sigma \in \mathscr{E}(S)$ and is their greatest lower bound, while $(\rho \cup \sigma)^e$ is their least upper bound; if $\rho, \sigma \in \mathscr{C}(S)$ then

$\rho \cap \sigma \in \mathscr{C}(S)$ and is their greatest lower bound, while $(\rho \cup \sigma)^\#$ is their least upper bound. Notice now that if $\rho, \sigma \in \mathscr{C}(S)$ then, by Lemma 5.7,

$$(\rho \cup \sigma)^c = \rho^c \cup \sigma^c = \rho \cup \sigma;$$

hence, by Proposition 5.9, $(\rho \cup \sigma)^\# = (\rho \cup \sigma)^e$. That is, the join of ρ and σ in the lattice $\mathscr{C}(S)$ coincides with their join in the lattice $\mathscr{E}(S)$; we may therefore unambiguously write $\rho \vee \sigma$ for the join of ρ and σ either as equivalences or as congruences.

We shall give further consideration to lattices of equivalences and congruences in Section 8. For the moment let us merely remark that in the foregoing analysis there is no difficulty in replacing the subset $\{\rho, \sigma\}$ by an arbitrary family of equivalences (or congruences), and so both the lattices $(\mathscr{E}(S), \subseteq, \cap, \vee)$ and $(\mathscr{C}(S), \subseteq, \cap, \vee)$ are complete. Both lattices have maximum element $S \times S$ and minimum element 1_S.

A specialization of Propositions 4.25 and 5.9 to the case where the relation $\mathbf{R}$ is the union $\rho \cup \sigma$ of two equivalences or congruences is worth stating separately:

Proposition 5.14. *Let ρ, σ be equivalences on a set S [congruences on a semigroup S]. If $a, b \in S$, then $(a, b) \in \rho \vee \sigma$ if and only if for some n in $\mathbf{N}$ there exist elements $x_1, x_2, \ldots, x_{2n-1}$ in S such that*

$$(a, x_1) \in \rho, (x_1, x_2) \in \sigma, (x_2, x_3) \in \rho, \ldots, (x_{2n-1}, b) \in \sigma.$$

Proof. The result says effectively that

$$\rho \vee \sigma = (\rho \circ \sigma)^\infty.$$

To see that this is so, notice that the general approach in Propositions 4.25 and 5.9 gives that

$$\rho \vee \sigma = \mathbf{R}^\infty,$$

where

$$\begin{aligned} \mathbf{R} &= (\rho \cup \sigma) \cup (\rho \cup \sigma)^{-1} \cup 1_S \\ &= \rho \cup \sigma \cup \rho^{-1} \cup \sigma^{-1} \cup 1_S \\ &= \rho \cup \sigma. \end{aligned}$$

(The last equality follows from the assumption that ρ and σ are equivalences.) Now $\rho \subseteq \rho \cup \sigma$ and $\sigma \subseteq \rho \cup \sigma$; hence $\rho \circ \sigma \subseteq (\rho \cup \sigma)^2$. Hence $(\rho \circ \sigma)^n \subseteq (\rho \cup \sigma)^{2n}$ for every n in $\mathbf{N}$, and so

$$(\rho \circ \sigma)^\infty \subseteq (\rho \cup \sigma)^\infty.$$

Conversely, $\rho \subseteq \rho \circ \sigma$ and $\sigma \subseteq \rho \circ \sigma$ since ρ and σ are equivalences, and so $\rho \cup \sigma \subseteq \rho \circ \sigma$. It follows that $(\rho \cup \sigma)^\infty \subseteq (\rho \circ \sigma)^\infty$. ■

An extremely useful specialization of this result is as follows:

Corollary 5.15. *If ρ and σ are equivalences on a set S [congruences on a semigroup S] such that $\rho \circ \sigma = \sigma \circ \rho$, then $\rho \vee \sigma = \rho \circ \sigma$.*

Proof. If $\rho \circ \sigma = \sigma \circ \rho$ then

$$(\rho \circ \sigma)^2 = \rho \circ (\sigma \circ \rho) \circ \sigma = (\rho \circ \rho) \circ (\sigma \circ \sigma) = \rho \circ \sigma,$$

and more generally $(\rho \circ \sigma)^n = \rho \circ \sigma$ for every n in $\mathbf{N}$. Hence $(\rho \circ \sigma)^\infty = \rho \circ \sigma$ and the result follows from Proposition 5.14. ■

6. FREE SEMIGROUPS

If A is a non-empty set, let us denote by F_A the set of all non-empty finite words $a_1 a_2 \ldots a_m$ in the "alphabet" A. A binary operation is defined on F_A by juxtaposition:

$$(a_1 a_2 \ldots a_m)(b_1 b_2 \ldots b_n) = a_1 a_2 \ldots a_m b_1 b_2 \ldots b_n.$$

With respect to this operation F_A is a semigroup, called the *free semigroup* on A. The set A is called the *generating set* of A. By contrast with the corresponding situation in group theory (see Kurosh 1956) the generating set is uniquely determined by F_A, since $A = F_A \backslash F_A^2$. In making this statement we are tacitly identifying each element a of A with the one-letter word a in F_A. This is certainly a reasonable identification and we shall almost always want to make it. At other times we shall refer to the mapping $\alpha: A \to F_A$ that associates each element of A with the corresponding one-letter word in F_A as the *standard embedding* of A in F_A.

If $A = \{a\}$ then $F_A = \{a, a^2, a^3, \ldots\}$ and is simply the infinite monogenic semigroup $\langle a \rangle$ encountered in Section 2. If $|A| > 1$ then F_A is not commutative.

The crucial property of free semigroups is given by the following theorem:

Theorem 6.1. *Let A be a non-empty set and let S be a semigroup. If $\phi: A \to S$ is an arbitrary mapping then there exists a unique homomorphism $\psi: F_A \to S$ such that $\psi | A = \phi$.*

Proof. Define $\psi: F_A \to S$ by

$$(a_1 a_2 \ldots a_m)\psi = (a_1 \phi)(a_2 \phi) \ldots (a_m \phi).$$

Then it is evident that ψ is a homomorphism and that $\psi|A = \phi$. Moreover ψ is unique, for if $\chi: F_A \to S$ is to have the required properties then

$$(a_1a_2\dots a_m)\chi = (a_1\chi)(a_2\chi)\dots(a_m\chi) = (a_1\phi)(a_2\phi)\dots(a_m\phi) = (a_1a_2\dots a_m)\psi$$

for every $a_1a_2\dots a_m$ in F_A, and so $\chi = \psi$. ■

If S is a semigroup and A is a generating set for S, then the theorem gives us a homomorphism ψ from F_A onto S. Hence $S \simeq F_A/\ker\psi$. We can always find a generating set for S (since S itself will do) and so we deduce that every semigroup can be expressed up to isomorphism as a quotient of a free semigroup by a congruence. The expression is of course not unique.

If $A = \{a_1, a_2, \dots, a_n\}$ is finite and if the congruence ρ on F_A can be generated by a finite set

$$\mathbf{R} = \{(w_1, z_1), \dots, (w_r, z_r)\}$$

of elements (w_i, z_i) in $F_A \times F_A$, we say that F_A/ρ is *finitely presented* and often write

$$F_A/\rho = \langle a_1, a_2, \dots, a_n | w_1 = z_1, \dots, w_r = z_r\rangle;$$

we say that F_A/ρ has *generators* $a_1, \dots, a_n$ and *defining relations* $w_1 = z_1, \dots, w_r = z_r$.

7. IDEALS AND REES CONGRUENCES

The reader will probably be familiar with the way in which homomorphisms in ring theory are closely connected with ideals. The corresponding link in semigroup theory is, as we shall see, less satisfactory, but there is one type of homomorphism, called here a *Rees homomorphism*, that does correspond very closely to an ideal of the semigroup. (See Rees (1940).)

First, if I is an ideal of a semigroup S, then

$$\rho_I = (I \times I) \cup 1_S$$

is a congruence on S. To see this, notice that $(x, y) \in \rho_I$ if and only if either $x = y$ or x and y both belong to I. It is then easy to verify that ρ_I is reflexive, symmetric, transitive and compatible. The quotient semigroup is

$$S/\rho_I = \{I\} \cup \{\{x\}: x \in S\backslash I\},$$

which it is convenient to regard as consisting of I together with the members of $S\backslash I$. In S/ρ_I the product of two elements of $S\backslash I$ is the same as their product in S if this lies in $S\backslash I$; otherwise the product is I. Since the element I of S/ρ_I

is the zero element of the semigroup, another useful way of thinking of S/ρ_I is as $(S\backslash I) \cup \{0\}$, where all products not falling in $S\backslash I$ are zero.

We shall call a congruence of this type a *Rees congruence*. If $\phi: S \to T$ is a homomorphism we shall say that ϕ is a *Rees homomorphism* if the congruence $\ker \phi$ is a Rees congruence. We shall normally write S/I rather than S/ρ_I; also, when we talk of the *kernel* of a Rees homomorphism we shall mean the ideal I rather than the congruence ρ_I.

It is important to note that not every semigroup homomorphism is of this type. Groups are certainly semigroups in good standing, but a homomorphism $\phi: G \to H$ between two non-trivial groups cannot possibly be a Rees homomorphism, since H has no zero element.

The next result requires only routine verification, and the proof is omitted:

Proposition 7.1. *Let I be an ideal of a semigroup S. Let $\mathscr{A}$ be the set of ideals of S containing I and let $\mathscr{B}$ be the set of ideals of S/I. Then the mapping θ: $J \mapsto J/I$ $(J \in \mathscr{A})$ is an inclusion-preserving bijection from $\mathscr{A}$ onto $\mathscr{B}$.* ■

8*. LATTICES OF EQUIVALENCES AND CONGRUENCES

We have seen that while the study of lattices of equivalences and congruences is in general hampered by the rather complicated way (Proposition 5.14) in which the join of two equivalences is formed, a useful simplification takes place (Corollary 5.15) if the equivalences commute. For this reason it is of interest to record the following result:

Proposition 8.1. *If G is a group, then $\rho \circ \sigma = \sigma \circ \rho$ for any two congruences ρ, σ on G.*

Proof. Let $(a, b) \in \rho \circ \sigma$; thus there exists c in G such that $(a, c) \in \rho$, $(c, b) \in \sigma$. Then

$$a = cc^{-1}a \equiv bc^{-1}a \pmod{\sigma}$$

and

$$bc^{-1}a \equiv bc^{-1}c = b \pmod{\rho},$$

giving $(a, b) \in \sigma \circ \rho$. Thus $\rho \circ \sigma \subseteq \sigma \circ \rho$, and similarly $\sigma \circ \rho \subseteq \rho \circ \sigma$. ■

This result can of course be proved in a more traditionally group-theoretic way. The connection between the general approach *via* congruences and the traditional group-theoretic approch *via* normal subgroups is outlined in the following proposition, whose proof is left to the reader:

Proposition 8.2. *Let G be a group with identity element e.*
(i) *If N is a normal subgroup of G, then*

$$\rho_N = \{(a, b) \in G \times G : ab^{-1} \in N\}$$

is a congruence on G. For each g in G the ρ_N-class $g\rho_N$ coincides with the coset Ng.
(ii) *If ρ is a congruence on G then $N = e\rho$ is a normal subgroup of G and $\rho = \rho_N$.*
(iii) *If M, N are normal subgroups of G, then*

$$\rho_M \cap \rho_N = \rho_{M \cap N}, \qquad \rho_M \circ \rho_N = \rho_{MN}. \blacksquare$$

The commuting of congruences in a group can therefore be derived as a consequence of the well-known group theoretic result—see, e.g., Hall (1959)—that normal subgroups commute.

A lattice $(L, \leqslant, \wedge, \vee)$ is called *modular* if $(\forall a, b, c \in L)$

$$a \leqslant c \Rightarrow (a \vee b) \wedge c = a \vee (b \wedge c). \tag{8.3}$$

Notice that in any lattice whatever, if $a \leqslant c$, then

$$a \leqslant a \vee b \quad \text{and} \quad a \leqslant c,$$

giving $a \leqslant (a \vee b) \wedge c$; also

$$b \wedge c \leqslant b \leqslant a \vee b \quad \text{and} \quad b \wedge c \leqslant c,$$

giving $b \wedge c \leqslant (a \vee b) \wedge c$; hence

$$a \vee (b \wedge c) \leqslant (a \vee b) \wedge c.$$

We therefore establish that a lattice is modular if we merely prove that $(\forall a, b, c \in L)$

$$a \leqslant c \Rightarrow (a \vee b) \wedge c \leqslant a \vee (b \wedge c). \tag{8.4}$$

Proposition 8.5. *Let $\mathscr{K}$ be a sublattice of the lattice $(\mathscr{C}(S), \subseteq, \cap, \vee)$ of congruences of a semigroup S, and suppose that $\rho \circ \sigma = \sigma \circ \rho$ for all ρ, σ in $\mathscr{K}$. Then $\mathscr{K}$ is a modular lattice.*

Proof. Let $\alpha, \beta, \gamma \in \mathscr{K}$ be such that $\alpha \subseteq \gamma$, and let $(x, y) \in (\alpha \vee \beta) \cap \gamma = (\alpha \circ \beta) \cap \gamma$. Then $(x, y) \in \gamma$ and there exists z in S such that $(x, z) \in \alpha$, $(z, y) \in \beta$. Since $\alpha \subseteq \gamma$ and since γ is a congruence, we have that $(z, x) \in \gamma$; hence (since $(x, y) \in \gamma$) it follows that $(z, y) \in \gamma$. We thus have $(x, z) \in \alpha$, $(z, y) \in \beta \cap \gamma$, and so $(x, y) \in \alpha \circ (\beta \cap \gamma) = \alpha \vee (\beta \cap \gamma)$. We have shown that $(\alpha \vee \beta) \cap \gamma \subseteq \alpha \vee (\beta \cap \gamma)$, and so $\mathscr{K}$ is modular as required. ■

Corollary 8.6. *The lattice of congruences on a group is modular.* ■

This result breaks down even for relatively small semigroups. In Exercise 13 below an example is given of a semigroup with four elements having a non-modular lattice of congruences. Indeed the lattice of congruences of that semigroup is not even *semimodular*, a concept that we now explain.

First, if a and b are elements of a lattice L, we say that *a covers b* in L if $a > b$ and if there is no x in L such that $a > x > b$. We write $a \succ b$. A lattice L is said to be (*upper*) *semimodular* if $(\forall a, b \in L)$

$$a \succ a \wedge b \text{ and } b \succ a \wedge b \;\Rightarrow\; a \vee b \succ a \text{ and } a \vee b \succ b.$$

Proposition 8.7. *A modular lattice is semimodular.*

Proof. If $a \succ a \wedge b$ and $b \succ a \wedge b$ in a modular lattice L, suppose that $x \in L$ is such that $a \leqslant x < a \vee b$. Then

$$a \wedge b \leqslant x \wedge b \leqslant (a \vee b) \wedge b = b,$$

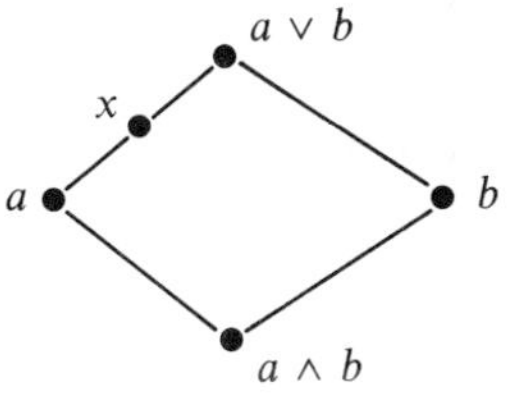

and $x \wedge b \neq b$, for otherwise $x \geqslant a, x \geqslant b$ and so $x \geqslant a \vee b$—which is not so. Hence $x \wedge b = a \wedge b$, since $b \succ a \wedge b$. It now follows that

$$\begin{aligned} x = (a \vee b) \wedge x &= a \vee (b \wedge x) \quad (\text{since } a \leqslant x) \\ &= a \vee (a \wedge b) = a. \end{aligned}$$

Thus $a \vee b \succ a$, and similarly $a \vee b \succ b$. ■

For further lattice-theoretic results, see, e.g., Birkhoff (1948). We shall here pursue only those results that have a direct bearing on the aspects of semigroup theory to be investigated. The concept of semimodularity is important for us because of the following result (from Dubreil-Jacotin *et al.* (1953)):

Proposition 8.8. *The lattice* $(\mathscr{E}(X), \subseteq, \cap, \vee)$ *of equivalences on a set* X *is semimodular.*

Proof. If ρ and σ are equivalences on X suppose that the $(\rho \cap \sigma)$-classes are

$$A_1, A_2, A_3, \ldots .$$

If $\rho \succ \rho \cap \sigma$ then there is exactly one ρ-class that is the union of two $(\rho \cap \sigma)$-classes; the other ρ-classes coincide with $(\rho \cap \sigma)$-classes. There is no loss of generality in supposing that the ρ-classes are

$$A_1 \cup A_2, A_3, \ldots.$$

The same argument now applies to σ, but $\sigma \neq \rho$ and so the two $(\rho \cap \sigma)$-classes that come together in σ are not the same two as come together in ρ. Essentially two possibilities arise: either the σ-classes are

$$A_1, A_2, A_3 \cup A_4, A_5, \ldots$$

or they are

$$A_1, A_2 \cup A_3, A_4, \ldots.$$

In the first case the $(\rho \vee \sigma)$-classes are

$$A_1 \cup A_2, A_3 \cup A_4, A_5, \ldots.$$

while in the second case they are

$$A_1 \cup A_2 \cup A_3, A_4, A_5, \ldots.$$

In either case $\rho \vee \sigma \succ \rho$ and $\rho \vee \sigma \succ \sigma$. ■

If $|X| \geqslant 5$ the lattice $\mathscr{E}(X)$ is not modular—see Exercise 14.

If $L_1, L_2, \ldots, L_n$ are lattices, the Cartesian product $L_1 \times L_2 \times \ldots \times L_n$ of $L_1, \ldots, L_n$ becomes a partially ordered set if we define

$$(x_1, x_2, \ldots, x_n) \leqslant (y_1, y_2, \ldots, y_n)$$

if and only if $x_1 \leqslant y_1, x_2 \leqslant y_2, \ldots, x_n \leqslant y_n$. It is in fact a lattice, with

$$(x_1, x_2, \ldots, x_n) \wedge (y_1, y_2, \ldots, y_n) = (x_1 \wedge y_1, x_2 \wedge y_2, \ldots, x_n \wedge y_n),$$

$$(x_1, x_2, \ldots, x_n) \vee (y_1, y_2, \ldots, y_n) = (x_1 \vee y_1, x_2 \vee y_2, \ldots, x_n \vee y_n).$$

We call it the *direct product* of $L_1, L_2, \ldots, L_n$. It is easy to see that

$$(a_1, a_2, \ldots, a_n) \succ (b_1, b_2, \ldots, b_n)$$

in $L_1 \times L_2 \times \ldots \times L_n$ if and only if $a_i \succ b_i$ for some i in $\{1, 2, \ldots, n\}$ and $a_j = b_j$ for all $j \neq i$. It is now not hard to deduce

Proposition 8.9. *The direct product $L_1 \times L_2 \times \ldots \times L_n$ of semimodular lattices $L_1, L_2, \ldots, L_n$ is semimodular.*

Proof. Let $x \wedge y = (z_1, \ldots, z_n)$, and suppose that $x \succ x \wedge y$, $y \succ x \wedge y$. Then $x = (x_1, \ldots, x_n)$, where $x_i \succ z_i$ in L_i for some i in $\{1, \ldots, n\}$, and $x_k = z_k$

if $k \neq i$. Equally, $y = (y_1, \ldots, y_n)$, where $y_j \succ z_j$ in L_j for some j in $\{1, \ldots, n\}$, and $y_k = z_k$ if $k \neq j$.

If $i \neq j$ then $x \vee y = (t_1, \ldots, t_n)$, where

$$t_i = x_i, t_j = y_j \quad \text{and} \quad t_k = z_k \ (k \notin \{i, j\});$$

in this case it is immediate that $x \vee y \succ x$, $x \vee y \succ y$. If $i = j$, then $x \vee y = (u_1, \ldots, u_n)$ where

$$u_i = x_i \vee y_i, \qquad u_k = z_k \qquad (k \neq i).$$

Since L_i is semimodular we have $u_i \succ x_i$, $u_i \succ y_i$ in L_i. Hence $x \vee y \succ x$ and $x \vee y \succ y$ in L in this case also. ∎

EXERCISES

1. An element e of a semigroup S is called a *left identity* if $ex = x$ for every x in S, and a *right identity* if $xe = x$ for every x in S. An element z of S is called a *left zero* if $zx = z$ for every x in S, and a *right zero* if $xz = z$ for every x in S.
   (i) If S has a left identity e and a right identity f, show that $e = f$ and that e is the unique (two-sided) identity of S.
   (ii) If S has a left zero z and a right zero u, show that $z = u$ and that z is the unique (two-sided) zero of S.
   (iii) Give an example of a semigroup having two (at least) distinct left identity elements and two (at least) distinct right zero elements.

2. Show that the definitions (1.3) and (1.4) of a group are equivalent.

3. (Baer and Levi 1932). Let X be a countably infinite set and let S be the set of one–one mappings $\alpha: X \to X$ with the property that $X \backslash X\alpha$ is infinite.
   (i) Show that S is a subsemigroup of $\mathscr{T}(X)$.
   (ii) Show that for any α in S there is a one–one correspondence between $X \backslash X\alpha$ and $X\alpha \backslash X\alpha^2$.
   (iii) Deduce that S contains no idempotent elements.

4. Let $\phi: S \to T$ be a homomorphism, where S, T are semigroups.
   (i) Show that the image under ϕ of an idempotent of S is an idempotent of T.
   (ii) Show that the image under ϕ of a subsemigroup of S is a subsemigroup of T.
   (iii) Show that if ϕ is onto then the image under ϕ of a right ideal [left ideal, ideal] of S is a right ideal [left ideal, ideal] of T.
   (iv) Show that if ϕ is onto then the image under ϕ of the identity [zero] of S is the identity [zero] of T.
   (v) Show that in (iii) and (iv) the hypothesis that ϕ is onto cannot be removed.

5. Show that if S is a semigroup without identity then the (unextended) right regular representation $\rho: S \to \mathscr{T}(S)$ may or may not be faithful. More specifically, show that it is faithful if S is a right zero semigroup but is not faithful if S is a left zero semigroup.

6. A semigroup may be isomorphic to a semigroup of mappings in more than one way. For example, for the semigroup $S = \{e, a, x, y\}$ with Cayley table

	e	a	x	y
e	e	a	x	y
a	a	e	x	y
x	x	y	x	y
y	y	x	x	y

show that $\phi: S \to \mathscr{T}(\{1, 2\})$, defined by

$$e\phi = \begin{pmatrix} 1 & 2 \\ 1 & 2 \end{pmatrix}, \quad a\phi = \begin{pmatrix} 1 & 2 \\ 2 & 1 \end{pmatrix}, \quad x\phi = \begin{pmatrix} 1 & 2 \\ 1 & 1 \end{pmatrix}, \quad y\phi = \begin{pmatrix} 1 & 2 \\ 2 & 2 \end{pmatrix},$$

is a monomorphism. Show also that $\psi: S \to \mathscr{T}(\{e, a, x, y\})$, defined by

$$e\psi = \begin{pmatrix} e & a & x & y \\ e & a & x & y \end{pmatrix}, \quad a\psi = \begin{pmatrix} e & a & x & y \\ a & e & y & x \end{pmatrix},$$

$$x\psi = \begin{pmatrix} e & a & x & y \\ x & x & x & x \end{pmatrix}, \quad y\psi = \begin{pmatrix} e & a & x & y \\ y & y & y & y \end{pmatrix},$$

is a monomorphism. (Notice that ψ is the right regular representation of S.)

7. Let $S = \langle a \rangle = M(m, r)$, where $m > 1$. Show that $S \backslash S^2 = \{a\}$. Deduce that (by contrast with group theory) the generator a of a finite monogenic semigroup S is uniquely determined by S unless S is a group.

9. For any set X, show that

$$|\mathscr{B}(X)| = 2^{n^2}, \qquad |\mathscr{PT}(X)| = \sum_{r=0}^{n} \binom{n}{r} n^r.$$

9. For any set, X, show that

$$\mathscr{E}(X) \cap \mathscr{PT}(X) = \{1_X\}.$$

10. Let S be a semigroup, and let R, C^l, C^r, C and T denote respectively the sets of reflexive, left compatible, right compatible, compatible and transitive relations in S.

(i) Show that

$$C \cap R \subseteq (C^l \cup C^r) \cap R, \qquad (C^l \cap C^r) \cap T \subseteq C \cap T.$$

(ii) Deduce that

$$C \cap (R \cap T) = (C^l \cup C^r) \cap (R \cap T),$$

i.e. that a reflexive and transitive relation is compatible if and only if it is left and right compatible.

(iii) Show that each of the inclusions in (i) can be strict. [Hint. Let $S = \{1, 2\} \times \{1, 2\}$ be a rectangular band. Write $x = (1, 1)$, $y = (1, 2)$, $z = (2, 1)$, $t = (2, 2)$ and consider the relations

$$\mathbf{R} = 1_S \cup \{(x, y), (x, z), (y, t), (z, t)\}, \qquad \mathbf{S} = \{(x, y)\}$$

on S.]

11. Let I, J be ideals of a semigroup S such that $I \subseteq J$. Show that

$$(S/J) \simeq (S/I)/(J/I).$$

12. Let I, J be ideals of a semigroup S. Show that $I \cap J$, $I \cup J$ are ideals of S. (Note that $IJ \subseteq I \cap J$ and so $I \cap J \neq \varnothing$.) Show also that

$$(I \cup J)/J \simeq I/(I \cap J).$$

13. Let $S = \{e, a, f, b\}$ be the semigroup with multiplication table

	e	a	f	b
e	e	a	f	b
a	a	e	b	f
f	f	b	f	b
b	b	f	b	f

Verify that the congruences on S are as follows:

1_S, with classes $\{e\}, \{a\}, \{f\}, \{b\}$;

σ, with classes $\{e, f\}, \{a, b\}$;

μ, with classes $\{e, a\}, \{f, b\}$;

ν, with classes $\{e\}, \{a\}, \{f, b\}$;

$S \times S$, with class $\{e, a, f, b\}$.

Draw a Hasse diagram for the lattice $(\mathscr{C}(S), \subseteq, \cap, \vee)$, and show that $\mathscr{C}(S)$ is not semimodular.

14. In Proposition 8.8 it is shown that the lattice $\mathscr{E}(X)$ of equivalences on a set X is semimodular. If $X = \{x_1, x_2, x_3, \ldots\}$ with $|X| \geqslant 5$, let α, β, γ be equivalences on X with classes as follows:

α: $\{x_1, x_2\}, \{x_3, x_4\}, \{x_5\}, \{x_6\}, \ldots$;

β: $\{x_1, x_3\}, \{x_2, x_5\}, \{x_4\}, \{x_6\}, \ldots$;

γ: $\{x_1, x_2\}, \{x_3, x_4, x_5\}, \{x_6\}, \ldots$.

Show that $\alpha \subseteq \gamma$ but $(\alpha \vee \beta) \cap \gamma \neq \alpha \vee (\beta \cap \gamma)$, and deduce that $\mathscr{E}(X)$ is not modular.

Chapter II

Green's Equivalences

INTRODUCTION

The notion of *ideal* mentioned in the last chapter leads naturally to the consideration of certain equivalence relations on a semigroup. These equivalences, first studied by J. A. Green (1951), have played a fundamental role in the development of semigroup theory. They are concerned with mutual divisibility of various kinds and all of them reduce to the universal equivalence in a group.

Green's equivalences are especially significant in the study of regular semigroups. The chapter ends with some general results concerning such semigroups that will be of use in later chapters.

1. THE EQUIVALENCES $\mathscr{L}$, $\mathscr{R}$, $\mathscr{H}$, $\mathscr{J}$ AND $\mathscr{D}$

If a is an element of a semigroup S, the smallest left ideal containing a is $Sa \cup \{a\}$, which we may conveniently write as S^1a, and which we shall call the *principal left ideal generated by a.* An equivalence relation $\mathscr{L}$ on S is then defined by the rule that $a \mathscr{L} b$ if and only if a and b generate the same principal left ideal, i.e. if and only if $S^1a = S^1b$.

Similarly, we define $\mathscr{R}$ by the rule that $a \mathscr{R} b$ if and only if a and b generate the same principal right ideal, i.e. if and only if $aS^1 = bS^1$.

An alternative characterization, making the "mutual divisibility" aspect of these equivalences more explicit, is given in the following lemma:

Lemma 1.1. *Let a, b be elements of a semigroup S. Then $a \mathscr{L} b$ if and only if there exist x, y in S^1 such that $xa = b$, $yb = a$. Also, $a \mathscr{R} b$ if and only if there exist u, v in S^1 such that $au = b$, $bv = a$.*■

Another immediate property of $\mathscr{L}$ and $\mathscr{R}$ is as follows (see Section I.5):

Lemma 1.2. *$\mathscr{L}$ is a right congruence and $\mathscr{R}$ is a left congruence.*■

We have seen (Section I.4) that the intersection of two equivalences is again an equivalence. Since the intersection of $\mathscr{L}$ and $\mathscr{R}$ is of great importance in the development of the theory, we reserve for it the letter $\mathscr{H}$. The join $\mathscr{L} \vee \mathscr{R}$ is also of great importance, and is denoted by $\mathscr{D}$. As we saw in Section I.4, the join of two equivalences can be rather hard to describe, but we are saved from these difficulties here by the following fortunate occurrence:

Proposition 1.3. *The relations $\mathscr{L}$ and $\mathscr{R}$ commute.*

Proof. If $(a, b) \in \mathscr{L} \circ \mathscr{R}$ then there exists $c \in S$ such that $a\mathscr{L}c$, $c\mathscr{R}b$. That is, there exist $x, y, u, v \in S^1$ such that

$$xa = c, \qquad cu = b;$$

$$yc = a, \qquad bv = c.$$

If we write d for the element ycu of S, then

$$au = ycu = d,$$

and

$$dv = ycuv = ybv = yc = a,$$

from which it follows that $a \mathscr{R} d$. Also,

$$yb = ycu = d$$

and

$$xd = xycu = xau = cu = b,$$

and so $d \mathscr{L} b$. We deduce that $(a, b) \in \mathscr{R} \circ \mathscr{L}$. Thus $\mathscr{L} \circ \mathscr{R} \subseteq \mathscr{R} \circ \mathscr{L}$, and we can equally well obtain the reverse inclusion. ■

By Proposition I.8.1 it now follows that

$$\mathscr{L} \circ \mathscr{R} = \mathscr{R} \circ \mathscr{L} = \mathscr{L} \vee \mathscr{R} = \mathscr{D}$$

—which makes $\mathscr{D}$ much easier to handle than one might have expected.

Our final equivalence is the two-sided analogue of $\mathscr{L}$ and $\mathscr{R}$. The principal two-sided ideal generated by an element a of S is S^1aS^1 $(= SaS \cup aS \cup Sa \cup \{a\})$, and we write $a \mathscr{J} b$ if $S^1aS^1 = S^1bS^1$, i.e. if there exist x, y, u, v in S^1 for which

$$xay = b, \qquad ubv = a.$$

It is immediate that $\mathscr{L} \subseteq \mathscr{J}$, $\mathscr{R} \subseteq \mathscr{J}$; hence, since $\mathscr{D}$ is the smallest equivalence containing $\mathscr{L}$ and $\mathscr{R}$, it follows that

$$\mathscr{D} \subseteq \mathscr{J}. \tag{1.4}$$

An example showing that this inclusion may be strict was given by Green (1951); a more striking example due to Andersen (1952) is given as Exercise 1 below. In certain classes of semigroups we do have equality. Certainly in commutative semigroups we have

$$\mathscr{H} = \mathscr{L} = \mathscr{R} = \mathscr{D} = \mathscr{J}.$$

Less trivially, however, we have a result which implies in particular that $\mathscr{D} = \mathscr{J}$ in any finite semigroup:

Proposition 1.5. *If S is a periodic semigroup, then* $\mathscr{D} = \mathscr{J}$.

Proof. Suppose that a, b in S are such that $a \mathscr{J} b$. Thus there exist x, y, u, v in S^1 such that

$$xay = b, \qquad ubv = a. \tag{1.6}$$

To prove the required result we require to show that there exists c in S for which $a \mathscr{L} c$, $c \mathscr{R} b$. Now, it follows easily from the equations (1.6) that

$$a = (ux)a(yv) = (ux)^2a(yv)^2 = (ux)^3a(yv)^3 = \dots,$$

$$b = (xu)b(vy) = (xu)^2b(vy)^2 = (xu)^3b(vy)^3 = \dots.$$

Since S is periodic we can by Proposition I.2.7 find an m for which $(ux)^m$ is idempotent. Then if we let $c = xa$, we have

$$a = (ux)^m a(yv)^m = (ux)^m(ux)^m a(yv)^m = (ux)^m a = (ux)^{m-1}uc,$$

and so $a \mathscr{L} c$. Also, $cy = xay = b$, and if we choose n so that $(vy)^n$ is idempotent, we have

$$\begin{aligned} c = xa &= x(ux)^{n+1}a(yv)^{n+1} = (xu)^{n+1}xay(vy)^n v \\ &= (xu)^{n+1}b(vy)^{2n}v = (xu)^{n+1}b(vy)^{n+1}(vy)^{n-1}v \\ &= b(vy)^{n-1}v. \end{aligned}$$

Hence $c \mathscr{R} b$ as required.∎

Some preliminaries are necessary before we can describe another important class of semigroups in which $\mathscr{D} = \mathscr{J}$. First, since the Green equivalences occur so often, it is worth making an exception to our general notational convention (Section I.4) about equivalence classes: the $\mathscr{L}$-class ($\mathscr{R}$-class, $\mathscr{H}$-class, $\mathscr{D}$-class, $\mathscr{J}$-class) containing the element a will be written $L_a(R_a, H_a, D_a, J_a)$. It is easy to see that in a semigroup with zero,

$$L_0 = R_0 = H_0 = D_0 = J_0 = \{0\}.$$

Next, notice that $\mathscr{L}$, $\mathscr{R}$ and $\mathscr{J}$ are defined in terms of principal ideals.

Thus the inclusion order among the principal ideals induces a corresponding order among the equivalence classes as follows:

$$L_a \leqslant L_b \quad \text{if} \quad S^1 a \subseteq S^1 b, \tag{1.7}$$

$$R_a \leqslant R_b \quad \text{if} \quad aS^1 \subseteq bS^1, \tag{1.8}$$

and

$$J_a \leqslant J_b \quad \text{if} \quad S^1 aS^1 \subseteq S^1 bS^1. \tag{1.9}$$

We thus have a partial order relation on each of the sets $S/\mathscr{L}$, $S/\mathscr{R}$, $S/\mathscr{J}$. Notice that

$$L_{xa} \leqslant L_a, \qquad R_{ax} \leqslant R_a, \qquad J_{xay} \leqslant J_a \tag{1.10}$$

for every a in S and for every x, y in S^1. Notice also that either of $L_a \leqslant L_b$ and $R_a \leqslant R_b$ implies that $J_a \leqslant J_b$.

We shall say that a semigroup S *satisfies the condition* $\min_L$, $\min_R$ or $\min_J$ according as the partially ordered set $S/\mathscr{L}$, $S/\mathscr{R}$, $S/\mathscr{J}$ satisfies the minimal condition (see Section I.3). These conditions are of course equivalent respectively to the minimal conditions on principal left ideals, on principal right ideals and on principal two-sided ideals, and are weaker than the corresponding conditions on (not necessarily principal) ideals. See (Clifford and Preston 1967, Section 6.6) for an exploration of the relationships among the various minimal conditions. Our immediate purpose in introducing these definitions is so that we can establish

Proposition 1.11. *If S is a semigroup satisfying* $\min_L$ *and* $\min_R$ *then* $\mathscr{D} = \mathscr{J}$.

Proof. If S satisfies $\min_L$ and $\min_R$ then so does S^1, for S^1 has exactly the same set of principal left and right ideals as S except for S^1 itself. We may thus assume that S has an identity element.

If, then, we have $a \mathscr{J} b$, we may assert that there exist p, q, r, s in S such that

$$paq = b, \qquad rbs = a.$$

It follows that the set

$$X = \{x \in S : (\exists y \in S) xay = b\}$$

is non-empty, and hence so also is the subset

$$\Lambda = \{L_x : x \in X\}$$

of $S/\mathscr{L}$. The condition $\min_L$ allows us to select a minimal element L_u in Λ; thus $uav = b$ for some element v of S.

Now $uruavsv = b$ and so $L_{uru} \in \Lambda$. Since by (1.10) we have $L_{uru} \leqslant L_u$, it follows by the minimality of L_u in Λ that $L_{uru} = L_u$. Hence we have

$$L_u = L_{uru} \leqslant L_{ru} \leqslant L_u$$

and so $ru \mathscr{L} u$. By Lemma 1.2 it follows that $ruav \mathscr{L} uav$, i.e. that $rb \mathscr{L} b$.

A similar argument establishes that $bs \mathscr{R} b$. Hence by Lemma 1.2 it follows that $rbs \mathscr{R} rb$, i.e. that $a \mathscr{R} rb$. Since $a \mathscr{R} rb$ and $rb \mathscr{L} b$, we conclude that $a \mathscr{D} b$ as required.■

2. THE STRUCTURE OF $\mathscr{D}$-CLASSES

Each $\mathscr{D}$-class in a semigroup is a union of $\mathscr{L}$-classes and also a union of $\mathscr{R}$-classes. The intersection of an $\mathscr{L}$-class and an $\mathscr{R}$-class is either empty or is an $\mathscr{H}$-class. In fact, by the very definition of $\mathscr{D}$,

$$a \mathscr{D} b \Leftrightarrow R_a \cap L_b \neq \varnothing \Leftrightarrow L_a \cap R_b \neq \varnothing.$$

Hence it is convenient to visualize a $\mathscr{D}$-class as what Clifford and Preston (1961) have called an "eggbox" in which each row represents an $\mathscr{R}$-class, each column an $\mathscr{L}$-class and each cell an $\mathscr{H}$-class. (It is of course possible for the "eggbox" to contain a single row or a single column of cells, or even for it to contain only one cell. Also, it may well be an infinite eggbox.)

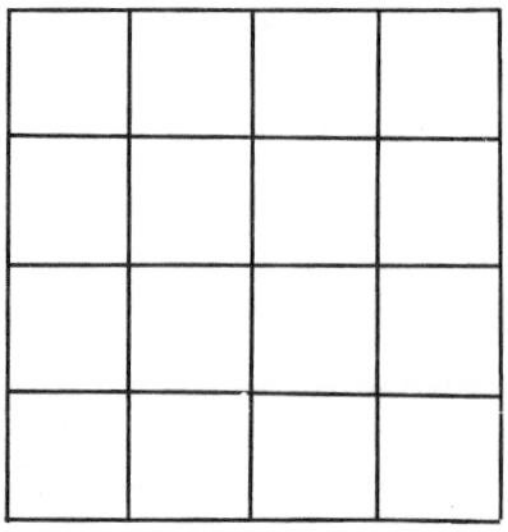

If D is an arbitrary $\mathscr{D}$-class in a semigroup S and if $a, b \in D$ are such that $a \mathscr{R} b$ (so that a and b lie in the same row of the eggbox), then by definition of $\mathscr{R}$ there exist $s, s' \in S^1$ such that

$$as = b, \qquad bs' = a.$$

The right translation $\rho_s\colon S \to S$ thus maps a to b. In fact it maps L_a into L_b; for if $x \in L_a$ then $xs \mathscr{L} as$ by Lemma 1.2; that is, $x\rho_s \in L_{as} = L_b$. Now it is just as easy to show that $\rho_{s'}$ maps L_b into L_a, and if we investigate the composition $\rho_s\rho_{s'}\colon L_a \to L_a$ we find that for any $x = ua \in L_a$,

$$x\rho_s\rho_{s'} = uass' = ubs' = ua = x.$$

Thus $\rho_s\rho_{s'}$ is the identity map on L_a. We may similarly show that $\rho_{s'}\rho_s$ is the identity map on L_b, and the conclusion is then that $\rho_s|L_a$ and $\rho_{s'}|L_b$ are mutually inverse bijections from L_a onto L_b and L_b onto L_a respectively.

We can say even more about these maps: if $x \in L_a$ then the element $y = x\rho_s$ of L_b has the property that

$$y = xs, \qquad ys' = x.$$

Thus $y \mathscr{R} x$ and so the map ρ_s is *$\mathscr{R}$-class preserving*. It maps each $\mathscr{H}$-class in L_a in a one–one manner onto the corresponding (i.e. $\mathscr{R}$-equivalent) $\mathscr{H}$-class in L_b. Similar remarks apply to $\rho_{s'}$.

To summarize, we have proved:

Lemma 2.1. (Green's Lemma). *Let a, b be $\mathscr{R}$-equivalent elements in a semigroup S and let s, s' in S^1 be such that*

$$as = b, \qquad bs' = a.$$

Then the right translations $\rho_s|L_a$, $\rho_{s'}|L_b$ are mutually inverse $\mathscr{R}$-class preserving bijections from L_a onto L_b and from L_b onto L_a respectively.■

The left–right dual is proved in an analogous way:

Lemma 2.2. (Green's Lemma). *Let a, b be $\mathscr{L}$-equivalent elements in a semigroup S and let t, t' in S^1 be such that*

$$ta = b, \qquad t'b = a.$$

Then the left translations $\lambda_t|R_a$, $\lambda_{t'}|R_b$ are mutually inverse $\mathscr{L}$-class preserving bijections from R_a onto R_b and from R_b onto R_a respectively.■

The combined effect of these two lemmas is as follows:

Lemma 2.3. *If a, b are $\mathscr{D}$-equivalent elements in a semigroup S, then $|H_a| = |H_b|$.*

Proof. If c is such that $a \mathscr{R} c$, $c \mathscr{L} b$, and if s, t, s', t' in S^1 are such that

$$as = c, \qquad cs' = a, \qquad tc = b, \qquad t'b = c,$$

then by the preceding lemmas $\rho_s|H_a$ is a bijection onto H_c and $\lambda_t|H_c$ is a bijection onto H_b. Thus $\rho_s\lambda_t\colon x \mapsto txs$ is a bijection from H_a onto H_b (with inverse $\lambda_{t'}\rho_{s'}\colon y \mapsto t'ys'$), and it follows that $|H_a| = |H_b|$.■

A more striking consequence of Lemmas 2.1 and 2.2 is concerned with the multiplicative properties of an $\mathscr{H}$-class. A preliminary lemma, which is just a specialization of Lemmas 2.1 and 2.2, is useful:

Lemma 2.4. If $x, y \in S$ *are such that* $xy \in H_x$ *then* $\rho_y | H_x$ *is a bijection of* H_x *onto itself. If* $xy \in H_y$ *then* $\lambda_x | H_y$ *is a bijection of* H_y *onto itself.*■

Now we have

Theorem 2.5. (Green's Theorem). *If* H *is an* $\mathscr{H}$*-class in a semigroup* S *then either* $H^2 \cap H = \varnothing$ *or* $H^2 = H$ *and* H *is a subgroup of* S.

Proof. Suppose that $H^2 \cap H \neq \varnothing$, so that there exist a, b in H such that $ab \in H$. By the lemma, ρ_b and λ_a are bijections of H onto itself. Hence $hb \in H$ and $ah \in H$ for every h in H. Again by the lemma it follows that λ_h and ρ_h are bijections of H onto itself. Hence $Hh = hH = H$ for every h in H. Hence certainly $H^2 = H$, and moreover by (I.1.8) it follows that H is a subgroup of S.■

We now immediately deduce

Corollary 2.6. *If* e *is an idempotent in a semigroup* S, *then* H_e *is a subgroup of* S. *No* $\mathscr{H}$*-class in* S *can contain more than one idempotent.*■

3. REGULAR $\mathscr{D}$-CLASSES

The results in this section are due to Miller and Clifford (1956).

An element a of semigroup S is called *regular* if there exists x in S such that $axa = a$. The semigroup S is called *regular* if all its elements are regular. Groups are of course regular semigroups, but we shall see that the class of regular semigroups is vastly more extensive than the class of groups. The concept of regularity was introduced by von Neumann (1936) in ring theory, where it has played an important role.

Our first observation is that regularity is a property of $\mathscr{D}$-classes rather than of elements:

Proposition 3.1. *If* a *is a regular element of a semigroup* S *then every element of* D_a *is regular.*

Proof. If $b \in D_a$ then $a \mathscr{L} c, c \mathscr{R} b$ for some c in S, and so there exist u, v, z, t in S^1 such that

$$ua = c, \qquad vc = a, \qquad cz = b, \qquad bt = c.$$

If x is such that $axa = a$, then

$$b(txv)b = cxvcz = uaxaz = uaz = cz = b,$$

and so b is a regular element.■

If, then, D is a $\mathscr{D}$-class then either every element of D is regular or no element of D is regular; we call the $\mathscr{D}$-class *regular* if all its elements are regular, and *irregular* otherwise. This dichotomy does not in general apply to $\mathscr{J}$-classes, and of course a single semigroup may contain both regular and irregular $\mathscr{D}$-classes. (See Exercise V.22.)

Since idempotents are regular ($eee = e$) it follows that a $\mathscr{D}$-class containing an idempotent is regular. Conversely, we can show that a regular $\mathscr{D}$-class must contain at least one idempotent. Indeed we can show more:

Proposition 3.2. *In a regular $\mathscr{D}$-class each $\mathscr{L}$-class and each $\mathscr{R}$-class contains at least one idempotent.*

Proof. If a is a member of a regular $\mathscr{D}$-class D in a semigroup S, and if $x \in S$ is such that $axa = a$, then xa is idempotent and is such that $a \mathscr{L} xa$. Similarly, ax is idempotent and $a \mathscr{R} ax$.■

Proposition 3.3. *Every idempotent e in a semigroup S is a left identity for R_e and a right identity for L_e.*

Proof. If $a \in R_e$ then $a = ex$ for some x in S^1 and so $ea = e(ex) = e^2x = a$. Similarly $be = b$ for every b in L_e.■

An idea of great importance in semigroup theory is that of an *inverse* of an element. If a is an element of a semigroup S we say that a' is an *inverse* of a if

$$aa'a = a, \qquad a'aa' = a'. \tag{3.4}$$

Notice that an element with an inverse is necessarily regular. Less obviously, every regular element has an inverse; for if $axa = a$ we need only define $a' = xax$ and verify that equations (3.4) are satisfied.

An element a may well have more than one inverse: in a *rectangular band* (Section I.1) every element is an inverse of every other element. This example shows also that the idea of inverse involved here is more general than the group-theoretic notion. (Of course a group inverse *is* an inverse in the sense of (3.4).)

The eggbox picture of a $\mathscr{D}$-class is exceedingly useful in the location of inverses. First, notice that if a is an element in a regular $\mathscr{D}$-class D then any inverse a' of a must also lie in D; for $a \mathscr{R} aa'$ (as in the proof of Proposition 3.2), and equally trivially $aa' \mathscr{L} a'$. We have proved part of

Theorem 3.5. *Let a be an element of a regular $\mathscr{D}$-class D in a semigroup S.*

(i) *If a' is an inverse of a then $a' \in D$ and the two $\mathscr{H}$-classes $R_a \cap L_{a'}$, $L_a \cap R_{a'}$ contain respectively the idempotents aa' and $a'a$.*

(ii) *If $b \in D$ is such that $R_a \cap L_b$, $L_a \cap R_b$ contains idempotents e, f respectively, then H_b contains an inverse a^* of a such that $aa^* = e$, $a^*a = f$.*

(iii) *No $\mathscr{H}$-class contains more than one inverse of a.*

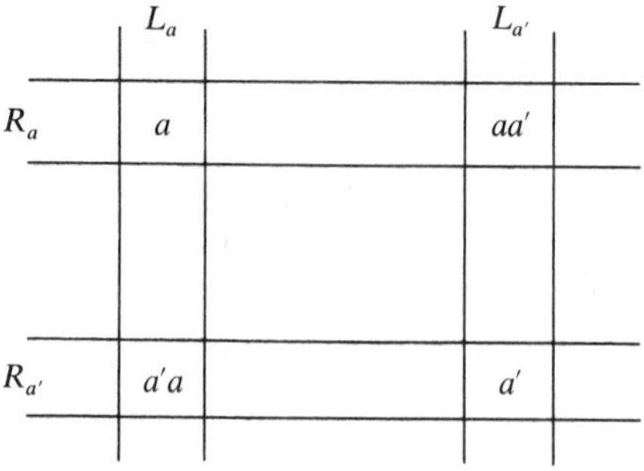

Proof. It remains to establish the second and third parts. To prove (ii), notice first that since $a \mathscr{R} e$ it follows by Proposition 3.3 that $ea = a$. Similarly, from $a \mathscr{L} f$ we deduce that $af = a$. Again from $a \mathscr{R} e$ it follows that there exists x in S^1 such that $ax = e$. If we denote fxe by a^*, we easily see that

$$aa^*a = (af)x(ea) = axa = ea = a,$$

and

$$a^*aa^* = fx(eaf)xe = fx(ax)e = fxe^2 = fxe = a^*;$$

thus a^* is an inverse of a. Moreover

$$aa^* = (af)xe = (ax)e = e^2 = e.$$

Further, since $a \mathscr{L} f$ we have an element y in S^1 for which $ya = f$; hence

$$a^*a = fxea = fxa = yaxa = yea = ya = f.$$

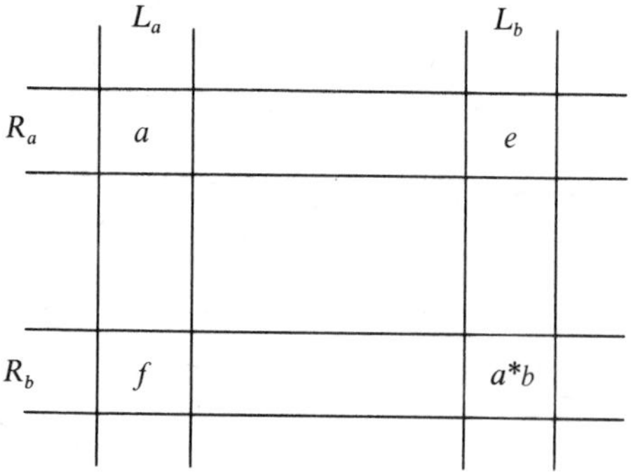

It now follows easily that

$$a^* \in L_e \cap R_f = L_b \cap R_b = H_b.$$

To prove part (iii) of the theorem, suppose that a' and a^* are both inverses of a inside the single $\mathscr{H}$-class H_b. It then follows that aa' and aa^* are both idempotents in the $\mathscr{H}$-class $R_a \cap L_b$; hence $aa' = aa^*$ by Corollary 2.6. Similarly $a'a = a^*a$, since both are idempotents in the $\mathscr{H}$-class $L_a \cap R_b$. Hence

$$a' = a'aa' = a'aa^* = a^*aa^* = a^*. \blacksquare$$

Notice that this theorem allows us to locate the inverses of a regular element provided that we know where the idempotents are. For example, in a finite semigroup, we can say immediately that the number of inverses of a regular element a is the number of idempotents in L_a multiplied by the number of idempotents in R_a. This idea of deducing facts about the semigroup from facts about its idempotents is a recurring theme in the study of regular semigroups and we shall have occasion to mention it several times.

The following easy consequence of Theorem 3.5 will be of considerable use in later chapters:

Proposition 3.6. *Let e, f be idempotents in a semigroup S. Then $(e, f) \in \mathscr{D}$ if and only if there exists an element a in S and an inverse a' of a such that $aa' = e, a'a = f$.*

Proof. If $(e, f) \in \mathscr{D}$ then e and f are members of the same regular $\mathscr{D}$-class. If a is any element of $R_e \cap L_f$ then by Theorem 3.5(ii) there is an inverse a' of a in $R_f \cap L_e$. By Theorem 3.5(i) it then follows that $aa' = e, a'a = f$.

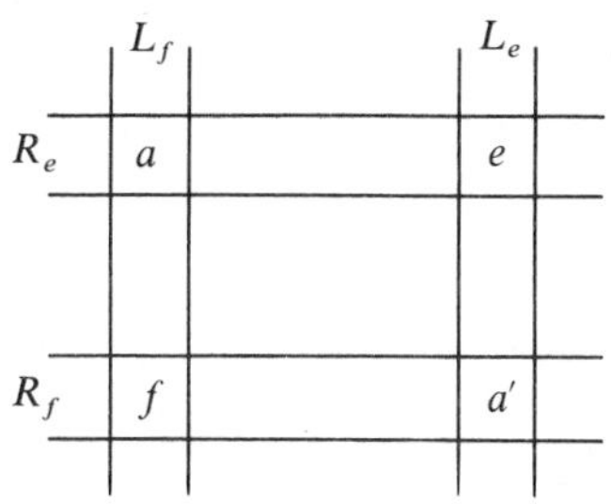

Conversely, if there exist mutually inverse elements a, a' such that $aa' = e$ and $a'a = f$, it follows from Theorem 3.5(i) that $e \mathscr{R} a, a \mathscr{L} f$. Hence $e \mathscr{D} f$ as required. $\blacksquare$

A final result for this section will shortly be of use:

Proposition 3.7. *If H and K are two group $\mathscr{H}$-classes in the same (regular) $\mathscr{D}$-class, then H and K are isomorphic.*

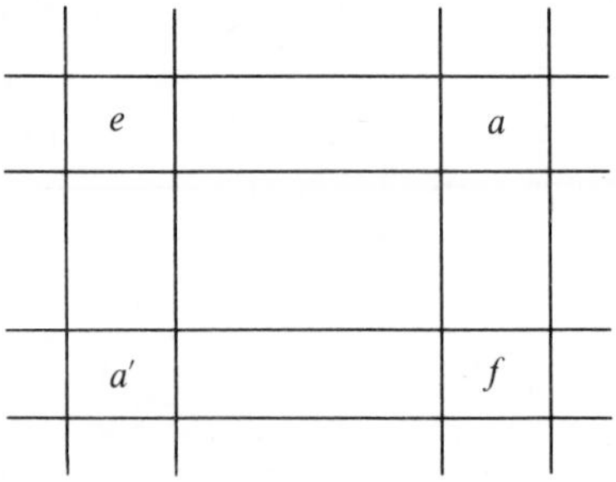

Proof. The method of proof is essentially that used in the proof of Lemma 2.3, with a careful (yet obvious) choice of translation maps. We have that $H = H_e$ and $K = H_f$ where e and f are idempotents and are respectively the identity elements of the groups H_e and H_f. If $a \in R_e \cap L_f$ (which is non-empty since $e \mathscr{D} f$ by assumption) then by Theorem 3.5 there exists a unique inverse a' of a in $L_e \cap R_f$, and we have

$$aa' = e, \qquad a'a = f, \qquad ea = af = a, \qquad a'e = fa' = a'.$$

By Green's Lemmas (2.1 and 2.2) it follows that $\rho_a | H_e$ is a bijection from H_e onto H_a and that $\lambda_{a'} | H_a$ is a bijection from H_a onto H_f. Thus the map $\phi = \rho_a \lambda_{a'}$ given by

$$x\phi = a'xa \qquad (x \in H_e)$$

is a bijection from H_e onto H_f, with inverse given by

$$y\phi^{-1} = aya' \qquad (y \in H_f).$$

The bijection ϕ is even an isomorphism, for if $x_1, x_2 \in H_e$, then

$$(x_1\phi)(x_2\phi) = a'x_1aa'x_2a = a'x_1ex_2a$$
$$= a'x_1x_2a = (x_1x_2)\phi,$$

since e is the identity element of the group H_e.■

4. REGULAR SEMIGROUPS

In this section are presented some results about regular semigroups and Green's equivalences that will be of use later, particularly though not exclusively in Chapter VI.

A first, very elementary remark is that in a regular semigroup $a = aa \in Sa$, and similarly $a \in aS$, $a \in SaS$. Thus in considering Green's equivalences on a regular semigroup we can drop all reference to S^1, and assert more simply that

$$a \mathscr{L} b \quad \text{if and only if} \quad Sa = Sb,$$
$$a \mathscr{R} b \quad \text{if and only if} \quad aS = bS,$$
$$a \mathscr{J} b \quad \text{if and only if} \quad SaS = SbS.$$

In the following characterizations of the equivalences $\mathscr{L}$, $\mathscr{R}$ and $\mathscr{H}$ we introduce the notation $V(a)$ for the set of all inverses of an element a. A regular semigroup S is characterized by the property that $V(a) \neq \varnothing$ for every a in S.

Proposition 4.1. *Let a, b be elements of a regular semigroup S. Then*
(i) *$(a, b) \in \mathscr{L}$ if and only if there exist $a' \in V(a)$, $b' \in V(b)$ such that $a'a = b'b$;*
(ii) *$(a, b) \in \mathscr{R}$ if and only if there exist $a' \in V(a)$, $b' \in V(b)$ such that $aa' = bb'$;*
(iii) *$(a, b) \in \mathscr{H}$ if and only if there exist $a' \in V(a)$, $b' \in V(b)$ such that $a'a = b'b$ and $aa' = bb'$.*

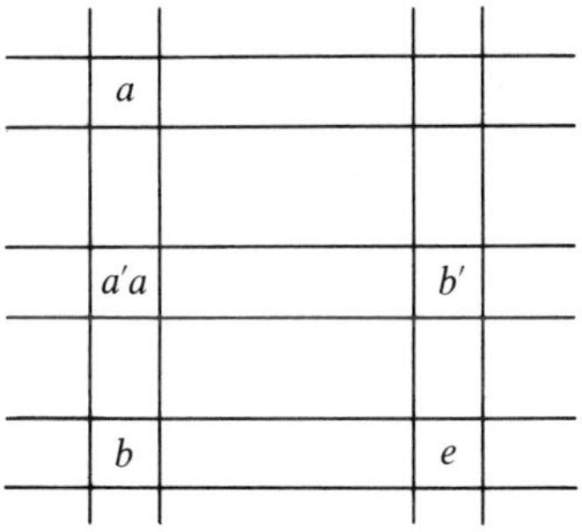

Proof. If $(a, b) \in \mathscr{L}$ and $a' \in V(a)$ then $a'a$ is an idempotent in $L_a = L_b$. The $\mathscr{R}$-class R_b contains at least one idempotent e by Proposition 3.2, and then by Theorem 3.5(ii) the $\mathscr{H}$-class $R_{a'a} \cap L_e$ contains an inverse b' of b with the property that $b'b = a'a$ (and $bb' = e$). Notice that we have in fact proved the stronger implication

$$(a, b) \in \mathscr{L} \Rightarrow (\forall a' \in V(a))(\exists b' \in V(b))\ aa' = bb'. \tag{4.2}$$

Conversely, if $a'a = b'b$ for some $a' \in V(a)$ and $b' \in V(b)$ then $a \mathscr{L} a'a$, $b'b \mathscr{L} b$ by Theorem 3.5(i), and so $a \mathscr{L} b$.

Part (ii) is similar, and once again we can prove

$$(a, b) \in \mathscr{R} \Rightarrow (\forall a' \in V(a))(\exists b' \in V(b))a'a = b'b. \tag{4.3}$$

To show part (iii), suppose that $a \mathscr{H} b$ and that $a' \in V(a)$. Then $aa' \in R_a = R_b$ and $a'a \in L_a = L_b$. Hence by Theorem 3.5(ii) the $\mathscr{H}$-class $L_{aa'} \cap R_{a'a}$ contains an inverse b' of b such that $bb' = aa'$ and $b'b = a'a$. Once again we have the implication

$$(a, b) \in \mathscr{H} \Rightarrow (\forall a' \in V(a))(\exists b' \in V(b))\; a'a = b'b \text{ and } aa' = bb'. \tag{4.4}$$

Conversely, if for some $a' \in V(a)$ and $b' \in V(b)$ we have $a'a = b'b$ and $aa' = bb'$, it is evident that $a \mathscr{L} b$ and $a \mathscr{R} b$, i.e. that $a \mathscr{H} b$.■

If U is a subsemigroup of a (not necessarily regular) semigroup S, and if $a, b \in U$, there can be some ambiguity about the meaning of (say) $a \mathscr{L} b$, since $\mathscr{L}$ may stand for the appropriate Green equivalence either in S or in U. When confusion of this sort is likely to arise we shall put superscripts on the symbols to distinguish between the two equivalences. Thus $(a, b) \in \mathscr{L}^U$ means that there exist u, v in U^1 such that $ua = b$, $a = vb$, while $(a, b) \in \mathscr{L}^S$ means that there exist s, t in S^1 such that $sa = b$, $a = tb$. Also, if $a \in U$,

$$L_a^U = \{u \in U : (a, u) \in \mathscr{L}^U\},$$

while

$$L_a^S = \{s \in S: \ (a, s) \in \mathscr{L}^S\}.$$

Clearly

$$\mathscr{L}^U \subseteq \mathscr{L}^S \cap (U \times U),$$

and also, with the obvious notations,

$$\mathscr{R}^U \subseteq \mathscr{R}^S \cap (U \times U), \qquad \mathscr{H}^U \subseteq \mathscr{H}^S \cap (U \times U),$$

$$\mathscr{D}^U \subseteq \mathscr{D}^S \cap (U \times U), \qquad \mathscr{J}^U \subseteq \mathscr{J}^S \cap (U \times U).$$

These inclusions may well be proper: e.g. if S is the infinite cyclic group generated by a and if $U = \langle a \rangle$ is the infinite monogenic subsemigroup of S consisting of the positive powers of a, then

$$\mathscr{L}^U = \mathscr{R}^U = \mathscr{H}^U = \mathscr{D}^U = \mathscr{J}^U = 1_U,$$

while

$$\mathscr{L}^S \cap (U \times U) = \mathscr{R}^S \cap (U \times U) = \mathscr{H}^S \cap (U \times U) = \mathscr{D}^S \cap (U \times U)$$
$$= \mathscr{J}^S \cap (U \times U) = U \times U.$$

However, as pointed out by T. E. Hall (1972), we do have

Proposition 4.5. *If U is a regular subsemigroup of a semigroup S, then*

$$\mathscr{L}^U = \mathscr{L}^S \cap (U \times U), \qquad \mathscr{R}^U = \mathscr{R}^S \cap (U \times U), \qquad \mathscr{H}^U = \mathscr{H}^S \cap (U \times U).$$

Proof. Suppose that $(a, b) \in \mathscr{L}^S \cap (U \times U)$, and that a', b' are inverses of a, b respectively in U. Then $(a'a, a) \in \mathscr{L}^U \subseteq \mathscr{L}^S$, $(a, b) \in \mathscr{L}^S$, and $(b, b'b) \in \mathscr{L}^U \subseteq \mathscr{L}^S$; hence, by transitivity, $(a'a, b'b) \in \mathscr{L}^S$. By Proposition 3.3 each of $a'a$, $b'b$ is a right identity for $L_{a'a}^S = L_{b'b}^S$; hence in particular

$$a'ab'b = a'a, \qquad b'ba'a = b'b.$$

These equations involve only elements of U and may be interpreted as implying that $(a'a, b'b) \in \mathscr{L}^U$. But we now have

$$(a, a'a), (a'a, b'b), (b'b, b) \in \mathscr{L}^U$$

and so $(a, b) \in \mathscr{L}^U$ as required.

The proof that $\mathscr{R}^U = \mathscr{R}^S \cap (U \times U)$ is similar. The corresponding statement for $\mathscr{H}$ is a simple set-theoretic consequence of the statements for $\mathscr{L}$ and $\mathscr{R}$. ■

The corresponding assertions for $\mathscr{D}$ and $\mathscr{J}$ are not true, even if S also is regular. For example, consider the semigroup $S = \{0, e, f, a, b\}$, where

$$0 = \begin{bmatrix} 0 & 0 \\ 0 & 0 \end{bmatrix}, \quad e = \begin{bmatrix} 1 & 0 \\ 0 & 0 \end{bmatrix}, \quad f = \begin{bmatrix} 0 & 0 \\ 0 & 1 \end{bmatrix},$$

$$a = \begin{bmatrix} 0 & 1 \\ 0 & 0 \end{bmatrix}, \quad b = \begin{bmatrix} 0 & 0 \\ 1 & 0 \end{bmatrix}$$

and where the operation is matrix multiplication. The Cayley table of S is

	0	e	f	a	b
0	0	0	0	0	0
e	0	e	0	a	0
f	0	0	f	0	b
a	0	a	0	0	e
b	0	0	b	f	0

and it is evident that S is regular. Then $U = \{0, e, f\}$ is a regular subsemigroup of S. Since U is a semilattice we have

$$\mathscr{L}^U = \mathscr{R}^U = \mathscr{H}^U = \mathscr{D}^U = \mathscr{J}^U = 1_U.$$

On the other hand it is easy to check that the $\mathscr{R}^S$-classes are $\{0\}$, $\{e, a\}$, $\{f, b\}$, that the $\mathscr{L}^S$-classes are $\{0\}$, $\{e, b\}$, $\{f, a\}$, and hence that the $\mathscr{D}^S$-classes are $\{0\}$, $\{e, f, a, b\}$. Now $\mathscr{J}^S = \mathscr{D}^S$ since S is finite (Proposition 1.5), and so

$$(e, f) \in \mathscr{D}^S \cap (U \times U) = \mathscr{J}^S \cap (U \times U).$$

If S is a regular semigroup and ρ is a congruence on S then S/ρ is a regular semigroup. Indeed if $a' \in V(a)$ in S then $a'\rho \in V(a\rho)$ in S/ρ, since

$$(a\rho)(a'\rho)(a\rho) = (aa'a)\rho = a\rho,$$
$$(a'\rho)(a\rho)(a'\rho) = (a'aa')\rho = a'\rho.$$

The following lemma, due to Lallement (1966), is crucial in the study of regular semigroups.

Lemma 4.6. (Lallement's Lemma). *Let ρ be a congruence on a regular semigroup S. If $a\rho$ is an idempotent in S/ρ then there exists an idempotent e in S such that $a\rho = e\rho$. Moreover, e can be chosen so that $R_e \leqslant R_a$, $L_e \leqslant L_a$.*

Proof. If $a\rho$ is an idempotent in S/ρ then $(a, a^2) \in \rho$. Let x be an inverse in S of a^2:

$$a^2xa^2 = a^2, \qquad xa^2x = x.$$

If $e = axa$, then

$$e^2 = axa^2xa = axa = e$$

and so e is idempotent. Also

$$e = axa \equiv a^2xa^2 \pmod{\rho}$$
$$= a^2 \equiv a \pmod{\rho}$$

and so $a\rho = e\rho$. It is clear from (1.10) that the idempotent $e = axa$ has the property that $R_e \leqslant R_a$, $L_e \leqslant L_a$.■

The close correspondence between congruences and homomorphisms discussed in Section I.5 enables us to obtain the following alternative version of Lallement's Lemma:

Lemma 4.7. *Let $\phi : S \to T$ be a homomorphism from a regular semigroup S into a semigroup T. Then $S\phi$ is regular. If f is an idempotent in $S\phi$ then there exists an idempotent e in S such that $e\phi = f$.*■

If S is a semigroup and E is the set of idempotents of S, we shall say that an equivalence relation ρ on S is *idempotent-separating* if

$$\rho \cap (E \times E) = 1_E,$$

i.e. if no ρ-class contains more than one idempotent. By Corollary 2.6 we have that $\mathscr{H}$ is an idempotent-separating equivalence on any semigroup S; hence any congruence contained in $\mathscr{H}$ is an idempotent-separating congruence. In fact, as Lallement (1966) showed:

Proposition 4.8. *If S is regular, then a congruence ρ on S is idempotent-separating if and only if $\rho \subseteq \mathscr{H}$. Hence $\mathscr{H}^\flat$ (as defined by (I.5.10)) is the maximum idempotent-separating congruence on S.*

Proof. We have seen that one half of this result holds in any semigroup whatever. It remains to show that if S is regular then any idempotent-separating congruence is contained in $\mathscr{H}$. Accordingly, suppose that $(a, b) \in \rho$, where ρ is an idempotent-separating congruence. Then $(aa', ba') \in \rho$ and so $(ba')\rho = (aa')\rho$, an idempotent of S/ρ. By Lallement's Lemma (4.6) there is an idempotent e in S such that $e\rho = (ba')\rho$ and $R_e \leqslant R_{na'}$. But then $e\rho = (aa')\rho$ and so $e = aa'$ since ρ is idempotent-separating. Hence

$$R_a = R_{aa'} \leqslant R_{ba'} \leqslant R_b.$$

Similarly $R_b \leqslant R_a$ and so $a \mathscr{R} b$. But then one can use a closely similar argument to show that $a \mathscr{L} b$. Hence $(a, b) \in \mathscr{H}$ as required.■

Remark. The hypothesis that S is regular cannot be removed in the above proposition. If $S = \{x, 0\}$ is the two-element null semigroup in which

$$x^2 = x0 = 0x = 00 = 0,$$

then 0 is the only idempotent and $\mathscr{H} = 1_S$. The universal congruence $S \times S$ is then idempotent-separating but is not contained in $\mathscr{H}$.

EXERCISES

1. (Andersen 1952) (i) Let C be a cancellative semigroup (i.e. $ca = cb \Rightarrow a = b$, $ac = bc \Rightarrow a = b$) without identity. Show that there cannot be any pair of elements e, a in C for which $ea = a$ or for which $ae = a$. Deduce that $\mathscr{L} = \mathscr{R} = \mathscr{D} = 1_C$ in such a semigroup.

(ii) Show that, with respect to matrix multiplication,

$$S = \left\{ \begin{bmatrix} a & 0 \\ b & 1 \end{bmatrix} : a, b \in \mathbf{R},\ a, b > 0 \right\}$$

is a cancellative semigroup without identity.

(iii) Show that $\mathscr{J} = S \times S$. Thus $\mathscr{D}$ is properly contained in $\mathscr{J}$.

(iv) Observe that S is not periodic, since (e.g.)

$$\begin{bmatrix} 1 & 0 \\ 1 & 1 \end{bmatrix}^n = \begin{bmatrix} 1 & 0 \\ n & 1 \end{bmatrix}.$$

(v) Observe also that S does not satisfy $\min_L$. More precisely, denote $\begin{bmatrix} 1 & 0 \\ n & 1 \end{bmatrix}$ by s_n and observe that

$$S^1 s_1 \supset S^1 s_2 \supset S^1 s_3 \supset \dots .$$

2. If e, f are idempotents in a semigroup S, show that

(i) $e \mathscr{L} f$ if and only if $ef = e$, $fe = f$;

(ii) $e \mathscr{R} f$ if and only if $ef = f$, $fe = e$.

3. (Lallement 1966) Let S be a semigroup. If λ is a right congruence on S contained in $\mathscr{L}$ and ρ is a left congruence on S contained in $\mathscr{R}$, show that $\lambda \circ \rho = \rho \circ \lambda$.

Deduce, using Proposition I.8.5, that the sublattice $[1_S, \mathscr{H}^\flat]$ of congruences on S contained in $\mathscr{H}$ is modular.

4. A semigroup S is called *right simple* [*left simple*] if $\mathscr{R} = S \times S$ [$\mathscr{L} = S \times S$]. It is called *right cancellative* [*left cancellative*] if (for all a, b, c in S) $ac = bc \Rightarrow a = b$ [$ca = cb \Rightarrow a = b$].

(i) Show that a left zero semigroup is left simple and right cancellative, but is neither right simple nor left cancellative.

(ii) Show that the Baer–Levi semigroup (Exercise I.3) is right simple and right cancellative, but is neither left simple nor left cancellative.

(iii) Show that a semigroup is right simple and left simple if and only if it is a group.

(iv) Show that a finite semigroup is left and right cancellative if and only if it is a group.

(v) Show that the word "finite" cannot be removed from the statement of (iv); i.e. give an example of a semigroup that is both right and left cancellative but is not a group.

5. A semigroup that is right simple and left cancellative is called a *right group*. If G is a group and E is a right zero semigroup, show that the direct product $G \times E$ is a right group.

Conversely, let S be a right group.

(i) Show that the set E of idempotents of S is non-empty.

(ii) Show that E is a right zero subsemigroup of S.

(iii) Show that $eb = b$ for every b in S.

(iv) Show that if $e \in E$ then Se is a subgroup of S.

(v) If f is a fixed element of E and G is the group Sf, show that the mapping $\phi: G \times E \to S$ defined by

$$(a, e)\phi = ae$$

is an isomorphism.

(vi) Deduce that a semigroup is a right group if and only if it is isomorphic to a direct product of a group and a right zero semigroup.

6. (Hall 1972) Let S be an arbitrary semigroup and let ρ be a congruence on S such that $\rho \subseteq \mathscr{L}$. Show that $a \mathscr{L} b$ in S if and only if $a\rho \mathscr{L} b\rho$ in S/ρ.

7. Show that a regular semigroup has exactly one idempotent if and only if it is a group.

8. If L is a left ideal and R a right ideal of a semigroup S, show that $R \cap L \supseteq RL$. Show that equality holds if S is regular.

9. Show that the full transformation semigroup $\mathscr{T}(X)$ is regular.

10. Show that, in $\mathscr{T}(X)$,

(i) $(\alpha, \beta) \in \mathscr{L}$ if and only if $\text{ran}(\alpha) = \text{ran}(\beta)$;

(ii) $(\alpha, \beta) \in \mathscr{R}$ if and only if $\alpha \circ \alpha^{-1} = \beta \circ \beta^{-1}$;

(iii) $(\alpha, \beta) \in \mathscr{D}$ if and only if $|\text{ran}(\alpha)| = |\text{ran}(\beta)|$;

(iv) $\mathscr{D} = \mathscr{J}$.

11. In general, in a regular semigroup S there will exist for each a in S elements x for which $axa = a$ but $xax \neq x$. For example, in the regular semigroup $\mathscr{T}(X)$, with $X = \{1, 2, 3\}$, show that if

$$\alpha = \begin{pmatrix} 1 & 2 & 3 \\ 2 & 1 & 1 \end{pmatrix} \quad \text{and} \quad \xi = \begin{pmatrix} 1 & 2 & 3 \\ 2 & 1 & 3 \end{pmatrix}$$

then $\alpha\xi\alpha = \alpha$ and $\xi\alpha\xi \neq \xi$. Check, however, that $\alpha' = \xi\alpha\xi$ is an inverse of α. (See also Exercise III.11.)

12. (Howie and Lallement 1966) A subset A of a semigroup S is called *right unitary* if

$$(\forall a \in A)\,(\forall s \in S)\; sa \in A \Rightarrow s \in A,$$

left unitary if

$$(\forall a \in A)\,(\forall s \in S)\; as \in A \Rightarrow s \in A,$$

and *unitary* if it is both left and right unitary. (These concepts were introduced by Dubreil (1941).)

Let E be the set of idempotents of a regular semigroup. Show that E is a right unitary subset of S if and only if it is a unitary subsemigroup of S. [Hint: show first that if $es \in E$ with e in E then $sess' \in E$ for every s' in $V(s)$.]

13. (Howie and Lallement 1966) Define a regular semigroup S to be *E-unitary* if the set E of idempotents is a unitary subsemigroup of S. Show that if S is *E-unitary*, then (for all a, b in S)

$$ab \in E \Rightarrow ba \in E.$$

[Hint: show first that $babb' \in E$.]

14. (Hall 1972) Let S be a regular semigroup and let ϕ be a homomorphism from S onto T. If c, d are mutually inverse elements of T, show that it is possible to choose a, b in S such that $a\phi = c$, $b\phi = d$ and a, b are mutually inverse elements of S. (This generalizes Lemma 4.7.) [Hint: if $x\phi = c$ and $y\phi = d$, consider

$$a = xy(xyxy)'xyx, \qquad b = y(xyxy)'xy,$$

where $(xyxy)' \in V(xyxy)$.]

15. (Fitz-Gerald 1972) (i) In a regular semigroup, show that any inverse of an idempotent is expressible as a product of two idempotents.

(ii) More generally, if E is the set of idempotents in a regular semigroup S, show that

$$V(E^n) = E^{n+1}.$$

[Hint: if $x = e_1 e_2 \dots e_n \in E^n$ and $y \in V(x)$, let

$$f_j = e_j \dots e_n y e_1 \dots e_{j-1};$$

then show that $f_j^2 = f_j$ and that $y = yx.f_n \dots f_2.xy \in E^{n+1}$. Conversely, if $x = e_1 e_2 \dots e_{n+1} \in E^{n+1}$, let

$$g_j = e_{j+1} \dots e_{n+1} y e_1 \dots e_j,$$

where $y \in V(x)$; then show that $g_j^2 = g_j$ and that the element $z = g_n \dots g_1$ in E^n has the property that $x \in V(z)$.]

Chapter III

0-simple Semigroups

INTRODUCTION

The chapter begins with some elementary results on simple and 0-simple semigroups and a decomposition theorem for semigroups in general that indicates why an understanding of simple and 0-simple semigroups is important. The main result of the chapter is a structure theorem due to Rees (1940) which applies to 0-simple semigroups satisfying both the minimal conditions $\min_L$ and $\min_R$—what are called *completely* 0-*simple semigroups*. (Rees himself used a different definition, but we shall see below (Theorem 3.1) that the two definitions are equivalent.)

Perhaps unfortunately, the word "simple" as used in semigroup theory does not have the same import as in group theory or ring theory, where it implies the total absence of non-trivial homomorphic images. By contrast with ring theory, not every congruence on a semigroup is associated with an ideal, and so it is normally the case that a simple (or 0-simple) semigroup has non-trivial congruences. Using the Rees structure theorem we describe a classification of the congruences on a completely 0-simple semigroup. A result on the nature of the lattice of congruences readily follows, and so also does a classification of the finite *congruence-free* (i.e. truly simple) semigroups.

1. SIMPLE AND 0-SIMPLE SEMIGROUPS; PRINCIPAL FACTORS

A semigroup S without zero is called *simple* if it has no proper ideals. A semigroup S with zero is called 0-*simple* if (i) $\{0\}$ and S are the only ideals and (ii) $S^2 \neq \{0\}$. The latter condition serves only to exclude the two-element null semigroup from the class of 0-simple semigroups, since any larger null semigroup fails to qualify on the grounds of having proper ideals.

It is easy to see that S is simple if and only if $\mathcal{J} = S \times S$. The corresponding criterion for 0-simplicity is that $S^2 \neq \{0\}$ and that $\{0\}$, $S \backslash \{0\}$ are the only $\mathcal{J}$-classes.

A simple semigroup can be made into a 0-simple semigroup by merely adjoining a zero element. Not all 0-simple semigroups arise from simple semigroups in this way, however; the zero element of a 0-simple semigroup S can be removed to leave behind a simple semigroup only if it is a "prime" element in the sense that

$$ab = 0 \Rightarrow a = 0 \text{ or } b = 0, \tag{1.1}$$

and this is not always the case, as will be clear by the end of the chapter. If the implication (1.1) does hold, we say that *S has no proper zero-divisors.* It will always be possible to deduce a theorem about simple semigroups from one about 0-simple semigroups, and for this reason we shall turn our attention primarily towards the 0-simple case.

Proposition 1.2. *A semigroup S is 0-simple if and only if $SaS = S$ for every a in $S\backslash\{0\}$, i.e. if and only if for every a, b in $S\backslash\{0\}$ there exist x, y in S such that $xay = b$.*

Proof. Suppose first that S is 0-simple. Then S^2 is an ideal of S and hence, since $S^2 \neq \{0\}$, we must have $S^2 = S$. Hence $S^3 = S^2.S = S.S = S$. Now, for any a in $S\backslash\{0\}$ the subset SaS of S is an ideal; hence either $SaS = S$ or $SaS = \{0\}$. If $SaS = \{0\}$ then the set $I = \{x \in S : SxS = \{0\}\}$ contains an element other than 0. Since I is easily seen to be an ideal of S it follows that $I = S$, i.e. that $SxS = \{0\}$ for every x in S. Thus $S^3 = \{0\}$, in contradiction to the previously noted fact that $S^3 = S$. Hence $SaS = S$ as required.

Conversely, if $SaS = S$ for every a in $S\backslash\{0\}$ then certainly $S^2 \neq \{0\}$. Also, if A is a non-zero ideal of S, containing a non-zero element a (say), then

$$A \supseteq SAS \supseteq SaS = S$$

and so $A = S$. ■

As a corollary, we have the following corresponding result for simple semigroups:

Corollary 1.3. *A semigroup S is simple if and only if $SaS = S$ for every a in S, i.e. if and only if for every a, b in S there exist x, y in S such that $xay = b$.* ■

By a *0-minimal* ideal M in a semigroup S with zero we mean an ideal minimal in the set of non-zero ideals. The next result shows that 0-simple semigroups may occur inside more general semigroups.

Proposition 1.4. *If M is a 0-minimal ideal of S then either $M^2 = \{0\}$ or M is a 0-simple semigroup.*

Proof. M^2 is an ideal of S contained in M and so either $M^2 = \{0\}$ or $M^2 = M$. In the latter case we deduce that also $M^3 = M$. If $a \in M\backslash\{0\}$ then S^1aS^1 is an ideal of S contained in M. It is certainly non-zero, since it contains a, and so $S^1aS^1 = M$. Thus $MaM \subseteq S^1aS^1 = M = M^3 = M(S^1aS^1)M = (MS^1)a(S^1M) \subseteq MaM$, and so $MaM = M$. By Proposition 1.2, it now follows that M is 0-simple. ■

The specialization of this result to the case of semigroups without zero is of special interest, since a semigroup can have at most one *minimal* ideal. To see this, suppose that M and N are both minimal ideals of S. Then MN, being an ideal contained in both M and N, must be *equal* to both M and N, which must therefore be equal to one another. Either, then, there are no minimal ideals at all in S (which can happen—see Exercise 1) or there is a unique *minimum* ideal, which we call the *kernel* $K = K(S)$ of S. A simplified version of the argument used to establish Proposition 1.4 now gives

Proposition 1.5. *If a semigroup S has a kernel K, then K is a simple semigroup.* ■

It is worth remarking that a semigroup with zero does always have a kernel, since $\{0\}$ is clearly the unique minimum ideal. In this case, however, the notion is not particularly useful. More significantly, every finite semigroup has a kernel, since the alternative to having a kernel is to have infinite strictly descending chains of ideals, and this alternative is not open to a finite semigroup.

The next result describes another context in which 0-simple semigroups can occur:

Proposition 1.6. *If I, J are ideals of a semigroup S such that $I \subset J$ and such that there is no ideal B of S for which $I \subset B \subset J$, then J/I is either* 0*-simple or null.*

Proof. By virtue of Proposition I.7.1, J/I is a 0-minimal ideal of S/I, and so the result is a direct consequence of Proposition 1.4. ■

This result, while not of any great depth, is the basis of an important decomposition method for any arbitrary semigroup S. Let us begin by recalling the natural partial order (II.1.9) among the $\mathcal{J}$-classes of S whereby $J_x < J_y$ if and only if $S^1xS^1 \subset S^1yS^1$. Let us now write $J(a)$ for the principal (two-sided) ideal S^1aS^1 generated by the element a of S. If J_a is minimal among the $\mathcal{J}$-classes of S then $J(a)$ is a minimal ideal of S and so is the unique minimum ideal $K(S)$ of S. In this case $S^1bS^1 = J(a)$ for every b in $J(a)$ and so

$$J(a) = J_a = K(S) \tag{1.7}$$

(where of course $K(S) = \{0\}$ in the case where S has a zero element).

If J_a is not minimal in $S/\mathscr{J}$ then the set

$$I(a) = \{b \in J(a) \colon J_b < J_a\}$$

is non-empty. In fact $I(a)$ is an ideal of S, since if $b \in I(a)$ and $u, v \in S$ then

$$J_{ubv} \leqslant J_b < J_a$$

and so $ubv \in I(a)$. Notice that

$$J(a) = J_a \cup I(a) \tag{1.8}$$

and that the union is disjoint.

Suppose now that B is an ideal of S such that

$$I(a) \subseteq B \subset J(a).$$

If $b \in B$ then $J(b) \subseteq B$, since $J(b)$ is the smallest ideal containing b. Hence $J(b) \subset J(q)$ and so $J_n < J_a$. It follows that $b \in I(a)$ and so $B = I(a)$. We have shown that $I(a)$ and $J(a)$ satisfy the criteria for Proposition 1.6; hence the Rees quotient $J(a)/I(a)$ is either 0-simple or null. The semigroups $K(S)$, $J(a)/I(a)$ $(a \in S)$ are called the *principal factors of S*.

Obviously, then, if we can find out something about 0-simple semigroups, we shall have gone some way to understanding the "local" structure of a semigroup. By virtue of (1.8) we can think of the principal factor $J(a)/I(a)$ as the $\mathscr{J}$-class J_a with a zero adjoined. If the factor is null, then the product of two elements in J_a always falls into a lower $\mathscr{J}$-class. If the factor is 0-simple then the product may lie in J_a or may fall into a lower $\mathscr{J}$-class.

We summarize these observations in a theorem as follows:

Theorem 1.9. *If J is a $\mathscr{J}$-class in a semigroup S, then either* (i) *J is the kernel of S, or* (ii) *the set $I = \{x \in S \colon J_x < J\}$ is non-empty and is an ideal of J such that J/I is either 0-simple or null.* ■

2. REES'S THEOREM

As was mentioned in the introduction to this chapter, we shall call a 0-simple semigroup *completely 0-simple* if it satisfies the conditions $\min_L$ and $\min_R$, that is to say, if every non-empty set either of $\mathscr{L}$-classes or of $\mathscr{R}$-classes possesses a minimal member.

Rees (1940) provided the following recipe for manufacturing completely 0-simple semigroups. Let G be a group with identity element e, and let I, Λ be non-empty sets. Let $P = (p_{\lambda i})$ be a $\Lambda \times I$ matrix with entries in the 0-group G° $(= G \cup \{0\})$, and suppose that P is *regular*, in the sense that no row or column consists entirely of zeros: formally,

$$\left.\begin{aligned}(\forall i \in I)(\exists \lambda \in \Lambda) \qquad p_{\lambda i} \neq 0,\\ (\forall \lambda \in \Lambda)(\exists i \in I) \qquad p_{\lambda i} \neq 0.\end{aligned}\right\} \tag{2.1}$$

Let $S = (G \times I \times \Lambda) \cup \{0\}$ and define a composition on S by

$$\left.\begin{aligned}(a, i, \lambda)(b, j, \mu) &= \begin{cases}(ap_{\lambda j}b, i, \mu) & \text{if} \quad p_{\lambda j} \neq 0,\\ 0 \quad \text{if} \quad p_{\lambda j} = 0.\end{cases}\\ (a, i, \lambda)0 &= 0(a, i, \lambda) = 00 = 0,\end{aligned}\right\} \tag{2.2}$$

Then we have

Lemma 2.3. *S is a completely 0-simple semigroup.*

Proof. It is fairly easy to verify directly the associativity of the composition (2.2), but it is probably more illuminating (and recalls Rees's own approach) to observe that $S\backslash\{0\}$ is in one–one correspondence with the set of $I \times \Lambda$ matrices $(a)_{i\lambda}$ $(a \in G)$, where $(a)_{i\lambda}$ denotes the matrix with entry a in the (i, λ) position and zeros elsewhere. Next, notice that $(0)_{i\lambda}$ is independent of i and λ and may be written simply as 0. Thus the correspondence extends to a one–one correspondence between S and the set

$$T = \{(a)_{i\lambda} : i \in I, \lambda \in \Lambda, a \in G^{\circ}\}.$$

It is a routine matter to verify that

$$(a)_{i\lambda}P(b)_{j\mu} = (ap_{\lambda j}b)_{i\mu},$$

where the juxtaposition on the left denotes matrix multiplication in the usual sense. (The low density of non-zero entries in the matrices means that no additions become necessary and so no trouble arises from the fact that the matrix entries lie in a 0-group rather than the more usual ring.) Thus the composition (2.2) in S corresponds in T to the evidently associative composition $\circ$ given by

$$(a)_{i\lambda} \circ (b)_{j\mu} = (a)_{i\lambda}P(b)_{j\mu}$$

and so is itself associative.

To verify that S is 0-simple, note that for any two non-zero elements (a, i, λ) and (b, j, μ) of S we may by the regularity of P choose $\nu \in \Lambda$ and $k \in I$ such that $p_{\nu i} \neq 0$, $p_{\lambda k} \neq 0$, and then easily show that

$$(a^{-1}p_{\nu i}^{-1}, j, \nu)(a, i, \lambda)(p_{\lambda k}^{-1}b, k, \mu) = (b, j, \mu).$$

Hence, by Proposition 1.2, S is 0-simple.

To complete the proof that S is completely 0-simple we must first obtain the form of the relations $\mathscr{L}$ and $\mathscr{R}$ in S.

Lemma 2.4. *In S,*

$$(a, i, \lambda)\,\mathscr{L}(b, j, \mu) \text{ if and only if } \lambda = \mu,$$

$$(a, i, \lambda)\,\mathscr{R}(b, j, \mu) \text{ if and only if } i = j.$$

Proof. It is easy to verify that if (a, i, λ) and (b, j, μ) are $\mathscr{L}$-equivalent, then $\lambda = \mu$. Conversely, if we choose $\nu \in \Lambda$ for which $p_{\nu i} \neq 0$ and choose $\kappa \in \Lambda$ for which $p_{\kappa j} \neq 0$, then

$$(ba^{-1}p_{\nu i}^{-1}, j, \nu)(a, i, \lambda) = (b, j, \lambda)$$

and

$$(ab^{-1}p_{\kappa j}^{-1}, i, \kappa)\,(b, j, \lambda) = (a, i, \lambda);$$

hence $(a, i, \lambda)\,\mathscr{L}(b, j, \lambda)$. This establishes the characterization of $\mathscr{L}$. The corresponding characterization of $\mathscr{R}$ follows similarly. ■

It follows now that the set of non-zero $\mathscr{L}$-classes in S is $\{L_\lambda : \lambda \in \Lambda\}$, where

$$L_\lambda = \{(a, i, \lambda) : a \in G, i \in I\},$$

and that the set of non-zero $\mathscr{R}$-classes in S is $\{R_i : i \in I\}$, where

$$R_i = \{(a, i, \lambda) : a \in G, \lambda \in \Lambda\}.$$

If $L_\lambda \leqslant L_\mu$ then there exists $(x, k, \nu) \in S$ such that

$$(e, i, \lambda) = (x, k, \nu)\,(e, i, \mu),$$

i.e. such that $(e, i, \lambda) = (xp_{\nu i}, k, \mu)$. This can happen only if $\lambda = \mu$. We thus have that

$$L_\lambda \leqslant L_\mu \Rightarrow L_\lambda = L_\mu$$

and so *every* non-zero $\mathscr{L}$-class is minimal in the set of non-zero $\mathscr{L}$-classes. Consequently the semigroup S satisfies the condition $\min_L$ and so, since by a similar argument it satisfies $\min_R$ also, S is completely 0-simple. ■

The semigroup S constructed in accordance with this recipe will, as in Clifford and Preston (1961), be denoted by $\mathscr{M}^\circ[G; I, \Lambda; P]$ and will be called the $I \times \Lambda$ *Rees matrix semigroup over the* 0-*group* G° *with the regular sandwich matrix* P.

The importance of Rees's recipe lies in its universality: Rees (1940) showed that *every* completely 0-simple semigroup is isomorphic to some

$\mathscr{M}^\circ[G; I, \Lambda; P]$. To spell it out formally, we have the following theorem, of which the easy half has been proved:

Theorem 2.5 (Rees's Theorem). *Let G° be a 0-group, let I, Λ be non-empty sets, and let $P = (p_{\lambda i})$ be a $\Lambda \times I$ matrix with entries in G°. Suppose that P is regular in the sense of* (2.1). *Let $S = (G \times I \times \Lambda) \cup \{0\}$ and define a binary operation on S by* (2.2), *Then S is a completely* 0-*simple semigroup.*

Conversely, every completely 0-*simple semigroup is isomorphic to one constructed in this way.*

Proof. As a first step towards proving the second half of the theorem we observe that since $\mathscr{D} = \mathscr{J}$ in a semigroup satisfying $\min_L$ and $\min_R$ (Proposition II.1.11), a completely 0-simple semigroup has only two $\mathscr{D}$-classes, namely $\{0\}$ and $S\backslash\{0\}$. Following the terminology now standard in the literature, we call such a semigroup 0-*bisimple*, reserving the term *bisimple* for a semigroup without zero in which $\mathscr{D}$ is the universal relation. Now, by virtue of Proposition II.3.1, a 0-bisimple semigroup either has no regular elements or consists entirely of regular elements. We now show that if S is completely 0-simple it must be the latter situation that obtains:

Lemma 2.6. *If S is completely* 0-*simple, then S is regular.*

Proof. We show first that there exists x in S such that $x^2 \neq 0$. If a is a non-zero element of S then, by Proposition 1.2,

$$a = xay = x^2ay^2$$

for some $x, y \in S$. If every element of S had zero square it would follow that $a = 0$, a contradiction.

We may thus assume that there is an x in $S\backslash\{0\}$ for which $x^2 \in S\backslash\{0\}$. By Proposition 1.2 there exist $u, v \in S$ such that

$$x = ux^2v.$$

We cannot say that u and v are unique with respect to this property, but we do have that the set

$$U_x = \{u \in S : (\exists v \in S)x = ux^2v\}$$

is non-empty. Hence so also is the set

$$U_x/\mathscr{L} = \{L_u : u \in U_x\}.$$

Using the condition $\min_L$ we may therefore suppose that u and v have been chosen in such a way that L_u is minimal in $U_x/\mathscr{L}$. Now

$$x = ux(ux^2v)v = (uxu)x^2v^2$$

and so, since clearly $L_{uxu} \leqslant L_u$, we must have $L_{uxu} = L_u$. Hence

$$L_u = L_{uxu} \leqslant L_{xu} \leqslant L_u$$

and so $xu \mathscr{L} u$. Since $\mathscr{L}$ is a right congruence (Proposition II.1.2) it therefore follows that $xux^2v \mathscr{L} ux^2v$, i.e. that $x^2 \mathscr{L} x$. Since we can equally well show that $x^2 \mathscr{R} x$ it follows that $x^2 \mathscr{H} x$ and hence, by Theorem II.2.5, that H_x is a group. The identity element of H_x, being idempotent, is thus a regular element of the $\mathscr{D}$-class $S\backslash\{0\}$, and it now follows that the entire semigroup is regular. ∎

It will be helpful shortly if we pause at this stage to record an interesting fact about the $\mathscr{H}$-classes of S. We have seen that either H_x is a group or $x^2 = 0$. In this latter case, if $y, z \in H_x$ then certainly there exist p, q in S^1 such that $y = px$ and $z = xq$. Hence

$$yz = px^2q = 0.$$

We have therefore established the following result:

Lemma 2.7. *If H is an $\mathscr{H}$-class in a completely 0-simple semigroup S, then either H is a group or $H^2 = \{0\}$.* ∎

To begin the process of shaping S into the form of a Rees matrix semigroup, let us denote the set of non-zero $\mathscr{R}$-classes of S by I and the set of non-zero $\mathscr{L}$-classes of S by Λ. As a matter of notation we shall treat I and Λ as index sets and write the $\mathscr{R}$-classes as R_i $(i \in I)$ and the $\mathscr{L}$-classes as L_λ $(\lambda \in \Lambda)$. The $\mathscr{H}$-class $R_i \cap L_\lambda$ will be written $H_{i\lambda}$.

Since $S\backslash\{0\}$ is a regular $\mathscr{D}$-class it follows by Proposition II.3.2 that each R_i contains at least one group $\mathscr{H}$-class $H_{i\lambda}$. Equally, each L_λ contains at least one group $\mathscr{H}$-class $H_{i\lambda}$. In an arbitrary way we choose and fix an i and a λ such that $H_{i\lambda}$ is a group. There is no loss of generality at all, and considerable notational simplification, if we write this chosen $\mathscr{H}$-class as H_{11}. We shall denote its identity element by e. Notice that the choice of one particular group $\mathscr{H}$-class rather than another does not affect the abstract properties of the group, for we have seen (Proposition II.3.7) that all the group $\mathscr{H}$-classes in a $\mathscr{D}$-class are isomorphic. The group H_{11} will turn out to be the group figuring in the Rees matrix semigroup we are looking for.

Now, again in a quite arbitrary way, we select for each i in I and λ in Λ an element r_i in H_{i1} and an element q_λ in $H_{1\lambda}$. Since $r_i \mathscr{L} e$ we have by Proposition II.3.3 that $r_i e = r_i$. Hence by Green's Lemma (II.2.2) it follows that $x \mapsto r_i x$ maps H_{11} onto H_{i1} in a one-one fashion. By a similar argument we can show that $eq_\lambda = q_\lambda$, and again by Green's Lemma (II.2.1) it follows that $y \mapsto yq_\lambda$ maps H_{i1} onto $H_{i\lambda}$ in a one–one fashion. Thus (once we have chosen $\{r_i : i \in I\}$

and $\{q_\lambda : \lambda \in \Lambda\}$) we have a unique expression $r_i a q_\lambda$ $(a \in H_{11})$ for each element of $H_{i\lambda}$. Since

$$S \backslash \{0\} = \bigcup \{H_{i\lambda} : i \in I, \lambda \in \Lambda\}$$

(and since the union is disjoint) it follows that there is a bijection $\phi : (H_{11} \times I \times \Lambda) \cup \{0\} \to S$ given by

$$(a, i, \lambda)\phi = r_i a q_\lambda, \qquad 0\phi = 0.$$

The final stage in the argument is to introduce a multiplication into $(H_{11} \times I \times \Lambda) \cup \{0\}$ so as to make it into a regular Rees matrix semigroup, and so that the bijection ϕ becomes an isomorphism. This amounts to defining a suitable sandwich matrix, and since

$$(r_i a q_\lambda)(r_j b q_\mu) = r_i(a q_\lambda r_j b) q_\mu,$$

it is natural to define

$$p_{\lambda i} = q_\lambda r_i \qquad (i \in I, \lambda \in \Lambda).$$

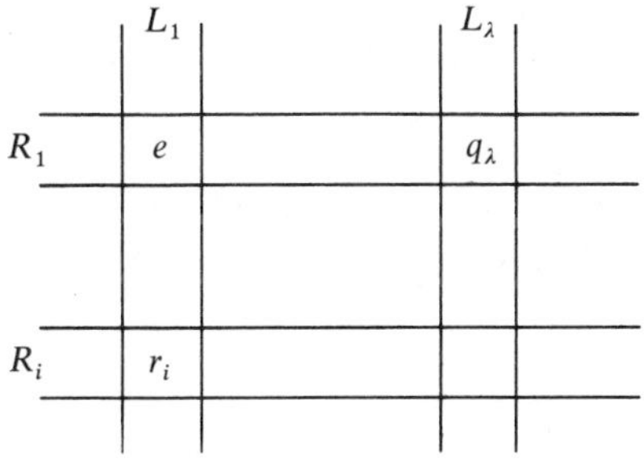

We now show that $p_{\lambda i} \in H_{11}^\circ$. If $H_{i\lambda}$ is a group, with identity f (say), then $fr_i = r_i$ by Proposition II.3.3 and so $x \mapsto xr_i$ is an $\mathscr{R}$-class-preserving bijection from L_λ onto L_1. It follows that $p_{\lambda i} = q_\lambda r_i \in H_{11}$. If, on the other hand, $H_{i\lambda}^2 = \{0\}$ (which is the only other possibility, by Lemma 2.7) then for any c in $H_{i\lambda}$ there exist $u, v \in S^1$ such that $q_\lambda = uc, r_i = cv$, and it follows immediately that

$$p_{\lambda i} = q_\lambda r_i = uc^2 v = 0.$$

Thus $p_{\lambda i} \in H_{11}^\circ$ as required.

We have shown, moreover, that $p_{\lambda i} = 0$ if and only if $H_{i\lambda}^2 = \{0\}$. From Proposition II.3.2 we know that every $\mathscr{L}$-class L_λ and every $\mathscr{R}$-class R_i of the regular $\mathscr{D}$-class $S \backslash \{0\}$ contains an idempotent. Hence for every i in I there exists at least one λ in Λ for which $H_{i\lambda}$ is a group, and equally for every λ in Λ there exists at least one i in I for which $H_{i\lambda}$ is a group. The regularity property of $P = (p_{\lambda i})$ (see (2.1)) now follows immediately.

It is now entirely routine to check that ϕ is an isomorphism from the Rees matrix semigroup $\mathscr{M}^\circ[H_{11}; I, \Lambda; P]$ onto S. ■

Of the structural elements obtained for S, the group H_{11} is unique up to isomorphism and the sets I and Λ are unique up to bijection, but the matrix $P = (p_{\lambda i})$, arising as it does from arbitrarily chosen elements q_λ in $H_{1\lambda}$ and r_i in H_{i1}, is subject to variation. The limits of this variability are set out in the following isomorphism theorem, also due to Rees (1940):

Theorem 2.8. *Two regular Rees matrix semigroups* $S = \mathscr{M}^\circ[G; I, \Lambda; P]$ *and* $T = \mathscr{M}^\circ[K; J, \mathrm{M}; Q]$ *are isomorphic if and only if there exists an isomorphism* $\theta\colon G \to K$, *bijections* $\psi: I \to J$, $\chi: \Lambda \to \mathrm{M}$ *and elements* u_i $(i \in I)$, v_λ $(\lambda \in \Lambda)$ *in* K *such that*

$$p_{\lambda i}\theta = v_\lambda q_{\lambda\chi, i\psi} u_i \tag{2.9}$$

for all $\lambda \in \Lambda$ *and* $i \in I$.

Proof. If $\phi\colon S \to T$ is an isomorphism, it maps the non-zero $\mathscr{R}$-classes R_i of S in a one-one fashion onto the non-zero $\mathscr{R}$-classes of T, and so there is a bijection $\psi\colon I \to J$ such that $(a, i, \lambda)\phi \in R_{i\psi}$. Similarly there is a bijection $\chi\colon \Lambda \to \mathrm{M}$ such that $(a, i, \lambda)\phi \in L_{\lambda\chi}$. Moreover, ϕ maps groups $\mathscr{H}$-classes to group $\mathscr{H}$-classes, and so $p_{\lambda i} \neq 0$ if and only if $q_{\lambda\chi, i\psi} \neq 0$.

Now we choose a group $\mathscr{H}$-class in S, and as in the proof of the Rees structure theorem we note that there is no harm at all in denoting it by H_{11}. The image $H_{11}\phi$ of H_{11} under the isomorphism ϕ is the non-null $\mathscr{H}$-class $H_{1\psi, 1\chi}$ of T. The element $(p_{11}^{-1}x, 1, 1)$ of H_{11} has image $(z, 1\psi, 1\chi)$ under ϕ, where z $(\in K)$ is uniquely determined by $x (\in G)$, and if we define $\theta\colon G \to K$ by the rule that

$$x\theta = q_{1\chi, 1\psi} z,$$

so that

$$(p_{11}^{-1}x, 1, 1)\phi = (q_{1\chi, 1\psi}^{-1}(x\theta), 1\psi, 1\chi),$$

we can see that θ is a homomorphism; for

$$[(p_{11}^{-1}x, 1, 1)(p_{11}^{-1}y, 1, 1)]\theta = (p_{11}^{-1}(xy), 1, 1)\phi = (q_{1\chi, 1\psi}^{-1}[(xy)\theta], 1\psi, 1\chi),$$

while

$$\begin{aligned}(p_{11}^{-1}x, 1, 1)\phi(p_{11}^{-1}y, 1, 1)\phi \\ &= (q_{1\chi, 1\psi}^{-1}(x\theta), 1\psi, 1\chi)(q_{1\chi, 1\psi}^{-1}(y\theta), 1\psi, 1\chi) \\ &= (q_{1\chi, 1\psi}^{-1}[(x\theta)(y\theta)], 1\psi, 1\chi).\end{aligned}$$

In fact, θ is an isomorphism; for

$$\begin{aligned} x\theta = y\theta &\Rightarrow (p_{11}^{-1}x, 1, 1)\phi = (p_{11}^{-1}y, 1, 1)\phi \\ &\Rightarrow (p_{11}^{-1}x, 1, 1) = (p_{11}^{-1}y, 1, 1) \\ &\Rightarrow x = y; \end{aligned}$$

and if $z \in K$ then $(q_{1\chi,1\psi}^{-1}z, 1\psi, 1\chi)\phi^{-1}$ belongs to H_{11} and so may be written as $(p_{11}^{-1}x, 1, 1)$ for some x in G; then it is easy to see that $x\theta = z$.

Next, observe that

$$(a, i, \lambda) = (e, i\ 1)(p_{11}^{-1}a, 1, 1)(p_{11}^{-1}, 1, \lambda);$$

hence, if we write

$$(e, i, 1)\phi = (u_i, i\psi, 1\chi), \qquad (p_{11}^{-1}, 1, \lambda) = (q_{1\chi,1\psi}^{-1}v_\lambda, 1\psi, \lambda\chi),$$

where $u_i, v_\lambda \in K$, then

$$(a, i, \lambda)\phi = (u_i(a\theta)v_\lambda, i\psi, \lambda\chi). \tag{2.10}$$

Hence, if $p_{\lambda i} \neq 0$,

$$\begin{aligned} (e, i, \lambda)\phi(e, i, \lambda)\phi &= (u_iv_\lambda, i\psi, \lambda\chi)(u_iv_\lambda, i\psi, \lambda\chi) \\ &= (u_iv_\lambda q_{\lambda\chi,i\psi}u_iv_\lambda, i\psi, \lambda\chi), \end{aligned}$$

while

$$[(e, i, \lambda)(e, i, \lambda)]\phi = (p_{\lambda i}, i, \lambda)\phi = (u_i(p_{\lambda i}\theta)v_\lambda, i\psi, \lambda\chi),$$

from which it follows that

$$p_{\lambda i}\theta = v_\lambda q_{\lambda\chi,i\psi}u_i$$

as required. If $p_{\lambda i} = 0$ then $q_{\lambda\chi,i\psi} = 0$ also and so the equality (2.9) holds also in this case.

Conversely, if we are given $\psi, \chi, \theta, \{u_i : i \in I\}$ and $\{v_\lambda : \lambda \in \Lambda\}$ with the stated properties, it is easy to check that (2.10), together with the rule that $0\phi = 0$, defines an isomorphism $\phi : S \to T$ whose inverse ϕ^{-1} is given by

$$\begin{aligned} (b, j, \mu)\phi^{-1} &= ((u_i^{-1}bv_\lambda^{-1})\theta^{-1}, i, \lambda), \\ 0\phi^{-1} &= 0, \end{aligned}$$

where $i = j\psi^{-1}$, $\lambda = \mu\chi^{-1}$. ■

A homomorphic image of a completely 0-simple semigroup is completely 0-simple (see Exercise 13). It is reasonable therefore to consider *homomorphisms* from a completely 0-simple semigroup $S = \mathscr{M}^\circ[G; I, \Lambda; P]$ onto a completely 0-simple semigroup $T = \mathscr{M}^\circ[K; J, M; Q]$. We shall not pursue this here, but refer the reader to Clifford and Preston (1961, Chapter 3.4) for a full treatment.

Rees's Theorem undergoes a good deal of simplification in the case where S has no zero and is completely *simple*. As we have already noted, this corresponds to the completely 0-simple case where there are no proper zero-divisors, i.e. to the case where the zero element can be removed to leave a completely simple semigroup. Now the Rees matrix semigroup $\mathcal{M}^\circ[G; I, \Lambda; P]$ has no proper zero-divisors precisely where $p_{\lambda i} \neq 0$ for *all* i in I and λ in Λ. If we remove the zero element in this case we are left with a semigroup, which we shall call $\mathcal{M}[G; I, \Lambda; P]$, consisting of triples (a, i, λ) in $G \times I \times \Lambda$ and with multiplication given by

$$(a, i, \lambda)(b, j, \mu) = (ap_{\lambda j}b, i, \mu).$$

To state our result formally, we have

Theorem 2.11. *Let G be a group, let I, Λ be non-empty sets, and let $P = (p_{\lambda i})$ be a $\Lambda \times I$ matrix with entries in G. Let $S = G \times I \times \Lambda$, and define a binary operation on S by the rule that*

$$(a, i, \lambda)(b, j, \mu) = (ap_{\lambda j}b, i, \mu).$$

Then S is a completely simple semigroup.

Conversely, any completely simple semigroup is isomorphic to one constructed in this manner. ∎

In essence this result is due to Suškevič (1928).

3. PRIMITIVE IDEMPOTENTS

An examination of the effect of various restrictions on the idempotents of a semigroup on the structure of the semigroup as a whole will, as mentioned in the foreword, be something of a general theme of this book. If e, f are idempotents of a semigroup S, we shall write $e \leqslant f$ if $ef = fe = e$. It is a routine matter to check that $\leqslant$ is a partial order relation on the set E of idempotents of S. Notice that if S has an identity element 1 then $e \leqslant 1$ for every e in E, and that if S has a zero element 0 then $0 \leqslant e$ for every e in E.

As in the theory of rings, an idempotent is called *primitive* if it is non-zero and is minimal in the set of non-zero idempotents (with respect to the order just described).

A Rees matrix semigroup $S = \mathcal{M}^\circ[G; I, \Lambda; P]$ contains one non-zero idempotent $(p_{\lambda i}^{-1}, i, \lambda)$ in each non-null $\mathcal{H}$-class $H_{i\lambda}$. If $e = (p_{\lambda i}^{-1}, i, \lambda)$ and

$f = (p_{\mu j}^{-1}, j, \mu)$ are two such idempotents, then $e \leqslant f$ if and only if $ef = fe = e$, i.e. if and only if

$$(p_{\lambda i}^{-1} p_{\lambda j} p_{\mu j}^{-1}, i, \mu) = (p_{\mu j}^{-1} p_{\mu i} p_{\lambda i}^{-1}, j, \lambda) = (p_{\lambda i}^{-1}, i, \lambda).$$

This can happen only if $i = j$ and $\lambda = \mu$, i.e. only if $e = f$. It follows that *every* non-zero idempotent of S is primitive, and so we have certainly established half of

Theorem 3.1. *A* 0*-simple semigroup is completely* 0*-simple if and only if it contains a primitive idempotent.*

Proof. To prove the remaining half, let us suppose that S is a 0-simple semigroup containing a primitive idempotent e. We begin by establishing

Lemma 3.2. *Se is a* 0*-minimal left ideal.*

Proof. Certainly Se is a left ideal, and is non-zero since $e = e^2 \in Se$. To show that it is 0-minimal, let A be a non-zero ideal of S contained in Se; we shall show that $A = Se$. If $a \in A \backslash \{0\}$ then $a = se$ for some s in S and so

$$ae = se^2 = se = a.$$

Since S is 0-simple, there exist x, y in S such that $xay = e$. In fact, if we write $z = ex$ and $t = eye$, we have that

$$zat = e, \qquad ez = z, \qquad et = te = t.$$

Now let $f = tza$. Then

$$f^2 = t(zat)za = teza = tza = f$$

and so f is idempotent. Also, it is easy to see that $ef = fe = f$ and so $f \leqslant e$. Since e is primitive it follows that $f = 0$ or $f = e$. But if $f = 0$ then

$$e = e^2 = za(tza)t = zaft = za0t = 0,$$

a contradiction. Hence $f = e$. But $f = tza \in A$, since A is a left ideal. Hence $e \in A$ and so $Se \subseteq A$ as required. ■

Our object is to show that S satisfies the conditions $\min_L$ and $\min_R$. To this end, we next note that the existence of a 0-minimal ideal $Se = L$ in S does ensure that there is a minimal non-zero $\mathscr{L}$-class in S.

Lemma 3.3. *If S is* 0*-simple and L is a* 0*-minimal left ideal of S, then $L \backslash \{0\}$ is a minimal non-zero $\mathscr{L}$-class in S.*

Proof. To prove that $L\backslash\{0\}$ is an $\mathscr{L}$-class we need only remark that

$$L\backslash\{0\} = \{a \in S: S^1 a = L\}.$$

It is evident that if $S^1 a = L$ then $a \in L\backslash\{0\}$. Conversely, if $a \in L\backslash\{0\}$ then $S^1 a$, being a non-zero ideal of S contained in L, must coincide with L.

To show that the $\mathscr{L}$-class $L\backslash\{0\}$ is minimal in the set of non-zero $\mathscr{L}$-classes, notice that

$$\begin{aligned} L_x \leqslant L\backslash\{0\} &\Rightarrow S^1 x \subseteq L \\ &\Rightarrow S^1 x = \{0\} \text{ or } S^1 x = L \\ &\Rightarrow L_x = \{0\} \text{ or } L_x = L\backslash\{0\}. \end{aligned}$$

The required result follows.■

The existence of a minimal non-zero $\mathscr{L}$-class in S is of course not quite enough to establish that S is completely 0-simple: we need to show that *every* non-empty set of $\mathscr{L}$-classes contains a minimal member (and the corresponding result for $\mathscr{R}$-classes). We shall in fact show that *every* non-zero $\mathscr{L}$-class is minimal in the set of non-zero $\mathscr{L}$-classes, which will imply that if L_1, Ls. are two such classes then

$$L_1 \leqslant L_2 \Rightarrow L_1 = L_2.$$

This will certainly suffice.

To prove this result, let us first notice that a "translate" Ls of a 0-minimal left ideal L must either be zero or must be a 0-minimal left ideal. That it is a left ideal is obvious, and if $Ls \neq \{0\}$ let B be a left ideal of S such that $\{0\} \neq B \subseteq Ls$. If we write

$$A = \{a \in L: as \in B\}$$

it is easy to see that A is a left ideal of S contained in L. Evidently $A \neq \{0\}$ and so $A = L$. Hence $B = Ls$ and so Ls is 0-minimal as required.

Now let M be the union of the collection (non-empty, since Se is one such) of all 0-minimal left ideals of S. Then $M \neq \{0\}$ and M is obviously a left ideal. In fact it is a right ideal also, for if $m \in M$ we must have $m \in L$ for some 0-minimal left ideal L. If $s \in S$ then $ms \in Ls$; hence either $ms = 0$ (which *certainly* is in M) or Ls is a 0-minimal left ideal of S, in which case $ms \in Ls \subseteq M$.

But S is 0-simple, and so $M = S$. If, then, a is an arbitrary element of $S\backslash\{0\}$, we must have that $a \in L\backslash\{0\}$ for some 0-minimal left ideal L, and so $L_a = L\backslash\{0\}$, a minimal non-zero $\mathscr{L}$-class of S. By a similar argument we may show that every non-zero $\mathscr{R}$-class in S is minimal in the set of non-zero $\mathscr{R}$-classes. Hence S is completely 0-simple as required.■

A simplified version of this argument establishes

Corollary 3.4. *A simple semigroup is completely simple if and only if it contains a primitive idempotent.*■

If $\{S_\alpha : \alpha \in A\}$ is a family of semigroups each with a zero element, we can combine these semigroups together to form a new semigroup in a very straightforward way as follows. First, as a matter of notational convenience we may denote all the zero elements by the same symbol 0. Then it is easy to verify that the set S which consists of 0 together with the disjoint union of all the sets $S_\alpha \backslash \{0\}$ becomes a semigroup if we define the product of x and y in S to be their product in S_α if they are from the same S_α and to be zero otherwise. Thus

$$S_\alpha S_\beta = \{0\}$$

if $\alpha \neq \beta$. In fact S, which, following Clifford and Preston (1961) we shall call the 0-*direct union* of the semigroups S_α, contains each of the semigroups S_α as a two-sided ideal, for

$$S S_\alpha S = S_\alpha^3 \subseteq S_\alpha.$$

Now, it is elementary to observe that S is a regular semigroup if every S_α is regular, and that any idempotent primitive in S_α is also primitive in S. Hence, since completely 0-simple semigroups are regular (Lemma 2.6) and since, as we remarked just before the statement of Theorem 3.1, every non-zero idempotent of a completely 0-simple semigroup is primitive, we may deduce that a 0-direct union of completely 0-simple semigroups is regular and has the property that every non-zero idempotent is primitive.

In fact, the converse also holds:

Theorem 3.5. *If S is a regular semigroup with zero in which every non-zero idempotent is primitive, then S is a 0-direct union of completely 0-simple semigroups.*

Proof. Since S is regular it follows by Proposition II.3.2 that every $\mathscr{D}$-class of S contains an idempotent. Hence certainly every $\mathscr{J}$-class contains an idempotent. If J_e is a typical non-zero $\mathscr{J}$-class, where e is a (primitive) idempotent, then in fact J_e is a *minimal* non-zero $\mathscr{J}$-class. To see this, suppose that $\{0\} \neq J_f \leqslant J_e$, where f is another (primitive) idempotent. Then $f = xey$ for some x, y in S^1 and the element $g = eyfxe$ has the properties

$$g^2 = eyfxe^2yfxe = eyf^3xe = eyfxe = g,$$

$$eg = ge = g, xgy = f.$$

Hence g is a non-zero idempotent and $g \leqslant e$. Thus $g = e$. Since $f = xgy$ and $g = eyfxe$ we have that $J_f = J_g = J_e$, and the minimality of J_e among the non-zero $\mathscr{J}$-classes is established.

It now follows that (in the notation of Section 1) $I(e) = \{0\}$ for each idempotent e, and so the principal factors of S are all of the form $J_e \cup \{0\}$ where e is a primitive idempotent and J_e is a minimal non-zero $\mathscr{J}$-class. Such principal factors cannot be null and so they are 0-simple by Theorem 1.9. By Theorem 3.1, since they contain primitive idempotents, they are completely 0-simple.

Finally, if $x \in J_e$ and $y \in J_f$, where J_e, J_f are distinct non-zero $\mathscr{J}$-classes of S, then, by (II.1.10).

$$J_{xy} \leqslant J_x = J_e, \qquad J_{xy} \leqslant J_y = J_f.$$

By minimality, this implies that

$$J_{xy} = \{0\} \quad \text{or} \quad J_{xy} = J_e, \quad \text{and} \quad J_{xy} = \{0\} \quad \text{or} \quad J_{xy} = J_f.$$

Since $J_e \neq J_f$ we conclude that $J_{xy} = \{0\}$. Hence $xy = 0$ and so S is the 0-direct union of the completely 0-simple semigroups $J_e \cup \{0\}$. ■

Theorem 3.5 simplifies considerably in the case where there is no zero element:

Corollary 3.6. *If S is a regular semigroup without zero in which every idempotent is primitive, then S is completely simple.*

Proof. If we follow the method of proof used for Theorem 3.5 we find that for each (primitive) idempotent e the $\mathscr{J}$-class J_e coincides with the kernel of S. Thus, if E is the set of idempotents of S,

$$S = \bigcup_{e \in E} J(e) = k(S).$$

That is, S coincides with its kernel. Hence, by Proposition 1.5, S is simple. But S contains a primitive idempotent and so by Corollary 3.4 is completely simple. ■

4.* CONGRUENCES ON COMPLETELY 0-SIMPLE SEMIGROUPS

There is often room for argument as to what constitutes a satisfactory structure theory in algebra. In the context of Abelian group theory Kaplansky (1954) proposed certain test questions that a truly satisfactory structure theory ought to enable one to answer. Kaplansky's questions do not seem very appropriate to semigroup theory, and the test question that seems to have been asked about semigroup structure theories is "Does the theory enable one to give an explicit description of the congruences?" The answer to the question often entails work of a tediously detailed nature, and in a book of modest size it will not be appropriate to investigate our test question in detail for every structure theorem we encounter. In most cases a reference to the literature will suffice. It seems reasonable, however, to make an exception in the case of the Rees structure theorem for completely 0-simple semigroups, partly because of its great importance in semigroup theory, but also because various authors (Gluskin (1956, 1957), Tamura (1960), Preston (1961) and Lallement (1967, 1974) have given differing accounts. Kapp and Schneider (1969) give a description of the congruences that does not depend explicitly on the Rees Theorem at all. The account given here has features in common with several of the published versions; it resembles most closely that of Lallement (1974).

If ρ is a congruence on a semigroup S with 0, then it is easy to verify that 0ρ, the ρ-class containing 0, is an ideal of S. If S is 0-simple, it follows that $0\rho = \{0\}$ or $0\rho = S$. In the latter case $\rho = S \times S$, the universal congruence, and so there will be little loss if we restrict attention for the moment to what we shall call *proper* congruences ρ, for which $0\rho = \{0\}$.

Let $S = \mathscr{M}^\circ[G; I, \Lambda; P]$ be a completely 0-simple semigroup. We define an equivalence relation $\mathscr{E}_I$ on I by the rule that

$$(i, j) \in \mathscr{E}_I \quad \text{if} \quad \{\lambda \in \Lambda : p_{\lambda i} = 0\} = \{\lambda \in \Lambda : p_{\lambda j} = 0\}, \tag{4.1}$$

and an equivalence relation $\mathscr{E}_\Lambda$ on Λ by the analogous rule that

$$(\lambda, \mu) \in \mathscr{E}_\Lambda \quad \text{if} \quad \{i \in I : p_{\lambda i} = 0\} = \{i \in I : p_{\mu i} = 0\}. \tag{4.2}$$

If ρ is a proper congruence on S, we define a relation ρ_I on I by the rule that $(i, j) \in \rho_I$ if $(i, j) \in \mathscr{E}_I$ and if

$$(p_{\lambda i}^{-1}, i, \lambda)\ \rho\ (p_{\lambda j}^{-1}, j, \lambda) \tag{4.3}$$

for every λ in Λ such that $p_{\lambda i}$ (and hence also $p_{\lambda j}$) is non-zero.

It is obvious that ρ_I is reflexive and symmetric. To show that it is transitive, note first that if $(i, j) \in \rho_I \subseteq \mathscr{E}_I$ and $(j, k) \in \rho_I \subseteq \mathscr{E}_I$, then certainly $(i, k) \in \mathscr{E}_I$. If λ is such that $p_{\lambda i} \neq 0$, then $p_{\lambda j}$ and $p_{\lambda k}$ are non-zero also, and

$$(p_{\lambda i}^{-1}, i, \lambda)\ \rho\ (p_{\lambda j}^{-1}, j, \lambda), \qquad (p_{\lambda i}^{-1}, j, \lambda)\ \rho\ (p_{\lambda k}^{-1}, k, \lambda).$$

The required result now follows by the transitivity of ρ.

By analogy, we have an equivalence relation ρ_Λ on Λ defined by the rule that $(\lambda, \mu) \in \rho_\Lambda$ if $(\lambda, \mu) \in \mathscr{E}_\Lambda$ and if

$$(p_{\lambda i}^{-1}, i, \lambda)\ \rho\ (p_{\mu i}^{-1}, i, \mu) \tag{4.4}$$

for every i in I such that $p_{\lambda i}$ (and hence also $p_{\mu i}$) is non-zero.

Once again, we select an arbitrary non-null $\mathscr{H}$-class of S and call it (without loss of generality) H_{11}. Let

$$N_\rho = \{a \in G : (a, 1, 1)\,\rho\,(e, 1, 1)\},$$

where e is the identity of G. Then $e \in N_\rho$ and so $N_\rho \neq \varnothing$. In fact,

Lemma 4.6. *N_ρ is a normal subgroup of G.*

Proof. If $a, b \in N_\rho$, so that

$$(a, 1, 1)\,\rho\,(e, 1, 1) \quad \text{and} \quad (b, 1, 1)\,\rho\,(e, 1, 1),$$

then

$$(a, 1, 1)(p_{11}^{-2}, 1, 1)(b, 1, 1)\,\rho\,(e, 1, 1)(p_{11}^{-2}, 1, 1)(e, 1, 1),$$

i.e.

$$(ab, 1, 1)\,\rho\,(e, 1, 1);$$

hence $ab \in N_\rho$. If $a \in N_\rho$, so that $(a, 1, 1)\,\rho\,(e, 1, 1)$, then

$$(a, 1, 1)(p_{11}^{-1}a^{-1}, 1, 1)\,\rho\,(e, 1, 1)(p_{11}^{-1}a^{-1}, 1, 1),$$

i.e.

$$(e, 1, 1)\,\rho\,(a^{-1}, 1, 1);$$

hence $a^{-1} \in N_\rho$ Finally, if $a \in N_\rho$ and $g \in G$, then

$$(g^{-1}p_{11}^{-1}, 1, 1)(a, 1, 1)(p_{11}^{-1}g, 1, 1)\,\rho\,(g^{-1}p_{11}^{-1}, 1, 1)(e, 1, 1)(p_{11}^{-1}g, 1, 1),$$

i.e.

$$(g^{-1}ag, 1, 1)\,\rho\,(e, 1, 1);$$

hence $g^{-1}ag \in N_\rho$. ■

Before moving on, let us record that (as may easily be verified)

$$(a, 1, 1)\,\rho\,(b, 1, 1) \quad \text{if and only if} \quad ab^{-1} \in N_\rho. \tag{4.7}$$

So far, then, we have seen that a proper congruence ρ determines equivalence relations $\rho_I(\subseteq \mathscr{E}_I)$ and $\rho_\Lambda(\subseteq \mathscr{E}_\Lambda)$ on I, Λ, respectively, and a normal subgroup N_ρ of G. These three objects are not independent, however, and to describe the precise nature of their interdependence we need to introduce a notion due to Lallement (1967). If $p_{\lambda i}, p_{\mu i}, p_{\lambda j}, p_{\mu j}$ are all non-zero, then $p_{\lambda i}p_{\mu i}^{-1}p_{\mu j}p_{\lambda j}^{-1}$ is an element of G. We call it an *extract* of the matrix P, and write

$$q_{\lambda\mu ij} = p_{\lambda i}p_{\mu i}^{-1}p_{\mu j}p_{\lambda j}^{-1}. \tag{4.8}$$

It is easy to verify the following identities, which will be required later:

$$q_{\lambda\mu ij}q_{\lambda\mu jk} = q_{\lambda\mu ik}, \tag{4.9}$$

and

$$p_{\lambda i}^{-1}q_{\lambda\mu ij}^{-1}p_{\lambda i} \cdot p_{\mu i}^{-1}q_{\mu\nu ij}^{-1}p_{\mu i} = p_{\lambda i}^{-1}q_{\lambda\nu ij}^{-1}p_{\lambda i}. \tag{4.10}$$

Also, we have

Lemma 4.11. *Let ρ be a proper congruence on a completely* 0*-simple semigroup $S = \mathscr{M}^\circ[G; I, \Lambda; P]$, and let i, j, λ, μ be such that $p_{\lambda i}, p_{\lambda j}, p_{\mu i}, p_{\mu j}$ are all non-zero. If either $(i, j) \in \rho_I$ or $(\lambda, \mu) \in \rho_\Lambda$ then $q_{\lambda\mu ij} \in N_\rho$.*

Proof. Let $(i, j) \in \rho_I$; then

$$(p_{\mu i}^{-1}, i, \mu)\, \rho\, (p_{\mu j}^{-1}, j, \mu)$$

and so

$$(e, 1, \lambda)(p_{\mu i}^{-1}, i, \mu)(p_{\lambda j}^{-1}, j, 1)\, \rho\, (e, 1, \lambda)(p_{\mu j}^{-1}, j, \mu)(p_{\lambda j}^{-1}, j, 1);$$

i.e.

$$(p_{\lambda i}p_{\mu i}^{-1}p_{\mu j}p_{\lambda j}^{-1}, 1, 1)\, \rho\, (e, 1, 1)$$

and hence $q_{\lambda\mu ij} \in N_\rho$. Similarly, if $(\lambda, \mu) \in \rho_\Lambda$, then

$$(e, 1, \lambda)(p_{\lambda i}^{-1}, i, \lambda)(p_{\lambda j}^{-1}, j, 1)\, \rho\, (e, 1, \lambda)(p_{\mu i}^{-1}, i, \mu)(p_{\lambda j}^{-1}, j, 1),$$

i.e.

$$(e, 1, 1)\, \rho\, (p_{\lambda i}p_{\mu i}^{-1}p_{\mu j}p_{\lambda j}^{-1}, 1, 1);$$

hence again $q_{\lambda\mu ij} \in N_\rho$. ■

This result motivates the following definition. A triple $(N, \mathscr{S}, \mathscr{T})$ consisting of a normal subgroup N of G, an equivalence relation $\mathscr{S}$ on I and an equivalence relation $\mathscr{T}$ on Λ will be called *admissible* if

(4.12) $\mathscr{S} \subseteq \mathscr{E}_I,\ \mathscr{T} \subseteq \mathscr{E}_\Lambda$;

(4.13) if $i, j \in I$ and $\lambda, \mu \in \Lambda$ are such that $p_{\lambda i}, p_{\lambda j}, p_{\mu i}, p_{\mu j}$ are all non-zero, then $q_{\lambda\mu ij} \in N$ whenever either $(i, j) \in \mathscr{S}$ or $(\lambda, \mu) \in \mathscr{T}$.

What we have shown thus far is that there exists a map Γ from the set of proper congruences of S into the set of admissible triples, associating ρ with the admissible triple $(N_\rho, \rho_I, \rho_\Lambda)$. In fact we shall see eventually that Γ is a bijection.

As a first step towards a full examination of the map Γ, we have

Lemma 4.14. *If the elements (a, i, λ) and (b, j, μ) of S are such that $(a, i, \lambda)\, \rho(b, j, \mu)$, then $(i, j) \in \rho_I$.*

Proof. We show first that $(i, j) \in \mathscr{E}_I$, i.e. that $(\forall \xi \in \Lambda)\ p_{\xi i} = 0$ if and only if $p_{\xi j} = 0$. To see this, notice that if $p_{\xi i} = 0$ then

$$(e, 1, \xi)(a, i, \lambda) = 0.$$

Since ρ is a proper congruence we must also have that

$$(e, 1, \xi)(b, j, \mu) = 0,$$

from which it follows that $p_{\xi j} = 0$. Thus

$$p_{\xi i} = 0 \Rightarrow p_{\xi j} = 0$$

and the converse implication clearly holds also. A similar argument shows that $(\lambda, \mu) \in \mathscr{E}_\Lambda$.

If now $k \in I$ is chosen so that $p_{\lambda k}$ (and hence also $p_{\mu k}$) is non-zero, then

$$(a, i, \lambda)(e, k, 1)\, \rho\, (b, j, \mu)(e, k, 1),$$

i.e.

$$(ap_{\lambda k}, i, 1)\, \rho\, (bp_{\mu k}, j, 1) \tag{4.15}$$

Hence, for any $\xi \in \Lambda$ such that $p_{\xi i} \neq 0$ (and so $p_{\xi j} \neq 0$ also),

$$(e, 1, \xi)(ap_{\lambda k}, i, 1)(e, 1, 1)\, \rho\, (e, 1, \xi)(bp_{\mu k}, j, 1)(e, 1, 1),$$

i.e.

$$(p_{\xi i} a p_{\lambda k} p_{11}, 1, 1)\, \rho\, (p_{\xi j} b p_{\mu k} p_{11}, 1, 1).$$

Writing this as $(c, 1, 1)\, \rho\, (d, 1, 1)$ for brevity, we deduce from (4.7) that $cd^{-1} \in N_\rho$. Hence $c^{-1}d = c^{-1}(cd^{-1})^{-1}c \in N_\rho$ and so, again by (4.7),

$$(c^{-1}, 1, 1)\, \rho\, (d^{-1}, 1, 1),$$

which in the original notation gives us that

$$(p_{11}^{-1} p_{\lambda k}^{-1} a^{-1} p_{\xi i}^{-1}, 1, 1)\ \rho\ (p_{11}^{-1} p_{\mu k}^{-1} b^{-1} p_{\xi j}^{-1}, 1, 1).$$

This, together with (4.15), implies that

$$(p_{\xi i}^{-1}, i, 1)\ \rho\ (p_{\xi j}^{-1}, j, 1),$$

and if we postmultiply this by $(p_{11}^{-1}, 1, \xi)$ we obtain, as required, that

$$(p_{\xi i}^{-1}, i, \xi)\ \rho\ (p_{\xi j}^{-1}, j, \xi)$$

for every ξ in Λ for which $p_{\xi i}$ (and hence also $p_{\xi j}$) is non-zero. ■

A similar argument establishes

Lemma 4.16. *If the elements (a, i, λ) and (b, j, μ) of S are such that (a, i, λ) $\rho(b, j, \mu)$, then $(\lambda, \mu) \in \rho_\Lambda$.* ■

The next result shows that the map Γ from the set of proper congruences on S into the set of admissible triples is order-preserving:

Lemma 4.17. *If ρ and σ are proper congruences on $S = \mathscr{M}^{\circ}[G; I, \Lambda; P]$, then $\rho \subseteq \sigma$ if and only if $\rho_I \subseteq \sigma_I$, $\rho_\Lambda \subseteq \sigma_\Lambda$ and $N_\rho \subseteq N_\sigma$.*

Proof. If $\rho \subseteq \sigma$ then it is clear from the definitions of ρ_I, ρ_Λ and N_ρ that $\rho_I \subseteq \sigma_I$, $\rho_\Lambda \subseteq \sigma_\Lambda$ and $N_\rho \subseteq N_\sigma$. Conversely, suppose that ρ and σ are congruences on S such that $\rho_I \subseteq \sigma_I$, $\rho_\Lambda \subseteq \sigma_\Lambda$ and $N_\rho \subseteq N_\sigma$, and suppose that $(a, i, \lambda)\, \rho\, (b, j, \mu)$. Then by Lemmas 4.14 and 4.16,

$$(i, j) \in \rho_I \subseteq \sigma_I, \qquad (\lambda, \mu) \in \rho_\Lambda \subseteq \sigma_\Lambda. \tag{4.18}$$

Also, if $\xi \in \Lambda$ and $x \in I$ are chosen so that $p_{\xi i} \neq 0$, $p_{\lambda x} \neq 0$ (and hence also so that $p_{\xi j} \neq 0$, $p_{\mu x} \neq 0$), then

$$(e, 1, \xi)(a, i, \lambda)(e, x, 1)\, \rho\, (e, 1, \xi)(b, j, \mu)(e, x, 1),$$

i.e.

$$(p_{\xi i} a p_{\lambda x}, 1, 1)\, \rho\, (p_{\xi j} b p_{\mu x}, 1, 1);$$

hence $p_{\xi i} a p_{\lambda x} p_{\mu x}^{-1} b^{-1} p_{\xi j}^{-1} \in N_\rho \subseteq N_\sigma$, and so

$$(p_{\xi i} a p_{\lambda x}, 1, 1)\, \sigma\, (p_{\xi j} b p_{\mu x}, 1, 1)$$

by (4.7). Now, it follows from (4.18) that

$$(p_{\xi i}^{-1}, i, \xi)\, \sigma\, (p_{\xi j}^{-1}, j, \xi);$$

hence, postmultiplying by $(p_{\xi i}^{-1} p_{11}^{-1}, i, 1)$ we obtain that

$$(p_{\xi i}^{-1} p_{11}^{-1}, i, 1)\, \sigma\, (p_{\xi j}^{-1} p_{11}^{-1}, j, 1).$$

Similarly,

$$(p_{11}^{-1} p_{\lambda x}^{-1}, 1, \lambda)\, \sigma\, (p_{11}^{-1} p_{\mu x}^{-1}, 1, \mu).$$

Since σ is a congruence, it now follows that

$$(p_{\xi i}^{-1} p_{11}^{-1}, i, 1)(p_{\xi i} a p_{\lambda x}, 1, 1)(p_{11}^{-1} p_{\lambda x}^{-1}, 1, \lambda)$$

and

$$(p_{\xi j}^{-1} p_{11}^{-1}, j, 1)(p_{\xi j} b p_{\mu x}, 1, 1)(p_{11}^{-1} p_{\mu x}^{-1}, 1, \mu)$$

are equivalent modulo σ, i.e. that $(a, i, \lambda)\, \sigma\, (b, j, \mu)$. ■

As a corollary to this result we have that if $\rho_I = \sigma_I$, $\rho_\Lambda = \sigma_\Lambda$ and $N_\rho = N_\sigma$, then $\rho = \sigma$; thus the map Γ is one–one. That Γ is onto is a consequence of the following result:

Lemma 4.19. *Let $(N, \mathscr{S}, \mathscr{T})$ be an admissible triple and let the relation ρ on $S \backslash \{0\}$ be defined by the rule that $(a, i, \lambda)\, \rho\, (b, j, \mu)$ if and only if*
(1) $(i, j) \in \mathscr{S}$, $(\lambda, \mu) \in \mathscr{T}$;

(2) $p_{\xi i}ap_{\lambda x}p_{\mu x}^{-1}b^{-1}p_{\xi j}^{-1} \in N$ *for some* $x \in I$ *and* $\xi \in \Lambda$ *such that* $p_{\xi i}, p_{\xi j}, p_{\lambda x}$ *and* $p_{\mu x}$ *are all non-zero.*

Then $\rho \cup \{(0, 0)\}$ *is a proper congruence on S such that* $\rho_I = \mathscr{S}$, $\rho_\Lambda = \mathscr{T}$ *and* $N_\rho = N$.

Proof. It is evident that ρ is reflexive. It is also symmetric, since $\mathscr{S}$ and $\mathscr{T}$ are symmetric, and since $p_{\xi i}ap_{\lambda x}p_{\mu x}^{-1}b^{-1}p_{\xi j}^{-1} \in N$ implies that

$$p_{\xi j}bp_{\mu x}p_{\lambda x}^{-1}a^{-1}p_{\xi i}^{-1} = (p_{\xi i}ap_{\lambda x}p_{\mu x}^{-1}b^{-1}p_{\xi j}^{-1})^{-1} \in N.$$

The verification of the transitivity of ρ presents us with some difficulties, and it pays first to develop a slightly different characterization of ρ in which the words "for some" are replaced by "for every":

Lemma 4.20. *If* ρ *is as defined in the statement of Lemma* 4.19, *then* $(a, i, \lambda)\ \rho\ (b, j, \mu)$ *if and only if*

(1) $(i, j) \in \mathscr{S}$, $(\lambda, \mu) \in \mathscr{T}$

(2) $p_{\xi i}ap_{\lambda x}p_{\mu x}^{-1}b^{-1}p_{\xi j}^{-1} \in N$ *for EVERY* $x \in I$ *and* $\xi \in \Lambda$ *such that* $p_{\xi i}, p_{\xi j}, p_{\lambda x}$ *and* $p_{\mu x}$ *are all non-zero.*

Proof. It is obvious that the relation defined in this new way is contained in ρ. To show the reverse inclusion we must show that if $p_{\xi i}ap_{\lambda x}p_{\mu x}^{-1}b^{-1}p_{\xi j}^{-1} \in N$ and if $y \in I$ and $\eta \in \Lambda$ are such that $p_{\eta i}, p_{\eta j}, p_{\lambda y}$ and $p_{\mu y}$ are all non-zero, then it is also the case that $p_{\eta i}ap_{\lambda y}p_{\mu y}^{-1}b^{-1}p_{\eta j}^{-1} \in N$. Now, a normal subgroup N of a group G has the "symmetric" property that for all c, d in G,

$$cd \in N \Rightarrow dc \in N$$

(for clearly $dc = c^{-1}(cd)c$). Using this, we deduce from the hypothesis that $p_{\xi i}ap_{\lambda x}p_{\mu x}^{-1}b^{-1}p_{\xi j}^{-1} \in N$ that

$$p_{\lambda x}p_{\mu x}^{-1}b^{-1}p_{\xi j}^{-1}p_{\xi i}a \in N.$$

Also, since $(\lambda, \mu) \in \mathscr{T}$ and since $(N, \mathscr{S}, \mathscr{T})$ is an admissible triple, we have that $q_{\lambda\mu yx} \in N$, i.e. that

$$p_{\lambda y}p_{\mu y}^{-1}p_{\mu x}p_{\lambda x}^{-1} \in N,$$

Multiplying these together, we find that

$$p_{\lambda y}p_{\mu y}^{-1}b^{-1}p_{\xi j}^{-1}p_{\xi i}a \in N.$$

Now, again by the symmetry property of N, we deduce that

$$ap_{\lambda y}p_{\mu y}^{-1}b^{-1}p_{\xi j}^{-1}p_{\xi i} \in N.$$

Also, since $(j, i) \in \mathscr{S}$ and $(N, \mathscr{S}, \mathscr{T})$ is an admissible triple, $p_{\xi i}^{-1}q_{\xi\eta ji}p_{\xi i} \in N$; i.e.

$$p_{\xi i}^{-1}p_{\xi j}p_{\eta j}^{-1}p_{\eta i} \in N.$$

Hence, by multiplying these two elements together, we obtain that

$$ap_{\lambda y}p_{\mu y}^{-1}b^{-1}p_{\eta j}^{-1}p_{\eta i} \in N,$$

from which it follows by the symmetry property that

$$p_{\eta i}ap_{\lambda y}p_{\mu y}^{-1}b^{-1}p_{\eta j}^{-1} \in N,$$

as required. ■

Returning now to the proof of Lemma 4.19, let us suppose that

$$(a, i, \lambda)\,\rho\,(b, j, \mu) \quad \text{and} \quad (b, j, \mu)\,\rho\,(c, k, \nu).$$

Then certainly $(i, k) \in \mathscr{S}$ and $(\lambda, \nu) \in \mathscr{T}$. Also, if $x \in I$ and $\xi \in \Lambda$ are such that $p_{\xi i}, p_{\xi k}, p_{\lambda x}$ and $p_{\nu x}$ are all non-zero then because $\mathscr{S} \subseteq \mathscr{E}_I$ and $\mathscr{T} \subseteq \mathscr{E}_\Lambda$ it follows that $p_{\xi j}$ and $p_{\mu x}$ are also non-zero. By the lemma just proved, we have

$$p_{\xi i}ap_{\lambda x}p_{\mu x}^{-1}b^{-1}p_{\xi j}^{-1} \in N \quad \text{and} \quad p_{\xi j}bp_{\mu x}p_{\nu x}^{-1}c^{-1}p_{\xi k}^{-1} \in N.$$

Multiplying these, we find that

$$p_{\xi i}ap_{\lambda x}p_{\nu x}^{-1}c^{-1}p_{\xi k}^{-1} \in N,$$

which is exactly what we require to show that $(a, i, \lambda)\,\rho(c, k, \nu)$.

We have shown that ρ is an equivalence relation. To show that it is a congruence, let $(a, i, \lambda)\,\rho\,(b, j, \mu)$ and let $(c, k, \nu) \in S$. Then either $p_{\nu i} = p_{\nu j} = 0$, in which case both products $(c, k, \nu)(a, i, \lambda)$ and $(c, k, \nu)(b, j, \mu)$ are zero and so certainly ρ-equivalent, or $p_{\nu i} \neq 0, p_{\nu j} \neq 0$, in which case we have to prove that

$$(cp_{\nu i}a, k, \lambda)\,\rho\,(cp_{\nu j}b, k, \mu).$$

This amounts to proving that for some $x \in I$, $\xi \in \Lambda$ such that $p_{\xi k}, p_{\lambda x}, p_{\mu x}$ are all non-zero,

$$p_{\xi k}cp_{\nu i}ap_{\lambda x}p_{\mu x}^{-1}b^{-1}p_{\nu j}^{-1}c^{-1}p_{\xi k}^{-1} \in N. \tag{4.21}$$

Now $(a, i, \lambda)\,\rho\,(b, j, \mu)$, and so by Lemma 4.20 we have that $p_{\xi i}ap_{\lambda x}p_{\mu x}^{-1}b^{-1}p_{\xi j}^{-1} \in N$ for *every* $x \in I$, $\xi \in \Lambda$ such that $p_{\xi i}, p_{\xi j}, p_{\lambda x}$ and $p_{\mu x}$ are all non-zero. Since we are presently assuming that $p_{\nu i}$ and $p_{\nu j}$ are non-zero we may substitute $\xi = \nu$ at this point and assert that

$$p_{\nu i}ap_{\lambda x}p_{\mu x}^{-1}b^{-1}p_{\nu j}^{-1} \in N$$

whenever $p_{\lambda x}$ and $p_{\mu x}$ are non-zero. The desired formula (4.21) follows immediately from this by conjugation. Thus ρ is a left congruence.

To show that ρ is a right congruence, once again let $(a, i, \lambda)\,\rho\,(b, j, \mu)$ and let $(c, k, \nu) \in S$. If $p_{\lambda k} = p_{\mu k} = 0$ then

$$(a, i, \lambda)(c, k, \nu) = (b, j, \mu)(c, k, \nu) = 0$$

and so the products are certainly ρ-equivalent. Since $(\lambda, \mu) \in \mathscr{T} \subseteq \mathscr{E}_\Lambda$, the only other possibility is that $p_{\lambda k}$ and $p_{\mu k}$ are both non-zero, and in this case we begin by noting that

$$p_{\xi i} a p_{\lambda k} p_{\mu k}^{-1} b^{-1} p_{\xi j}^{-1} \in N$$

for every $\xi \in \Lambda$ such that $p_{\xi i}$ and $p_{\xi j}$ are non-zero. (Here we have chosen $x = k$, as we may.) Hence, if $x \in I$ is such that $p_{\nu x} \neq 0$,

$$p_{\xi i} a p_{\lambda k} c p_{\nu x} p_{\nu x}^{-1} c^{-1} p_{\mu k}^{-1} b^{-1} p_{\xi j}^{-1} \in N,$$

from which it follows that

$$(a p_{\lambda k} c, i, \nu)\ \rho\ (b p_{\mu k} c, j, \nu)$$

as required. Thus $\rho \cup \{(0, 0)\}$ is a proper congruence on S.

It remains to show that $\rho_I = \mathscr{S}$, $\rho_\Lambda = \mathscr{T}$ and $N_\rho = N$. First, if $(i, j) \in \rho_I$ then

$$(p_{\lambda i}^{-1}, i, \lambda)\ \rho\ (p_{\lambda j}^{-1}, j, \lambda)$$

for every λ such that $p_{\lambda i} \neq 0$. Hence, by the definition of ρ we have that $(i, j) \in \mathscr{S}$. Conversely if $(i,j) \in \mathscr{S}$ and $p_{\lambda i} \neq 0$ then for every x in I and ξ in Λ such that $p_{\xi i}$, $p_{\xi j}$ and $p_{\lambda x}$ are all non-zero,

$$p_{\xi i} p_{\lambda i}^{-1} p_{\lambda x} p_{\lambda x}^{-1} p_{\lambda j} p_{\xi j}^{-1} = q_{\xi \lambda i j} \in N.$$

Hence $(p_{\lambda i}^{-1}, i, \lambda)\ \rho\ (p_{\lambda j}^{-1}, j, \lambda)$ and so $(i, j) \in \rho_I$. Thus $\rho_I = \mathscr{S}$ and similarly we may show that $\rho_\Lambda = \mathscr{T}$.

Finally, $a \in N_\rho$ if and only if $(a, 1, 1)\ \rho\ (e, 1, 1)$, i.e. if and only if for every x in I and ξ in Λ such that $p_{\xi 1}$ and p_{1x} are non-zero,

$$p_{\xi 1} a p_{1x} p_{1x}^{-1} e p_{\xi 1}^{-1} \in N,$$

i.e. if and only if $a \in N$. Thus $N_\rho = N$. ∎

We summarize the results of this section in a theorem as follows:

Theorem 4.22. *Let $S = \mathscr{M}^\circ[G; I, \Lambda; P]$ be a completely 0-simple semigroup. Then the mapping $\Gamma: \rho \mapsto (N_\rho, \rho_I, \rho_\Lambda)$ is an order-preserving bijection from the set of proper congruences on S onto the set of admissible triples.* ∎

If we study completely simple semigroups (without zero) many of the complications in the foregoing account disappear. In particular, we do not need to restrict to proper congruences, and the equivalences $\mathscr{E}_I$, $\mathscr{E}_\Lambda$ do not appear An *admissible* triple is then simply a triple $(N, \mathscr{S}, \mathscr{T})$ consisting of a normal subgroup of G and equivalence relations $\mathscr{S}, \mathscr{T}$ on I, Λ respectively, with the property that $(i, j) \in \mathscr{S}$ or $(\lambda, \mu) \in \mathscr{T}$ implies that $q_{\lambda \mu i j} \in N$. We then have

Theorem 4.23. *Let* $S = \mathcal{M}[G; I, \Lambda; P]$ *be a completely simple semigroup. Then the mapping* $\Gamma: \rho \mapsto (N, \rho_I, \rho_\Lambda)$ *is an order-preserving bijection from the set of congruences on S onto the set of admissible triples.* ■

5.* THE LATTICE OF CONGRUENCES ON A COMPLETELY 0-SIMPLE SEMIGROUP

In this section we use the correspondence between the proper congruences of a completely 0-simple semigroup and admissible triples to derive information on the nature of the lattice of congruences of the semigroup. If $(N, \mathcal{S}, \mathcal{T})$ is an admissible triple of $S = \mathcal{M}^\circ[G; I, \Lambda; P]$, we shall write the congruence $(N, \mathcal{S}, \mathcal{T})\Gamma^{-1}$ corresponding to the triple as $[N, \mathcal{S}, \mathcal{T}]$ (with square brackets). Thus $\rho = [N_\rho, \rho_I, \rho_\Lambda]$.

Lemma 5.1. *If* ρ *and* σ *are proper congruences on* $S = \mathcal{M}^\circ[G; I, \Lambda; P]$, *then*

$$\rho \cap \sigma = [N_\rho \cap N_\sigma, \rho_I \cap \sigma_I, \rho_\Lambda \cap \sigma_\Lambda],$$
$$\rho \vee \sigma = [N_\rho N_\sigma, \rho_I \vee \sigma_I, \rho_\Lambda \vee \sigma_\Lambda].$$

Proof. To prove the first of these statements, notice first that $(N_\rho \cap N_\sigma, \rho_I \cap \sigma_I, \rho_\Lambda \cap \sigma_\Lambda)$ is an admissible triple; for certainly $\rho_I \cap \sigma_I \subseteq \mathscr{E}_I$ and $\rho_\Lambda \cap \sigma_\Lambda \subseteq \mathscr{E}_\Lambda$; and if i, j, λ, μ are such that $p_{\lambda i}, p_{\lambda j}, p_{\mu i}$ and $p_{\mu j}$ are all non-zero, then

$$\begin{aligned}(i,j) \in \rho_I \cap \sigma_I &\Rightarrow (i,j) \in \rho_I \quad \text{and} \quad (i,j) \in \sigma_I \\ &\Rightarrow q_{\lambda\mu ij} \in N_\rho \quad \text{and} \quad q_{\lambda\mu ij} \in N_\sigma \\ &\Rightarrow q_{\lambda\mu ij} \in N_\rho \cap N_\sigma.\end{aligned}$$

Similarly,

$$(\lambda, \mu) \in \rho_\Lambda \cap \sigma_\Lambda \Rightarrow q_{\lambda\mu ij} \in N_\rho \cap N_\sigma.$$

Thus there *is* a congruence $[N_\rho \cap N_\sigma, \rho_I \cap \sigma_I, \rho_\Lambda \cap \sigma_\Lambda]$ which, by Lemma 4.17, is then necessarily contained in ρ and in σ, and is the largest congruence with these properties.

The same approach works for $\rho \vee \sigma$, except that there is more difficulty in establishing that $(N_\rho N_\sigma, \rho_I \vee \sigma_I, \rho_\Lambda \vee \sigma_\Lambda)$ is an admissible triple. Since $\rho_I, \sigma_I \subseteq \mathscr{E}_I$ we do have that $\rho_I \vee \sigma_I \subseteq \mathscr{E}_I$, and similarly $\rho_\Lambda \vee \sigma_\Lambda \subseteq \mathscr{E}_\Lambda$. Also, by elementary group-theoretic results (Hall 1959) $N_\rho N_\sigma$ is a normal subgroup of G and is the smallest normal subgroup containing N_ρ and N_σ. If $p_{\lambda i}, p_{\lambda j}, p_{\mu i}$ and $p_{\mu j}$ are all non-zero and if $(i, j) \in \rho_I \vee \sigma_I$ then by Proposition I.5.14 there exist $i_1, i_2, \ldots, i_{2n-1}$ in I such that

$$(i, i_1) \in \rho_I, (i_1, i_2) \in \sigma_I, \ldots, (i_{2n-2}, i_{2n-1}) \in \rho_I, (i_{2n-1}, j) \in \sigma_I.$$

Since ρ_I and σ_I are both contained in $\mathscr{E}_I$ we have that $p_{\lambda i_r}$ and $p_{\mu i_r}$ are non-zero for every r, and by the admissibility of the triples $(N_\rho, \rho_I, \rho_\Lambda)$ and $(N_\sigma, \sigma_I, \sigma_\Lambda)$ we have that

$$q_{\lambda\mu i i_1} \in N_\rho, q_{\lambda\mu i_1 i_2} \in N_\sigma, \ldots, q_{\lambda\mu i_{2n-2} i_{2n-1}} \in N_\rho, q_{\lambda\mu i_{2n-1} j} \in N_\sigma.$$

Hence the product

$$q_{\lambda\mu i i_1} q_{\lambda\mu i_1 i_2} \cdots q_{\lambda\mu i_{2n-2} i_{2n-1}} q_{\lambda\mu i_{2n-1} j}$$

is in $N_\rho N_\sigma$; i.e. $q_{\lambda\mu ij} \in N_\rho N_\sigma$, by virtue of (4.9).

If $(\lambda, \mu) \in \rho_\Lambda \vee \sigma_\Lambda$ then there exist $\lambda_1, \ldots, \lambda_{2n-1}$ in Λ such that

$$(\lambda, \lambda_1) \in \rho_\Lambda, (\lambda_1, \lambda_2) \in \sigma_\Lambda, \ldots, (\lambda_{2n-2}, \lambda_{2n-1}) \in \rho_\Lambda, (\lambda_{2n-1}, \mu) \in \sigma_\Lambda.$$

Again, if $i, j \in I$ are such that $p_{\lambda i}, p_{\lambda j}, p_{\mu i}$ and $p_{\mu j}$ are all non-zero, then $p_{\lambda_r i}$ and $p_{\lambda_r j}$ are non-zero for every r, and

$$q_{\lambda\lambda_1 ij} \in N_\rho, q_{\lambda_1\lambda_2 ij} \in N_\sigma, \ldots, q_{\lambda_{2n-1}\mu ij} \in N_\sigma.$$

Hence, since N_ρ and N_σ are normal subgroups,

$$p_{\lambda i}^{-1} q_{\lambda\lambda_1 ij}^{-1} p_{\lambda i} \in N_\rho, p_{\lambda_1 i}^{-1} q_{\lambda_1\lambda_2 ij}^{-1} p_{\lambda_1 i} \in N_\sigma, \ldots, p_{\lambda_{2n-1} i}^{-1} q_{\lambda_{2n-1}\mu ij}^{-1} p_{\lambda_{2n-1} i} \in N_\sigma.$$

Hence the product of these is in $N_\rho N_\sigma$. That is, by (4.10), $p_{\lambda i}^{-1} q_{\lambda\mu ij}^{-1} p_{\lambda i} \in N_\rho N_\sigma$, from which it follows immediately that $q_{\lambda\mu ij} \in N_\rho N_\sigma$ as required.

Thus $(N_\rho N_\sigma, \rho_I \vee \sigma_I, \rho_\Lambda \vee \sigma_\Lambda)$ is an admissible triple, and it now follows easily from Lemma 4.17 that $[N_\rho N_\sigma, \rho_I \vee \sigma_I, \rho_\Lambda \vee \sigma_\Lambda] = \rho \vee \sigma$. ■

We can now use these characterizations of the intersection and the join of two proper congruences on S to prove the following result:

Theorem 5.2. *The lattice of congruences on a completely* 0-*simple semigroup* S *is semimodular.*

Proof. Let $S = \mathscr{M}^\circ[G; I, \Lambda; P]$. Let $(\mathscr{N}, \cap, .)$ be the lattice of normal subgroups of G, let $(\mathscr{P}, \cap, \vee)$ be the lattice consisting of all equivalences on I contained in $\mathscr{E}_I$, and let $(\mathscr{Q}, \cap, \vee)$ be the lattice consisting of all equivalences on Λ contained in $\mathscr{E}_\Lambda$. Let $\mathscr{X} = \mathscr{N} \times \mathscr{P} \times \mathscr{Q}$ be the direct product of these three lattices. Then by Lemma 5.1 the subset $\mathscr{Y}$ of $\mathscr{X}$ consisting of all *admissible* triples $(N, \mathscr{S}, \mathscr{T})$ in $\mathscr{X}$ is a sublattice of $\mathscr{X}$. The effect of Theorem 4.22 and Lemma 5.1 is to give us an isomorphism Γ between the lattice $(\mathscr{K}, \cap, \vee)$ of proper congruences on S and the lattice $\mathscr{Y}$. Now $\mathscr{X}$ is semimodular by Corollary I.8.6 and Propositions I.8.7, I.8.8 and I.8.9. It will follow that $\mathscr{Y}$ is semimodular if we establish the following result:

Lemma 5.3. *If* $(N_1, \mathscr{S}_1, \mathscr{T}_1), (N_2, \mathscr{S}_2, \mathscr{T}_2) \in \mathscr{Y}$, *then* $(N_1, \mathscr{S}_1, \mathscr{T}_1)$ *covers* $(N_2, \mathscr{S}_2, \mathscr{T}_2)$ *in* $\mathscr{Y}$ *if and only if* $(N_1, \mathscr{S}_1, \mathscr{T}_1)$ *covers* $(N_2, \mathscr{S}_2, \mathscr{T}_2)$ *in* $\mathscr{X}$.

Proof. One way round, this is obvious. The difficulty arises because it is conceivable that $(N_1, \mathscr{S}_1, \mathscr{T}_1)$ could cover $(N_2, \mathscr{S}_2, \mathscr{T}_2)$ in $\mathscr{Y}$ but that there might be an element $(N, \mathscr{S}, \mathscr{T})$ in $\mathscr{X}\backslash\mathscr{Y}$ (an "inadmissible" triple) strictly in between. To see that this cannot happen, notice that if $N_1 \supset N_2$ then $(N_1, \mathscr{S}_2, \mathscr{T}_2)$ is an admissible triple such that

$$(N_1, \mathscr{S}_1, \mathscr{T}_1) \geqslant (N_1, \mathscr{S}_2, \mathscr{T}_2) > (N_2, \mathscr{S}_2, \mathscr{T}_2);$$

hence, by the covering property, $\mathscr{S}_1 = \mathscr{S}_2$ and $\mathscr{T}_1 = \mathscr{T}_2$.

Equally, if $\mathscr{S}_1 \supset \mathscr{S}_2$, then $(N_1, \mathscr{S}_2, \mathscr{T}_1)$ is an admissible triple such that

$$(N_1, \mathscr{S}_1, \mathscr{T}_1) > (N_1, \mathscr{S}_2, \mathscr{T}_1) \geqslant (N_2, \mathscr{S}_2, \mathscr{T}_2),$$

and hence $N_1 = N_2$, $\mathscr{T}_1 = \mathscr{T}_2$.

Applying a similar argument to the case where $\mathscr{T}_1 \supset \mathscr{T}_2$ we draw the general conclusion that $(N_1, \mathscr{S}_1, \mathscr{T}_1)$ covers $(N_2, \mathscr{S}_2, \mathscr{T}_2)$ in $\mathscr{Y}$ if and only if either

$$\text{(a) } N_1 \succ N_2 \quad \text{in } \mathscr{N},\ \mathscr{S}_1 = \mathscr{S}_2, \mathscr{T}_1 = \mathscr{T}_2;$$

or

$$\text{(b) } N_1 = N_2, \mathscr{S}_1 \succ \mathscr{S}_2 \quad \text{in } \mathscr{P},\ \mathscr{T}_1 = \mathscr{T}_2;$$

or

$$\text{(c) } N_1 = N_2, \mathscr{S}_1 = \mathscr{S}_2, \mathscr{T}_1 \succ \mathscr{T}_2 \quad \text{in } \mathscr{Q}.$$

It is now clear that there is no room at all for a strictly intermediate triple $(N, \mathscr{S}, \mathscr{T})$, even an inadmissible one. ■

We may now conclude that the lattice $\mathscr{K}$ of proper congruences on S is semimodular, since it is isomorphic to $\mathscr{Y}$. The lattice $\mathscr{C}(S)$ of *all* congruences on S is obtained from $\mathscr{K}$ by adjoining a single extra maximum element, the universal congruence $S \times S$. Since the maximum element can never figure either as a or as b in the hypothesis "if $a \succ a \wedge b$ and $b \succ a \wedge b$" appearing in the definition of semimodularity, the adjunction of this element does not destroy the semimodularity property of the lattice. The proof of Theorem 5.2 is thus complete. ■

6.* FINITE CONGRUENCE-FREE SEMIGROUPS

If a semigroup S has a proper ideal I, then the Rees quotient S/I is a proper homomorphic image of S. Thus, if S is to be *congruence-free*, i.e. is to have no congruences other than 1_S and $S \times S$, then it must in the first instance be simple or 0-simple. In the finite case this implies that S is completely simple or completely 0-simple. (This need not be so in the infinite case: see Munn (1972, 1974).)

If S has a zero, we may thus identify it with $\mathscr{M}^\circ[G; I, \Lambda; P]$. Now the triple $(G, 1_I, 1_\Lambda)$ is certainly admissible, and gives rise to a nonidentical congruence on S unless $G = \{e\}$. Thus $G = \{e\}$ if S is congruence-free. Hence every $p_{\lambda i}$ is equal either to e or to 0 and so every extract $q_{\lambda\mu ij} = e$. Thus $(\{e\}, \mathscr{E}_I, \mathscr{E}_\Lambda)$ is an admissible triple, and determines a non-identical congruence unless $\mathscr{E}_I = 1_I$, $\mathscr{E}_\Lambda = 1_\Lambda$.

Now, in the case we are considering, where $G = \{e\}$, the pair $(i, j) \in \mathscr{E}_I$ if and only if $p_{\lambda i} = e$ whenever $p_{\lambda j} = e$, i.e. if and only if column i and column j are identical. Thus to say that $\mathscr{E}_I = 1_I$ is to say that no two columns of P are identical. Equally, to say that $\mathscr{E}_\Lambda = 1_\Lambda$ is to say that no two rows of P are identical. We can summarize what we have found as follows:

Theorem 6.1. (Yamura 1956). *Let $I = \{1, \ldots, m\}$ and $\Lambda = \{1, \ldots, n\}$ be finite sets and let $P = (p_{\lambda i})$ be a regular $n \times m$ matrix of 1's and 0's such that no two rows are identical and no two columns are identical. Let $S = (I \times \Lambda) \cup \{0\}$ and let a binary operation on S be defined by*

$$(i, \lambda)(j, \mu) = \begin{cases} (i, \mu) & \text{if} \quad p_{\lambda j} = 1 \\ 0 & \text{if} \quad p_{\lambda j} = 0 \end{cases}$$

$$(i, \lambda)0 = 0(i, \lambda) = 00 = 0.$$

Then S is a congruence-free semigroup of order $mn + 1$.

Conversely, any finite congruence-free semigroup with zero is isomorphic to one of this kind. ■

This constitutes a fairly reasonable classification of finite congruence-free semigroups with zero. It is not possible to write down a general formula for the number of congruence-free semigroups of a given order, but for small orders it is not too hard to compute. (See Exercise 14.)

In the case where S has no zero, a simpler, more striking statement can be made:

Theorem 6.2. *If S is a finite congruence-free semigroup without 0 and if $|S| > 2$, then S is a simple group.*

Proof. First, let it be said that we are here including among the simple groups the cyclic groups of prime order; that is to say, we are including all non-trivial groups having no proper homomorphic images.

We may take $S = \mathscr{M}[G; I, \Lambda; P]$. If N is a proper normal subgroup of G, then $(N, 1_I, 1_\Lambda)$ is an admissible triple (since $q_{\lambda\lambda ij} = q_{\lambda\mu ii} = e$) and so $[N, 1_I, 1_\Lambda]$ is a congruence on S distinct from both 1_S $(= [\{e\}, 1_I, 1_\Lambda])$ and $S \times S$

$(= [G, I \times I, \Lambda \times \Lambda])$. Hence if S is congruence-free we must either have that G is simple or that $G = \{e\}$.

If $G = \{e\}$, then $|I \times \Lambda| = |S| > 2$ and so either $|I| = |\Lambda| = 2$ or at least one of I, Λ (say I) has more than two elements. In the first case we find admissible triples $(\{e\}, 1_I, \Lambda \times \Lambda)$ and $(\{e\}, I \times I, 1_\Lambda)$ giving rise to non-trivial congruences, while in the second case there exists an equivalence $\mathscr{S}$ on I such that $1_I \subset \mathscr{S} \subset I \times I$ and this gives rise to a non-trivial congruence $[\{e\}, \mathscr{S}, 1_\Lambda]$ on S. Hence if S is congruence-free we must have that $G \neq \{e\}$.

Thus G is a simple group. If either of I, Λ has more than one element, then $[G, 1_I, 1_\Lambda]$ is a congruence on S distinct both from 1_S and from $S \times S$. Hence if S is congruence-free we must have $|I| = |\Lambda| = 1$. Thus $S \simeq G$, a simple group. ■

The classification of finite congruence-free semigroups without zero is thus complete as far as semigroup theory is concerned, having been reduced to a group-theoretic problem. The group-theoretic problem in question, however, is known to be an extremely difficult one.

Remark. If in Theorem 6.2 we drop the proviso that $|S| > 2$ then it becomes impossible to dismiss altogether the case $G = \{e\}$. In fact we obtain a congruence-free semigroup of order 2 when $G = \{e\}$ if either $|I| = 2$ and $|\Lambda| = 1$, or $|I| = 1$ and $|\Lambda| = 2$. In the former case we obtain the left zero semigroup of order 2.

EXERCISES

1. In the infinite monogenic semigroup $S = \langle a \rangle$, show that every proper ideal is of the form Sa^m, and that

$$S \supset Sa \supset Sa^2 \supset \dots$$

Deduce that S has no minimal ideals.

2. If $S = \langle a \rangle = M(m, r)$, a monogenic semigroup with index m and period r, show that

$$K(S) = \{a^m, a^{m+1}, \dots, a^{m+r-1}\}.$$

3. (Green 1951) Show that if I, J are ideals of a semigroup S then $I \cap J$ and $I \cup J$ are ideals. Show also that $(I \cup J)/J \simeq I/(I \cap J)$.

Let A, B be ideals of a semigroup S such that $A \subset B$ and such that there does not exist any ideal C of S for which $A \subset C \subset B$. Let $b \in B \backslash A$.

(i) Show that $A \cup J(b) = B$.
(ii) Show that $I(b) = A \cap J(b)$.
(iii) Deduce that $B/A \simeq J(b)/I(b)$.

4. (Green 1951) Define

$$S_1 \supset S_2 \supset \ldots \supset S_m$$

to be a *principal series* of a semigroup S if
(i) each S_i is a (two-sided) ideal of S;
(ii) there is no ideal of S strictly between S_i and S_{i+1} $(i = 1, \ldots, m-1)$;
(iii) $S_1 = S$, $S_m = K(S)$.
Show that the factors $S_1/S_2, S_2/S_3, \ldots, S_{m-1}/S_m, S_m$ of the principal series are isomorphic, in some order, to the principal factors of the semigroup. Deduce that any two principal series have isomorphic factors.

5. (Munn 1955) A semigroup S is called *semisimple* if none of its principal factors is null. Show that S is semisimple if and only if $A^2 = A$ for every ideal of S.

6. Use the Rees theorem to show that if S is completely simple and $\mathscr{H} = 1_S$ then S is isomorphic to a rectangular band $I \times \Lambda$.

If S is completely 0-simple and $\mathscr{H} = 1_S$, show that S is isomorphic to a semigroup $(I \times \Lambda) \cup \{0\}$ whose multiplication is given in terms of a regular $\Lambda \times I$ matrix $P = (p_{\lambda i})$ with entries in $\{1, 0\}$ as follows:

$$(i, \lambda)(j, \mu) = \begin{cases} (i, \mu) & \text{if} \quad p_{\lambda j} = 1 \\ 0 & \text{if} \quad p_{\lambda j} = 0 \end{cases}$$

$$(i, \lambda)0 = 0(i, \lambda) = 00 = 0.$$

7. Let $S = \mathscr{M}[G; I, \Lambda; P]$ be a completely simple semigroup. Show that the idempotents of S form a subsemigroup if and only if

$$(\forall i, j \in I)(\forall \lambda, \mu \in \Lambda) \qquad q_{\lambda\mu ij} = e$$

(where e is the identity of G).

8. Let $S = \mathscr{M}[G; I, \Lambda; P]$ be a completely simple semigroup in which the idempotents form a subsemigroup. Choose a fixed i_0 in I and a fixed λ_0 in Λ, and for simplicity of notation write $i_0 = \lambda_0 = 1$. If E is the rectangular band $I \times \Lambda$ show that the mapping $\phi: G \times E \to S$ defined by

$$(g, (i, \lambda))\phi = (p_{1i}^{-1} g p_{11} p_{\lambda 1}^{-1}, i, \lambda)$$

is an isomorphism.

9. (See Exercise II.10) If $X = \{1, 2, 3, 4\}$ the semigroup $\mathscr{T}(X)$ has four $\mathscr{J}$-classes, $J^{(1)}$, $J^{(2)}$, $J^{(3)}$ and $J^{(4)}$, where $J^{(i)} = \{\alpha \in \mathscr{T}(X) : |\operatorname{ran} \alpha| = i\}$. Show that

$$|J^{(1)}| = 4, \qquad |J^{(2)}| = 84, \qquad |J^{(3)}| = 144, \qquad |J^{(4)}| = 24.$$

The $\mathscr{J}$-class $J^{(2)}$ can be enumerated in the egg box fashion as follows:

$\begin{pmatrix}1 & 234\\1 & 2\end{pmatrix}$ $\begin{pmatrix}1 & 234\\2 & 1\end{pmatrix}$	$\begin{pmatrix}1 & 234\\1 & 3\end{pmatrix}$ $\begin{pmatrix}1 & 234\\3 & 1\end{pmatrix}$	$\begin{pmatrix}1 & 234\\1 & 4\end{pmatrix}$ $\begin{pmatrix}1 & 234\\4 & 1\end{pmatrix}$	$\begin{pmatrix}1 & 234\\2 & 3\end{pmatrix}$ $\begin{pmatrix}1 & 234\\3 & 2\end{pmatrix}$	$\begin{pmatrix}1 & 234\\2 & 4\end{pmatrix}$ $\begin{pmatrix}1 & 234\\4 & 2\end{pmatrix}$	$\begin{pmatrix}1 & 234\\3 & 4\end{pmatrix}$ $\begin{pmatrix}1 & 234\\4 & 3\end{pmatrix}$
$\begin{pmatrix}2 & 134\\1 & 2\end{pmatrix}$ $\begin{pmatrix}2 & 134\\2 & 1\end{pmatrix}$	$\begin{pmatrix}2 & 134\\1 & 3\end{pmatrix}$ $\begin{pmatrix}2 & 134\\3 & 1\end{pmatrix}$	$\begin{pmatrix}2 & 134\\1 & 4\end{pmatrix}$ $\begin{pmatrix}2 & 134\\4 & 1\end{pmatrix}$	$\begin{pmatrix}2 & 134\\2 & 3\end{pmatrix}$ $\begin{pmatrix}2 & 134\\3 & 2\end{pmatrix}$	$\begin{pmatrix}2 & 134\\2 & 4\end{pmatrix}$ $\begin{pmatrix}2 & 134\\4 & 2\end{pmatrix}$	$\begin{pmatrix}2 & 134\\3 & 4\end{pmatrix}$ $\begin{pmatrix}2 & 134\\4 & 3\end{pmatrix}$
$\begin{pmatrix}3 & 124\\1 & 2\end{pmatrix}$ $\begin{pmatrix}3 & 124\\2 & 1\end{pmatrix}$	$\begin{pmatrix}3 & 124\\1 & 3\end{pmatrix}$ $\begin{pmatrix}3 & 124\\3 & 1\end{pmatrix}$	$\begin{pmatrix}3 & 124\\1 & 4\end{pmatrix}$ $\begin{pmatrix}3 & 124\\4 & 1\end{pmatrix}$	$\begin{pmatrix}3 & 124\\2 & 3\end{pmatrix}$ $\begin{pmatrix}3 & 124\\3 & 2\end{pmatrix}$	$\begin{pmatrix}3 & 124\\2 & 4\end{pmatrix}$ $\begin{pmatrix}3 & 124\\4 & 2\end{pmatrix}$	$\begin{pmatrix}3 & 124\\3 & 4\end{pmatrix}$ $\begin{pmatrix}3 & 124\\4 & 3\end{pmatrix}$
$\begin{pmatrix}4 & 123\\1 & 2\end{pmatrix}$ $\begin{pmatrix}4 & 123\\2 & 1\end{pmatrix}$	$\begin{pmatrix}4 & 123\\1 & 3\end{pmatrix}$ $\begin{pmatrix}4 & 123\\3 & 1\end{pmatrix}$	$\begin{pmatrix}4 & 123\\1 & 4\end{pmatrix}$ $\begin{pmatrix}4 & 123\\4 & 1\end{pmatrix}$	$\begin{pmatrix}4 & 123\\2 & 3\end{pmatrix}$ $\begin{pmatrix}4 & 123\\3 & 2\end{pmatrix}$	$\begin{pmatrix}4 & 123\\2 & 4\end{pmatrix}$ $\begin{pmatrix}4 & 123\\4 & 2\end{pmatrix}$	$\begin{pmatrix}4 & 123\\3 & 4\end{pmatrix}$ $\begin{pmatrix}4 & 123\\4 & 3\end{pmatrix}$
$\begin{pmatrix}12 & 34\\1 & 2\end{pmatrix}$ $\begin{pmatrix}12 & 34\\2 & 1\end{pmatrix}$	$\begin{pmatrix}12 & 34\\1 & 3\end{pmatrix}$ $\begin{pmatrix}12 & 34\\3 & 1\end{pmatrix}$	$\begin{pmatrix}12 & 34\\1 & 4\end{pmatrix}$ $\begin{pmatrix}12 & 34\\4 & 1\end{pmatrix}$	$\begin{pmatrix}12 & 34\\2 & 3\end{pmatrix}$ $\begin{pmatrix}12 & 34\\3 & 2\end{pmatrix}$	$\begin{pmatrix}12 & 34\\2 & 4\end{pmatrix}$ $\begin{pmatrix}12 & 34\\4 & 2\end{pmatrix}$	$\begin{pmatrix}12 & 34\\3 & 4\end{pmatrix}$ $\begin{pmatrix}12 & 34\\4 & 3\end{pmatrix}$
$\begin{pmatrix}13 & 24\\1 & 2\end{pmatrix}$ $\begin{pmatrix}13 & 24\\2 & 1\end{pmatrix}$	$\begin{pmatrix}13 & 24\\1 & 3\end{pmatrix}$ $\begin{pmatrix}13 & 24\\3 & 1\end{pmatrix}$	$\begin{pmatrix}13 & 24\\1 & 4\end{pmatrix}$ $\begin{pmatrix}13 & 24\\4 & 1\end{pmatrix}$	$\begin{pmatrix}13 & 24\\2 & 3\end{pmatrix}$ $\begin{pmatrix}13 & 24\\3 & 2\end{pmatrix}$	$\begin{pmatrix}13 & 24\\2 & 4\end{pmatrix}$ $\begin{pmatrix}13 & 24\\4 & 2\end{pmatrix}$	$\begin{pmatrix}13 & 24\\3 & 4\end{pmatrix}$ $\begin{pmatrix}13 & 24\\4 & 3\end{pmatrix}$
$\begin{pmatrix}14 & 23\\1 & 2\end{pmatrix}$ $\begin{pmatrix}14 & 23\\2 & 1\end{pmatrix}$	$\begin{pmatrix}14 & 23\\1 & 3\end{pmatrix}$ $\begin{pmatrix}14 & 23\\3 & 1\end{pmatrix}$	$\begin{pmatrix}14 & 23\\1 & 4\end{pmatrix}$ $\begin{pmatrix}14 & 23\\4 & 1\end{pmatrix}$	$\begin{pmatrix}14 & 23\\2 & 3\end{pmatrix}$ $\begin{pmatrix}14 & 23\\3 & 2\end{pmatrix}$	$\begin{pmatrix}14 & 23\\2 & 4\end{pmatrix}$ $\begin{pmatrix}14 & 23\\4 & 2\end{pmatrix}$	$\begin{pmatrix}14 & 23\\3 & 4\end{pmatrix}$ $\begin{pmatrix}14 & 23\\4 & 3\end{pmatrix}$

The principal factor $J^{(2)}/J^{(1)}$ is a 0-simple semigroup and so, being finite, is completely 0-simple. We may express it as a Rees matrix semigroup as follows. The $\mathscr{H}$-class in the top left-hand corner is a group $\mathscr{H}$-class; treat it as H_{11} and denote its elements by

$$e = \begin{pmatrix}1 & 234\\1 & 2\end{pmatrix}, \qquad a = \begin{pmatrix}1 & 234\\2 & 1\end{pmatrix}.$$

Call the $\mathscr{R}$-classes $R_1, \ldots, R_7$, reading from top to bottom, and the $\mathscr{L}$-classes $L_1, \ldots, L_6$, reading from left to right. For $\lambda = 1, \ldots, 6$, choose q_λ as the *first-named* element in the $\mathscr{H}$-class $H_{1\lambda}$. For $i = 1, \ldots, 7$, choose r_i as the *first-named* element in the $\mathscr{H}$-class H_{i1}. If $p_{\lambda i} = q_\lambda r_i$ (the product being in $J^{(2)}/J^{(1)}$), then $P = (p_{\lambda i})$ is the matrix

$$\begin{bmatrix} e & a & 0 & 0 & 0 & e & e \\ e & 0 & a & 0 & e & 0 & e \\ e & 0 & 0 & a & e & e & 0 \\ 0 & e & a & 0 & e & a & 0 \\ 0 & e & 0 & a & e & 0 & a \\ 0 & 0 & e & a & 0 & e & a \end{bmatrix}$$

and $J^{(2)}/J^{(1)} \simeq \mathscr{M}^\circ[H_{11}; \{1, \ldots, 7\}, \{1, \ldots, 6\}; P]$.

The principal factors corresponding to $J^{(1)}$ and $J^{(4)}$ are easily described, since the former is a right zero semigroup and the latter is a 0-group. The principal factor $J^{(3)}/(J^{(1)} \cup J^{(2)})$ is a completely 0-simple semigroup with 6 $\mathscr{R}$-classes, 4 $\mathscr{L}$-classes and in which each $\mathscr{H}$-class is of order 6. Find G, I, Λ and P such that $J^{(3)}/(J^{(1)} \cup J^{(2)}) \simeq \mathscr{M}^\circ[G; I, \Lambda; P]$.

10. Use the Rees theorem to show that a completely simple right simple semigroup is a right group. (See Exercise II.5.)

11. (Croisot 1953) Use the Rees theorem to show that a completely simple semigroup is regular and has the property

$$(\forall a, b \in S) \qquad aba = a \Rightarrow bab = b.$$

Conversely, show that a regular semigroup with this property cannot have a zero and must have the property that all its idempotents are primitive. Deduce that the semigroup is completely simple.

12. Let S be a regular semigroup in which the idempotents form a rectangular band. Show that S has no zero and that every idempotent of S is primitive. Deduce by Corollary 3.6 and Exercise 8 that S is isomorphic to the direct product of a group and a rectangular band.

13. Let ρ be a proper congruence on a completely 0-simple semigroup S, i.e. a congruence ρ for which $0\rho = \{0\}$.

(i) Use Proposition 1.2 to show that S/ρ is 0-simple.

(ii) If e is a primitive idempotent in S, show that $e\rho$ is a primitive idempotent in in S/ρ. [Hint: if $h \in S$ is such that $h\rho$ is an idempotent in S/ρ and $h\rho \leqslant e\rho$, show that $(ehe)\rho = h\rho$. Then by Lallement's Lemma (II.4.6) let g be an idempotent in S such that $g\rho = (ehe)\rho$, $L_g \leqslant L_{ehe}$, $R_g \leqslant R_{ehe}$.]

(iii) Deduce that S/ρ is completely 0-simple.

14. (i) If $K(n)$ denotes the number of non-isomorphic congruence-free semigroups with 0 having order n, show that $K(5) = 2$, $K(6) = 0$, $K(7) = 2$.

(ii) Show that the two congruence-free semigroups with zero having order 7 are anti-isomorphic.

(iii) Show that $K(p + 1) = 0$ for every prime p.

Chapter IV

Unions of Groups

INTRODUCTION

In the last chapter we saw how a completely 0-simple semigroup could be constructed as a Rees matrix semigroup over a 0-group. In this chapter we introduce another constructive device of major importance in semigroup theory. The crucial, extremely simple observation on which the construction depends is that if ϕ is a homomorphism from a semigroup S onto a semilattice Y, then each of the subsets $\alpha\phi^{-1}$ $(\alpha \in Y)$ is a subsemigroup of S. If we denote $\alpha\phi^{-1}$ by S_α we thus have that S is the disjoint union of the subsemigroups $S_\alpha(\alpha \in Y)$, and that

$$S_\alpha S_\beta \subseteq S_{\alpha\beta},$$

where $\alpha\beta$ is the product of α and β in the semilattice Y.

This can of course be done for any semigroup admitting a proper homomorphism onto a semilattice, but gives rise to a meaningful structure theorem only if the semigroups S_α are of some recognizable type $\mathscr{T}$ that is more special and better understood than the type to which S itself belongs. We then say that S is a *semilattice of semigroups of type* $\mathscr{T}$.

Such a structure theorem has defects, of course, for only what we might call the 'gross' structure is elucidated: we know that the product of an element x in S_α and an element y in S_β lies in $S_{\alpha\beta}$, but that is the limit of what we know. In certain cases, however, one can obtain much more complete information, by virtue of the construction method now described.

Let Y be a semilattice and let $\{S_\alpha : \alpha \in Y\}$ be a family of disjoint semigroups of type $\mathscr{T}$, indexed by Y. For each pair α, β of elements of Y such that $\alpha \geqslant \beta$, let $\phi_{\alpha,\beta} : S_\alpha \to S_\beta$ be a homomorphism, and suppose that

(a) $\phi_{\alpha,\alpha}$ is the identical automorphism of S_α for each α in Y,

(b) $\phi_{\alpha,\beta}\phi_{\beta,\gamma} = \phi_{\alpha,\gamma}$ for every α, β, γ in Y such that $\alpha \geqslant \beta \geqslant \gamma$.

Let $S = \bigcup\{S_\alpha : \alpha \in Y\}$, and define a multiplication $\circ$ on S by the rule that if $a_\alpha \in S_\alpha$ and $b_\beta \in S_\beta$,

$$a_\alpha \circ b_\beta = (a_\alpha \phi_{\alpha,\alpha\beta})(b_\beta \phi_{\beta,\alpha\beta}).$$

(Here $\alpha\beta$ denotes the product of α and β in the semilattice Y, while the multiplication of $a_\alpha\phi_{\alpha,\alpha\beta}$ and $b_\beta\phi_{\beta,\alpha\beta}$ takes place in the semigroup $S_{\alpha\beta}$.) Then $(S, \circ)$ is a semigroup, since if $a_\alpha, b_\beta, c_\gamma$ are in $S_\alpha, S_\beta, S_\gamma$ respectively, and if we write $\alpha\beta = \delta$, $\beta\gamma = \varepsilon$, $\alpha\beta\gamma = \eta$ in Y, then

$$\begin{aligned}(a_\alpha \circ b_\beta) \circ c_\gamma &= [(a_\alpha\phi_{\alpha,\delta})(b_\beta\phi_{\beta,\delta})] \circ c_\gamma \\ &= [(a_\alpha\phi_{\alpha,\delta})(b_\beta\phi_{\beta,\delta})]\phi_{\delta,\eta} \cdot c_\gamma\phi_{\gamma,\eta} \\ &= a_\alpha\phi_{\alpha,\delta}\phi_{\delta,\eta} \cdot b_\beta\phi_{\beta,\delta}\phi_{\delta\eta} \cdot c_\gamma\phi_{\gamma,\eta} \\ &= a_\alpha\phi_{\alpha,\eta} \cdot b_\beta\phi_{\beta,\eta} \cdot c_\gamma\phi_{\gamma,\eta},\end{aligned}$$

and similarly

$$a_\alpha \circ (b_\beta \circ c_\gamma) = a_\alpha\phi_{\alpha,\eta} \cdot b_\beta\phi_{\beta,\eta} \cdot c_\gamma\phi_{\gamma,\eta}.$$

Notice that each S_α is a subsemigroup of S, and that $S_\alpha S_\beta \subseteq S_{\alpha\beta}$ for all α, β in Y; thus S is certainly a semilattice of semigroups of type $\mathscr{T}$. The structure of S is, however, much more completely determined, and we give recognition to this fact by writing

$$S = \mathscr{S}(Y; \{S_\alpha : \alpha \in Y\}; \{\phi_{\alpha,\beta} : \alpha, \beta \in Y, \alpha \geqslant \beta\})$$

or more briefly $S = \mathscr{S}(Y; S_\alpha; \phi_{\alpha,\beta})$. We refer to S as a *strong semilattice of semigroups of type* $\mathscr{T}$.

In Sections 1 and 2 we obtain a structure theorem of the weak sort (Theorem 1.7) for semigroups in which every $\mathscr{H}$-class is a group and one of the strong sort (Theorem 2.1) for regular semigroups in which the idempotents are central. Both are due to Clifford (1941).

In Section 3 we examine the implication of the first of these theorems when specialized to *bands*, i.e. to semigroups consisting solely of idempotents, and then prove a theorem due to Petrich (1967) giving a (necessarily rather complicated) description of the "fine" structure of a band. Lallement (1967) considers a more general situation, but his results in their most general form are exceedingly complex and are beyond the scope of this book.

In Section 4 we consider free bands, as a preliminary to an examination in Section 5 of varieties of bands.

1. UNIONS OF GROUPS

A semigroup S is called a *union of groups* if each of its elements is contained in some subgroup of S. If a is an element of such a semigroup then $a \in G$, a subgroup of S. If we denote the identity of G by e then within the group G we have

$$ea = ae = a, \qquad aa^{-1} = a^{-1}a = e;$$

hence certainly a $\mathscr{H}e$ in S. Thus the $\mathscr{H}$-class H_a coincides with H_2, which by Corollary II.2.6 is a group.

In a union of groups, then, we have that every $\mathscr{H}$-class is a group. Such a semigroup is certainly regular, but by no means every regular semigroup is of this type: the full transformation semigroup is an obvious example. We have, however, already come across unions of groups in the shape of completely simple semigroups. What we shall now show is that any union of groups can be built up out of completely simple semigroups.

As a first step we have

Proposition 1.1. *If S is a union of groups and is simple, then it is completely simple.*

Proof. We shall show that every idempotent of S is primitive, which by Theorem III.3.1 is certainly enough, for clearly S contains at least one idempotent. Let e be an idempotent of S and suppose that $f \leqslant e$, i.e. that

$$ef = fe = f, \tag{1.2}$$

where f also is idempotent. Then, since S is simple, there exist $z, t \in S$ such that $e = zft$; hence, if we define $x = ezf$ and $y = fte$, we readily obtain that

$$xfy = ezf^3te = e(zft)e = e^3 = e;$$

also

$$ex = xf = x, \qquad fy = ye = y.$$

Now, since S is a union of groups, we must have that $x \in H_g$ for some idempotent g. Hence $gx = xg = x$, and there exists an element x^{-1} in H_g such that $xx^{-1} = x^{-1}x = g$.

From $xf = x$ it follows that $x^{-1}xf = x^{-1}x$; hence $gf = g$. But we also have

$$gf = gef = gxfyf = xfyf = ef = f$$

and so $g = f$. Hence

$$f = fe = ge = gxfy = xfy = e.$$

Thus e is primitive. ■

Consider now a semigroup S that is a union of groups and is *not* necessarily simple. Since $a\,\mathscr{H}a^2$ for every a in S it is certainly the case that

$$a\,\mathscr{J}a^2 \tag{1.3}$$

for every a. As a consequence, if a, b are arbitrary elements of S,

$$J_{ab} = J_{(ab)^2} = J_{a(ba)b} \leqslant J_{ba},$$

and similarly $J_{ba} \leqslant J_{ab}$; thus

$$J_{ab} = J_{ba}. \tag{1.4}$$

Next, if $J_a = J_b$ then $b = xay$ and $a = ubv$ for some x, y, u, v in S^1. If $c \in S$, then

$$J_{ca} = J_{cubv} \leqslant J_{cub} = J_{bcu} \qquad \text{(by (1.4))}$$
$$\leqslant J_{bc} = J_{cb},$$

and similarly $J_{cb} \leqslant J_{ca}$. Hence $J_{ca} = J_{cb}$ and it follows by (1.4) that $J_{ac} = J_{bc}$ also. Thus $\mathcal{J}$ is a congruence on S, and

$$J_a J_b \subseteq J_{ab} \tag{1.5}$$

for all a, b in S. Moreover, it follows from (1.3) and (1.4) that $S/\mathcal{J}$ is a semilattice.

Each J_a is a subsemigroup of S and so none of the principal factors of S is null. The kernel of S (if it exists) is simple, and every other principal factor is of the form $J_a \cup \{0\}$, a 0-simple semigroup in which J_a is a subsemigroup. Thus J_a is simple. Now each $\mathcal{J}$-class is a union of $\mathcal{H}$-classes, i.e. is a union of groups. Hence, by Proposition 1.1, each $\mathcal{J}$-class of S is a completely simple semigroup.

Let us now denote the semilattice $S/\mathcal{J}$ by Y and write $S_\alpha (\alpha \in Y)$ for the subset $\alpha(\mathcal{J}^\natural)^{-1}$ of S. Each S_α is a $\mathcal{J}$-class of S and is a completely simple semigroup. Thus S is the disjoint union of its subsemigroups S_α $(\alpha \in Y)$ and formula (1.5) becomes

$$S_\alpha S_\beta \subseteq S_{\alpha\beta}. \tag{1.6}$$

We have therefore proved

Theorem 1.7 (Clifford 1941). *If a semigroup S is a union of groups, then it is a semilattice of completely simple semigroups.* ■

At a first glance this does not look like progress at all, since completely simple semigroups are more complicated than groups. The progress is, however, real, and lies in the "gross" multiplication formula (1.6). Previously we had

$$S = \bigcup_{e \in E} H_e,$$

where E is the set of idempotents of S and each H_e is a group, but we had no idea at all where to look for the product of an element x in H_e and an element y in H_f, or even whether the product of H_e and H_f was contained in a single $\mathcal{H}$-class.

A number of authors have attempted to improve Theorem 1.7 by giving the "fine" structure of S. Clifford (1941) himself made a start on the problem, and more recently (1972) presented an improved version of a treatment by Fantham (1960) of the case where the idempotents of S form a subsemigroup. Lallement (1967) and Petrich (1974) have considered the general case, but it would be beyond the scope of this book to present their results in general, and we shall confine ourselves to a consideration (in Section 3) of the special case where S is a band.

Before moving on we pause to give a very brief further consideration to *semilattice congruences* on a semigroup S, i.e. congruences ρ for which S/ρ is a semilattice. There is always at least one such congruence, namely the universal congruence. If $\{\rho_i : i \in I\}$ is a non-empty family of semilattice congruences on S, and if $\rho = \bigcap\{\rho_i : i \in I\}$, then for all x, y in S and all i in I,

$$(x^2, x) \in \rho_i \quad \text{and} \quad (xy, yx) \in \rho_i;$$

hence $(x^2, x) \in \rho$, $(xy, yx) \in \rho$, and so ρ also is a semilattice congruence.

The intersection of *all* semilattice congruences on S is thus the unique minimum semilattice congruence on S. We denote it by η.

Proposition 1.8. *If S is a union of groups, then $\mathscr{J} = \eta$.*

Proof. We have already remarked that if S is a union of groups then $\mathscr{J}$ is a congruence and $S/\mathscr{J}$ is a semilattice. Hence $\eta \subseteq \mathscr{J}$. Conversely, if $(a, b) \in \mathscr{J}$ then there exist x, y, u, v in S such that $xay = b$, $ubv = a$. (Since S is regular we may, as remarked at the beginning of Section II.4, take x, y, u, v in S rather than in S^1.) Hence, in S/η,

$$(x\eta)(a\eta)(y\eta) = b\eta, \qquad (u\eta)(b\eta)(v\eta) = a\eta$$

and so $a\eta$, $b\eta$ are $\mathscr{J}$-equivalent in S/η. But in the semilattice S/η

$$\mathscr{J} = \mathscr{H} = 1_{S/\eta};$$

hence $a\eta = b\eta$. That is, $(a, b) \in \eta$, and so we have shown that $\mathscr{J} \subseteq \eta$. ■

For a more general result, to the effect that $\mathscr{J}^{\#} = \eta$ in any regular semigroup, and further results of this type, see (Howie and Lallement 1966).

2. SEMILATTICES OF GROUPS

One of the main themes of this book concerns the influence of the properties of the idempotents of a semigroup on the structure of the semigroup as a whole. In pursuance of this theme, let us define a *Clifford semigroup* to be a regular semigroup S in which the idempotents are *central*, i.e. in which $ex = xe$ for every idempotent e and every x in S.

Theorem 2.1. *If S is a semigroup with set E of idempotents, then the following statements are equivalent:*

(A) *S is a Clifford semigroup;*
(B) *S is regular and $\mathscr{D}^S \cap (E \times E) = 1_E$;*
(C) *S is a semilattice of groups;*
(D) *S is a strong semilattice of groups.*

Proof. (A) $\Rightarrow$ (B), Certainly a Clifford semigroup is regular. Also, if $(e, f) \in \mathscr{D}^S \cap (E \times E)$ then by Proposition II.3.6 there exist a in S and a' in $V(a)$ such that $aa' = e$, $a'a = f$. Hence, since idempotents are central,

$$e = e^2 = aa'aa' = afa' = faa' = fe = a'ae = a'ea = a'aa'a = f^2 = f.$$

Thus $\mathscr{D}^S \cap (E \times E) = 1_E$.

(B) $\Rightarrow$ (C). For every a in S and every a' in $V(a)$ we have that

$$(aa', a'a) \in \mathscr{D}^S \cap (E \times E);$$

hence $aa' = a'a$ ($=e$, say). Since $a\mathscr{R}aa'$ and $a\mathscr{L}a'a$, it follows that $a\,\mathscr{H}e$. Hence S is a union of groups and so by Theorem 1.7 is a semilattice of completely simple semigroups $S_\alpha(\alpha \in Y)$. Now a completely simple semigroup is certainly bisimple by Proposition II.1.11; hence any two idempotents e, f in any S_α are $\mathscr{D}$-equivalent in S_α and so certainly $\mathscr{D}$-equivalent in S. Hence $e = f$ and so S_α, being a completely simple semigroup with only one idempotent, is a group.

(C) $\Rightarrow$ (D). Let S be a semilattice Y of groups $G_\alpha(\alpha \in Y)$ and for each α in Y let e_α be the identity element of the group G_α. If $\alpha \geqslant \beta$ in Y then $\alpha\beta = \beta$ and so, for all a_α in G_α,

$$a_\alpha e_\beta \in G_\alpha G_\beta \subseteq G_{\alpha\beta} = G_\beta.$$

Hence we may define a mapping $\phi_{\alpha,\beta}: G_\alpha \to G_\beta$ by the rule that

$$a_\alpha \phi_{\alpha,\beta} = a_\alpha e_\beta \qquad (a_\alpha \in G_\alpha).$$

Notice that $\phi_{\alpha,\alpha}$ is the identical automorphism of G_α.

Next, notice that for all a_α, b_α in G_α.

$$(a_\alpha \phi_{\alpha,\beta})(b_\alpha \phi_{\alpha,\beta}) = a_\alpha e_\beta b_\alpha e_\beta = a_\alpha b_\alpha e_\beta,$$

since $b_\alpha e_\beta \in G_\beta$ and since e_β is the identity element of G_β. Thus

$$(a_\alpha \phi_{\alpha,\beta})(b_\alpha \phi_{\alpha,\beta}) = (a_\alpha b_\alpha)\phi_{\alpha,\beta}$$

and so $\phi_{\alpha,\beta}$ is a homomorphism.

Since a group homomorphism from G_α to G_β necessarily maps the identity of G_α to the identity of G_β, we may deduce that $e_\alpha e_\beta = e_\beta$ whenever $\alpha \geqslant \beta$. Hence, if $\alpha \geqslant \beta \geqslant \gamma$,

$$(a_\alpha\phi_{\alpha,\beta})\phi_{\beta,\gamma} = (a_\alpha e_\beta)e_\gamma = a_\alpha(e_\beta e_\gamma) = a_\alpha e_\gamma = a_\alpha\phi_{\alpha,\gamma};$$

thus $\phi_{\alpha,\beta}\phi_{\beta,\gamma} = \phi_{\alpha,\gamma}$.

Finally, if α, β are arbitrary elements of Y and if $\alpha\beta = \gamma$ in Y, then for all a_α in G_α and b_β in G_β,

$$\begin{aligned} a_\alpha b_\beta &= (a_\alpha b_\beta)e_\gamma \quad (\text{since } a_\alpha b_\beta \in G_\gamma) \\ &= a_\alpha(b_\beta e_\gamma) \\ &= a_\alpha e_\gamma b_\beta e_\gamma \quad (\text{since } b_\beta e_\gamma \in G_\gamma) \\ &= (a_\alpha\phi_{\alpha,\gamma})(b_\beta\phi_{\beta,\gamma}). \end{aligned}$$

Hence S is a strong semilattice $\mathscr{S}(Y; G_\alpha; \phi_{\alpha,\beta})$ of groups.

$(D) \Rightarrow (A)$. If S is a strong semilattice $\mathscr{S}(Y; G_\alpha; \phi_{\alpha,\beta})$ of groups, then it is certainly regular. The idempotents of S are the identity elements e_α of the groups G_α. If e_α is an idempotent and b_β ($\in G_\beta$) is an arbitrary element of S, then, writing γ for $\alpha\beta$, we have

$$\begin{aligned} e_\alpha b_\beta &= (e_\alpha\phi_{\alpha,\gamma})(b_\beta\phi_{\beta,\gamma}) = e_\gamma(b_\beta\phi_{\beta,\gamma}) \\ &= b_\beta\phi_{\beta,\gamma} = (b_\beta\phi_{\beta,\gamma})e_\gamma = (b_\beta\phi_{\beta,\gamma})(e_\alpha\phi_{\alpha,\gamma}) = b_\beta e_\alpha; \end{aligned}$$

thus S is a Clifford semigroup. ■

Notice that the idempotents of a *commutative* regular semigroup are necessarily central. Hence we have

Corollary 2.2. *A commutative semigroup is regular if and only if it is a strong semilattice of abelian groups.* ■

3. BANDS

By a *band* we mean a semigroup S in which every element is idempotent. Such a semigroup is certainly a union of groups, since each element constitutes a group all on its own. Hence Clifford's Theorem (1.7) applies, and S is a semilattice Y of completely simple semigroups S_α ($\alpha \in Y$). Each S_α, being a subsemigroup of the band S, is also a band, and it is not at all hard to see that a Rees matrix semigroup $\mathscr{M}[G; I, \Lambda; P]$ is a band if and only if G is the trivial group $\{e\}$ (which of course implies that $p_{\lambda i} = e$ for every i and λ). The multiplication in $\mathscr{M}[G; I, \Lambda; P]$ is thus given by

$$(e, i, \lambda)(e, j, \mu) = (e, i, \mu)$$

and so clearly $\mathscr{M}[G; I, \Lambda; P]$ is isomorphic to the rectangular band $I \times \Lambda$ with multiplication given (as in Section I.1) by

$$(i, \lambda)(j, \mu) = (i, \mu).$$

We have thus proved:

Theorem 3.1. (Clifford 1941; McLean 1952) *A band is a semilattice of rectangular bands.* ■

This is an exceedingly useful result, as we shall see, but it cannot be regarded as solving the problem of the structure of bands, since in general we do not obtain a *strong* semilattice. We shall eventually identify (in Theorem 5.14) the precise class of bands for which we do obtain a strong semilattice, but in this section our aim will be to describe a general structure theorem for bands due to Petrich (1967).

First, however, it is important to learn more about rectangular bands. The definition given of a rectangular band in Section I.1 is unsatisfactory, since it does not have the 'abstract algebra' property that any semigroup isomorphic to a semigroup satisfying the definition also satisfies the definition. From now on, therefore, we shall define a rectangular band as a semigroup *isomorphic* to a Cartesian product $I \times \Lambda$, with multiplication as described earlier.

Proposition 3.2. *If S is a semigroup then the following statements are equivalent:*

(A) *S is a rectangular band;*
(B) $(\forall a, b \in S)\ ab = ba \Rightarrow a = b$;
(C) $(\forall a, b \in S)\ aba = a$;
(D) $(\forall a \in S)\ a^2 = a$, *and* $(\forall a, b, c \in S)\ abc = ac$.

Proof. (A) ⇒ (B). If $S = I \times \Lambda$ is a rectangular band and if $a = (i, \lambda)$, $b = (j, \mu)$ then

$$ab = ba \Rightarrow (i, \mu) = (j, \lambda) \Rightarrow i = j, \lambda = \mu \Rightarrow a = b.$$

(B) ⇒ (C). Since $aa^2 = a^2a$ for every a in S, it follows that $a^2 = a$. Hence, since $a \,.\, aba = aba = aba \,.\, a$, we deduce that $aba = a$ for all a, b in S.

(C) ⇒ (D). Putting $b = a$ in (C) we deduce that $a^3 = a$ for every a in S. Hence $a^4 = a^2$ and so $a = a(a^2)a = a^4 = a^2$. Also, for all a, b, c in S,

$$\begin{aligned} abc &= ab(cac) && \text{(by (C))} \\ &= a(bc)a \,.\, c \\ &= ac && \text{(by (C)).} \end{aligned}$$

(D) ⇒ (A). Since, by (D),

$$a = a^2 = aba, \qquad b = b^2 = bab,$$

it follows that

$$a \,\mathcal{R}\, ab, \qquad ab \,\mathcal{L}\, b \tag{3.3}$$

and hence that $a \,\mathcal{D}\, b$ for all a, b in S. Hence $R \cap L$ is non-empty for every $\mathcal{R}$-class R and every $\mathcal{L}$-class L in S. But since S is a band the $\mathcal{H}$-class $R \cap L$

consists of a single element (Corollary II.2.6). Hence the mapping $\psi: S \to S/\mathscr{R} \times S/\mathscr{L}$ defined by

$$a\psi = (R_a, L_a)$$

is a bijection. If we furnish the Cartesian product $S/\mathscr{R} \times S/\mathscr{L}$ with the rectangular band multiplication we obtain that ψ is an isomorphism, since

$$\begin{aligned}(ab)\psi &= (R_{ab}, L_{ab}) = (R_a, L_b) \qquad \text{(by (3.3))}\\ &= (R_a, L_a)(R_b, L_b) = (a\psi)(b\psi). \blacksquare\end{aligned}$$

We remark that in condition (D) it is not possible to dispense with the requirement that $a^2 = a$ for every a in S. Exercise 9 below gives an example of a semigroup (containing six elements) in which $abc = ac$ for every a, b and c but which is not a band.

Next, we have

Proposition 3.4. *If ϕ is a homomorphism from a rectangular band $I_1 \times \Lambda_1$ into a rectangular band $I_2 \times \Lambda_2$ then there exist mappings $\phi^l: I_1 \to I_2$, $\phi^r: \Lambda_1 \to \Lambda_2$ such that*

$$(x_1, \xi_1)\phi = (x_1\phi^l, \xi_1\phi^r) \tag{3.5}$$

for every $(x_1, \xi_1) \in I_1 \times \Lambda_1$.

Conversely, for any mappings $\phi^l: I_1 \to I_2$, $\phi^r: \Lambda_1 \to \Lambda_2$, the formula (3.5) *defines a homomorphism ϕ from $I_1 \times \Lambda_1$ into $I_2 \times \Lambda_2$.*

Proof. Let $\phi: I_1 \times \Lambda_1 \to I_2 \times \Lambda_2$ be a homomorphism. Choose a fixed λ_1 in Λ_1 and for every x_1 in I_1 define $x_1\phi^l$ by

$$(x_1, \lambda_1)\phi = (x_1\phi^l, \lambda_2).$$

Similarly, choose a fixed i_1 in I_1 and for every ξ_1 in Λ_1 define $\xi_1\phi^r$ by

$$(i_1, \xi_1)\phi = (i_2, \xi_1\phi^r).$$

Then for all (x_1, ξ_1) in $I_1 \times \Lambda_1$,

$$\begin{aligned}(x_1, \xi_1)\phi &= [(x_1, \lambda_1)(i_1, \xi_1)]\phi\\ &= (x_1, \lambda_1)\phi \cdot (i_1, \xi_1)\phi\\ &= (x_1\phi^l, \lambda_2)(i_2, \xi_1\phi^r) = (x_1\phi^l, \xi_1\phi^r).\end{aligned}$$

Conversely, if ϕ is defined by (3.5), then for all (x_1, ξ_1), (y_1, η_1) in $I_1 \times \Lambda_1$,

$$\begin{aligned}&[(x_1, \xi_1)(y_1, \eta_1)]\phi = (x_1, \eta_1)\phi\\ &= (x_1\phi^l, \eta_1\phi^r) = (x_1\phi^l, \xi_1\phi^r)(y_1\phi^l, \eta_1\phi^r)\\ &= (x_1, \xi_1)\phi \cdot (y_1, \eta_1)\phi,\end{aligned}$$

and so ϕ is a homomorphism. $\blacksquare$

If, as in Proposition I.1.9 we choose to regard a rectangular band $I \times \Lambda$ as the direct product of a left zero semigroup I and a right zero semigroup Λ, then we may interpret the mappings $\phi^l\colon I_1 \to I_2$, $\phi^r\colon \Lambda_1 \to \Lambda_2$ as *homomorphisms* of the left zero and right zero semigroups respectively. There is no extra information involved in this version, since *every* mapping between left (or right) zero semigroups is easily seen to be a homomorphism. We state the result formally as follows:

Corollary 3.6. *Let L_1, L_2 be left zero semigroups and R_1, R_2 be right zero semigroups. If ϕ is a homomorphism from the rectangular band $L_1 \times R_1$ into the rectangular band $L_2 \times R_2$ then there exist homomorphisms $\phi^l\colon L_1 \to L_2$, $\phi^r\colon R_1 \to R_2$ such that*

$$(l_1, r_1)\phi = (l_1\phi^l, r_1\phi^r) \tag{3.7}$$

for all (l_1, r_1) in $L_1 \times R_1$.

Conversely, for any homomorphisms $\phi^l\colon L_1 \to L_2$, $\phi^r\colon R_1 \to R_2$, the formula (3.7) *defines a homomorphism ϕ from $L_1 \times R_1$ into $L_2 \times R_2$.* ■

The basic idea in Petrich's approach to the problem of the structure of bands lies in the idea of translations of a semigroup. We have already come across the maps λ_a and ρ_a associated with each element a of a semigroup S:

$$\lambda_a s = as, \qquad s\rho_a = sa \qquad (s \in S).$$

Here, for convenience, we are writing λ_a as a left mapping and ρ_a as a right mapping. We refer to λ_a and ρ_a as the *inner left and right translations* of S associated with a. Because of associativity, we have

$$\lambda_a(st) = (\lambda_a s)t, \qquad (st)\rho_a = s(t\rho_a), \qquad s(\lambda_a t) = (s\rho_a)t$$

for all a, s, t in S. This motivates the following definitions: the left mapping $\lambda\colon S \to S$ is called a *left translation* of S if

$$\lambda(st) = (\lambda s)t$$

for all s, t in S; the right mapping $\rho\colon S \to S$ is called a *right translation* of S if

$$(st)\rho = s(t\rho)$$

for all s, t in S; the left translation λ and the right translation ρ are said to be *linked* if

$$s(\lambda t) = (s\rho)t$$

for all s, t in S. The set of all linked pairs (λ, ρ) of left and right translations is called the *translational hull* $\Omega(S)$ of S. It is a semigroup under the obvious multiplication

$$(\lambda, \rho)(\lambda', \rho') = (\lambda\lambda', \rho\rho'),$$

where $\lambda\lambda'$ denotes the composition of the left mappings λ and λ' (i.e. first λ', then λ), while $\rho\rho'$ denotes the composition of the right mappings ρ and ρ' (i.e. first ρ, then ρ'). The proof of this assertion entails merely the routine verification that $\lambda\lambda'$ and $\rho\rho'$ are linked left and right translations; associativity is a consequence of the known associativity of the operation of composition of mappings (whether on the left or on the right).

Within $\Omega(S)$ we have a linked pair (λ_a, ρ_a) for each a in S, and it is easily verified that

$$(\lambda_a, \rho_a)(\lambda_b, \rho_b) = (\lambda_{ab}, \rho_{ab})$$

for all a, b in S. Thus we have a homomorphism $a \mapsto (\lambda_a, \rho_a)$ of S into $\Omega(S)$. In general this is not a monomorphism, for it can happen that $\lambda_a = \lambda_b$ and $\rho_a = \rho_b$ for distinct a, b in S. (See Exercise 6.) In the case where S is completely simple, however, we do get a monomorphism, by virtue of the following lemma:

Lemma 3.8. *If a and b are elements of a regular semigroup S, then*

$$\lambda_a = \lambda_b \text{ and } \rho_a = \rho_b \;\Rightarrow\; a = b.$$

Proof. Suppose that $\lambda_a = \lambda_b$ and $\rho_a = \rho_b$. If $a' \in V(a)$ and $b' \in V(b)$ then

$$a = aa'a = \lambda_a(a'a) = \lambda_b(a'a) = ba'a$$

and so $R_a \leqslant R_b$. Similar arguments show that $L_a \leqslant L_b$, $R_b \leqslant R_a$ and $L_b \leqslant L_a$; hence $a \mathscr{H} b$. By Proposition II.4.1 we may now suppose that a' and b' have been chosen so that $aa' = bb'$ and $a'a = b'b$. Hence

$$a = ba'a = bb'b = b. \blacksquare$$

Next, if a semigroup S is expressed as a semilattice Y of completely simple semigroups S_α and if α, β are two elements of Y with $\alpha \geqslant \beta$, then

$$S_\alpha S_\beta \subseteq S_\beta, \qquad S_\beta S_\alpha \subseteq S_\beta.$$

Hence if $c \in S_\alpha$ and $x \in S_\beta$ it follows that both cx and xc are in S_β. Thus c induces mappings λ_c and ρ_c from S_β into S_β in the obvious way:

$$\lambda_c x = cx, \qquad x\rho_c = xc.$$

By associativity of S, these mappings are evidently linked left and right translations, and so $c \mapsto (\lambda_c, \rho_c)$ defines a mapping (indeed a homomorphism) from S_α into $\Omega(S_\beta)$, the translational hull of S_β.

In the case considered in Section 2, where S is a Clifford semigroup and all the S_α are groups, the translational hull of S_β can be shown to be isomorphic to S_β itself (see Exercise 5) and the homomorphism $c \mapsto (\lambda_c, \rho_c)$ of S_α into $\Omega(S_\beta)$ essentially reduces to the homomorphism $\phi_{\alpha,\beta}\colon S_\alpha \to S_\beta$ already defined for this case. In general, however, the translational hull of a completely simple semigroup contains linked pairs that are not inner.

The problem of describing the translational hull of a Rees matrix semigroup $\mathscr{M}[G; I, \Lambda; P]$ has been tackled by Petrich (1968). We shall here confine ourselves to the case of a rectangular band $I \times \Lambda$. If λ is a left translation of such a semigroup, then

$$\begin{aligned}\lambda(i, \mu) &= \lambda[(i, \mu)(i, \mu)] = [\lambda(i, \mu)](i, \mu)\\ &= (i*, \mu*)(i, \mu) \qquad \text{(say)}\\ &= (i*, \mu).\end{aligned}$$

Moreover, for *any* ξ in Λ,

$$\begin{aligned}\lambda(i, \xi) &= \lambda[(i, \mu)(i, \xi)] = [\lambda(i, \mu)](i, \xi)\\ &= (i*, \mu)(i, \xi) = (i*, \xi).\end{aligned}$$

Thus λ determines a mapping $\phi\colon I \to I$ (which we shall find convenient to write as a left mapping) such that

$$\lambda(i, \xi) = (\phi i, \xi) \tag{3.9}$$

for every ξ in Λ. Conversely, for any left mapping $\phi\colon I \to I$ the formula (3.9) defines a left translation of $I \times \Lambda$.

A closely similar argument shows that every right translation ρ of the rectangular band $I \times \Lambda$ determines and is determined by a right mapping $\psi\colon \Lambda \to \Lambda$ such that

$$(x, \mu)\rho = (x, \mu\psi) \tag{3.10}$$

for every (x, μ) in $I \times \Lambda$.

Now notice that if λ, defined by (3.9), is a left translation of $I \times \Lambda$ and ρ, defined by (3.10), is a right translation, then, for all (i, μ), (j, ν) in $I \times \Lambda$,

$$(i, \mu)[\lambda(j, \nu)] = (i, \mu)(\phi j, \nu) = (i, \nu),$$

while

$$[(i, \mu)\rho](j, \nu) = (i, \mu\psi)(j, \nu) = (i, \nu).$$

Hence *every* pair (λ, ρ) is linked. It is now easily verified that the mapping $(\lambda, \rho) \mapsto (\phi, \psi)$ gives an isomorphism from the translational hull $\Omega(I \times \Lambda)$ onto the Cartesian product $\mathscr{T}^*(I) \times \mathscr{T}(\Lambda)$ of the semigroup $\mathscr{T}^*(I)$ of left mappings from I into I with the semigroup $\mathscr{T}(\Lambda)$ of right mappings from Λ into Λ.

Consider now a general band B, and let us suppose that we have expressed B as a semilattice of rectangular bands $E_\alpha (\alpha \in Y)$. Let us write $E_\alpha = I_\alpha \times \Lambda_\alpha$ for each α in Y. If $\alpha \geqslant \beta$ then, as we have seen, each a in E_α induces a linked pair (λ_a, ρ_a) of left and right translations of E_β. Moreover, $a \mapsto (\lambda_a, \rho_a)$ is a homomorphism from E_α *into* $\Omega(E_\beta)$. As a result of our investigations into the nature of the translational hull of a rectangular band, we can now be somewhat more explicit and assert that a induces a left mapping $\phi_\beta^a : I_\beta \to I_\beta$ and a right mapping $\psi_\beta^a : \Lambda_\beta \to \Lambda_\beta$ in accordance with the formulae

$$a(x_\beta, \xi_\beta) = (\phi_\beta^a x_\beta, \xi_\beta), \qquad (x_\beta, \xi_\beta)a = (x_\beta, \xi_\beta \psi_\beta^a). \tag{3.11}$$

To put it more globally, we have, whenever $\alpha \geqslant \beta$, a homomorphism

$$\Phi_{\alpha, \beta} : E_\alpha \to \mathscr{T}^*(I_\beta) \times \mathscr{T}(\Lambda_\beta)$$

given by

$$a\Phi_{\alpha, \beta} = (\phi_\beta^a, \psi_\beta^a) \qquad (a \in E_\alpha).$$

If $\beta = \alpha$, and if $a = (i, \mu)$, then

$$a(x_\alpha, \xi_\alpha) = (i, \xi_\alpha), \qquad (x_\alpha, \xi_\alpha)a = (x_\alpha, \mu).$$

Hence the mapping $\phi_\alpha^{(i,\mu)} : I_\alpha \to I_\alpha$ has the property that $\phi_\alpha^{(i,\mu)} x_\alpha = i$ for every x_α in I_α. Similarly, $\xi_\alpha \psi_\alpha^{(i,\mu)} = \mu$ for every ξ_α in Λ_α. Thus, if we adopt a notation whereby the constant value of a constant mapping χ is denoted by $\langle \chi \rangle$, we may write

$$\langle \phi_\alpha^{(i,\mu)} \rangle = i, \qquad \langle \psi_\alpha^{(i,\mu)} \rangle = \mu \tag{3.12}$$

whenever $(i, \mu) \in E_\alpha$.

Now let us consider a more general product in B. Specifically, suppose that $a \in E_\alpha$, $b \in E_\beta$ and that $z = (x_\gamma, \xi_\gamma)$ is an arbitrary element of E_γ, where $\gamma = \alpha\beta$. Then, if we write $ab = (i_\gamma, \mu_\gamma)$, it follows that

$$abz = (ab)z = (i_\gamma, \mu_\gamma)(x_\gamma, \xi_\gamma) = (i_\gamma, \xi_\gamma)$$

and also that

$$\begin{aligned} abz = a(bz) &= a[b(x_\gamma, \xi_\gamma)] \\ &= a(\phi_\gamma^b x_\gamma, \xi_\gamma) = (\phi_\gamma^a \phi_\gamma^b x_\gamma, \xi_\gamma). \end{aligned}$$

We deduce that the left mapping $\phi_\gamma^a \phi_\gamma^b$ of I_γ has the property that

$$\phi_\gamma^a \phi_\gamma^b x_\gamma = i_\gamma$$

for every x_γ in I_γ, i.e. that $\phi_\gamma^a \phi_\gamma^b$ is a constant mapping, with constant value i_γ. Dually, by considering zab in two different ways, we may show that the right mapping $\psi^a_\gamma\psi^b_\gamma$ of Λ_γ has the constant value μ_γ. We thus obtain the product of a and b in terms of the mappings ϕ^a_γ, ψ^a_γ, ϕ^b_γ, ψ^b_γ as follows:

$$ab = (\langle\phi^a_\gamma\phi^b_\gamma\rangle, \langle\psi^a_\gamma\psi^b_\gamma\rangle). \tag{3.13}$$

If, then, we think of the homomorphisms $\Phi_{\alpha,\beta}$ as "known", formula (3.13) shows how the product ab of two arbitrary elements of B is determined by these homomorphisms.

Consider now what happens when we multiply the product ab on the right by an element $d = (x_\delta, \xi_\delta)$ of E_δ, where $\delta \leqslant \alpha\beta$. On the one hand

$$abd = (ab)d = (\phi^{ab}_\delta x_\delta, \xi_\delta),$$

while on the other hand

$$abd = a(bd) = a(\phi^b_\delta x_\delta, \xi_\delta) = (\phi^a_\delta\phi^b_\delta x_\delta, \xi_\delta).$$

We deduce that

$$\phi^{ab}_\delta = \phi^a_\delta\phi^b_\delta, \tag{3.14}$$

and a similar argument based on the two computations of dab leads to the corresponding formula

$$\psi^{ab}_\delta = \psi^a_\delta\psi^b_\delta. \tag{3.15}$$

It is convenient now to state the theorem towards which we are working:

Theorem 3.16 (Petrich 1967). *Let Y be a semilattice and let $\{E_\alpha : \alpha \in Y\}$ be a family of pairwise disjoint rectangular bands indexed by Y. For each α, let $E_\alpha = I_\alpha \times \Lambda_\alpha$, and for each pair α, β of elements of Y such that $\alpha \geqslant \beta$ let let $\Phi_{\alpha,\beta} : E_\alpha \to \mathscr{T}^*(I_\beta) \times \mathscr{T}(\Lambda_\beta)$ be a homomorphism, where*

$$a\Phi_{\alpha,\beta} = (\phi^a_\beta, \psi^a_\beta) \qquad (a \in E_\alpha).$$

Suppose also that

(i) *if $a = (i, \mu) \in E_\alpha$, then ϕ^a_α, ψ^a_α are constant mappings, and*

$$\langle\phi^{(i,\mu)}_\alpha\rangle = i, \qquad \langle\psi^{(i,\mu)}_\alpha\rangle = \mu;$$

(ii) *if $a \in S_\alpha$, $b \in S_\beta$ and $\alpha\beta = \gamma$, then $\phi^a_\gamma\phi^b_\gamma$ and $\psi^a_\gamma\psi^b_\gamma$ are constant mappings;*

(iii) *if $\langle\phi^a_\gamma\phi^b_\gamma\rangle$ is denoted by j and $\langle\psi^a_\gamma\psi^b_\gamma\rangle$ by ν, then, for all $\delta \leqslant \gamma$,*

$$\phi^{(j,\nu)}_\beta = \phi^a_\delta\phi^b_\delta, \qquad \psi^{(j,\nu)}_\delta = \psi^a_\delta\psi^b_\delta.$$

Let $B = \bigcup\{E_\alpha : \alpha \in Y\}$ and define the product of a in E_α and b in E_β by

$$a * b = (\langle\phi^a_\gamma\phi^b_\gamma\rangle, \langle\psi^a_\gamma\psi^b_\gamma\rangle),$$

where $\gamma = \alpha\beta$. Then $(B, *)$ is a band, whose $\mathscr{J}$-classes are the rectangular bands E_α.

Conversely, every band is determined in this way by a semilattice Y, a family of rectangular bands $E_\alpha = I_\alpha \times \Lambda_\alpha$ *indexed by Y, and a family of homomorphisms* $\Phi_{\alpha,\beta}\colon E_\alpha \to \mathscr{T}^*(I_\beta) \times \mathscr{T}(\Lambda_\beta)$ $(\alpha, \beta \in Y, \alpha \geqslant \beta)$ *satisfying* (i), (ii) *and* (iii).

Proof. We have in fact already established the more difficult converse half of this result, by virtue of formulae (3.12) to (3.15). To prove the direct half, we first show that the given multiplication is associative. If $a \in E_\alpha$, $b \in E_\beta$ and $c \in E_\gamma$ are arbitrarily chosen elements of B, with $\alpha\beta = \delta$, $\beta\gamma = \varepsilon$ and $\alpha\beta\gamma = \zeta$, write

$$a * b = (\langle \phi_\delta^a \phi_\delta^b \rangle, \langle \psi_\delta^a \psi_\delta^b \rangle) = (j, \nu),\ b * c = (\langle \phi_\varepsilon^b \phi_\varepsilon^c \rangle, \langle \psi_\varepsilon^b \psi_\varepsilon^c \rangle) = (k, \pi).$$

Then

$$\begin{aligned}(a * b) * c &= (\langle \phi_\zeta^{(j,\nu)} \phi_\zeta^c \rangle, \langle \psi_\zeta^{(j,\nu)} \psi_\zeta^c \rangle) = (\langle \phi_\zeta^a \phi_\zeta^b \phi_\zeta^c \rangle, \langle \psi_\zeta^a \psi_\zeta^b \psi_\zeta^c \rangle)\\ &= (\langle \phi_\zeta^a \phi_\zeta^{(k,\pi)} \rangle, \langle \psi_\zeta^a \psi_\zeta^{(k,\pi)} \rangle) = a * (b * c).\end{aligned}$$

Next, note that if $a = (i, \mu)$ and $b = (j, \nu)$ both belong to E_α, then the multiplication formula gives

$$\begin{aligned}a * b &= (\langle \phi_\alpha^{(i,\mu)} \phi_\alpha^{(j,\nu)} \rangle, \langle \psi_\alpha^{(i,\mu)} \psi_\alpha^{(j,\nu)} \rangle)\\ &= (i, \nu),\end{aligned}$$

by property (i) and by the properties of constant left and right mappings. This coincides exactly with the product of a and b in the rectangular band E_α. In particular it follows that $a * a = a$ and so B is a band.

The multiplication formula implies that

$$E_\alpha * E_\beta \subseteq E_{\alpha\beta}$$

and so two elements of B can be $\mathscr{J}$-equivalent only if they fall in the same E_α. Since any two elements in the same E_α are easily seen to be $\mathscr{J}$-equivalent, we thus conclude that the $\mathscr{J}$-classes of B are the rectangular bands E_α. ■

In Exercises 10 and 11 below the foregoing analysis is applied to the free band on three generators. Bands of this type are the topic for the next section.

4. FREE BANDS

In Section I.6 we defined the *free semigroup* F_A on a non-empty set A, and established the crucial property (Theorem I.6.1) that if $\phi\colon A \to S$ is an arbitrary mapping from A into a semigroup S then there is a unique homomorphism $\psi\colon F_A \to S$ with the property that the diagram

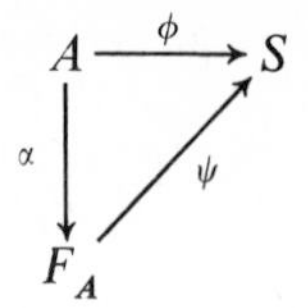

commutes, where α is the standard mapping embedding A in F_A.

Now let us consider the relation

$$\mathbf{B} = \{(x^2, x): x \in F_A\}$$

on F_A. The *free band B_A on the set A of generators* is defined to be F_A/β, where $\beta = \mathbf{B}^\sharp$, the smallest congruence on F_A containing $\mathbf{B}$. Certainly B_A is a band, since the very definition of β makes it clear that $(x\beta)^2 = x\beta$ for every $x\beta$ in B_A.

Suppose now that E is a band and that $\phi: A \to E$ is an arbitrary mapping. We certainly have a unique homomorphism ψ from the free *semigroup* F_A into E such that

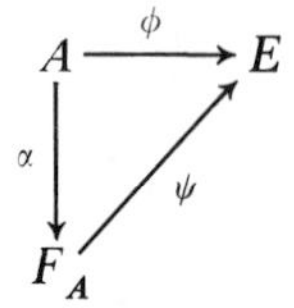

$$\begin{array}{ccc} A & \xrightarrow{\phi} & E \\ {\scriptstyle\alpha}\downarrow & \nearrow{\scriptstyle\psi} & \\ F_A & & \end{array} \qquad (4.1)$$

is a commutative diagram. For every x in F_A we have

$$x^2\psi = (x\psi)^2 = x\psi;$$

hence $(x^2, x) \in \psi \circ \psi^{-1}$ and so, since $\psi \circ \psi^{-1}$ is a congruence, it follows that β, being the *smallest* congruence containing $\mathbf{B}$, is contained in $\psi \circ \psi^{-1}$. By Theorem I.5.4 there exists a unique homomorphism $\chi: F_A/\beta \to E$ such that

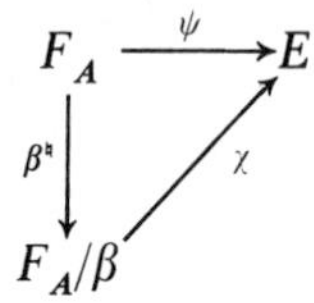

$$\begin{array}{ccc} F_A & \xrightarrow{\psi} & E \\ {\scriptstyle\beta^\sharp}\downarrow & \nearrow{\scriptstyle\chi} & \\ F_A/\beta & & \end{array} \qquad (4.2)$$

is a commutative diagram. Writing F_A/β as B_A and combining the diagrams (4.1) and (4.2), we obtain a commutative diagram

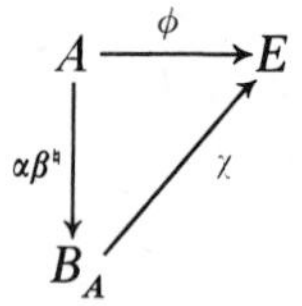

$$\begin{array}{ccc} A & \xrightarrow{\phi} & E \\ {\scriptstyle\alpha\beta^\sharp}\downarrow & \nearrow{\scriptstyle\chi} & \\ B_A & & \end{array} \qquad (4.3)$$

which justifies us in referring to B_A as a *free* band.

If $w \in F_A$ we define the *content* $C(w)$ of w to be the (necessarily finite) set of elements of A appearing in w. By Proposition I.5.10, any two distinct

β-equivalent elements of F_A are connected by a finite sequence of elementary **B**-transitions, i.e. by steps of one or other of the types

$$pxq \to px^2q, \qquad px^2q \to pxq \qquad (p, q \in F_A^1);$$

hence β-equivalent elements have the same content, and it is possible to talk in an unambiguous way about the *content* $C(w\beta)$ of an element $w\beta$ of B_A. For each non-empty finite subset P of A we denote by U_P the set of elements of B_A with content P. Each U_P is fairly obviously a sub-band of B_A, and $U_P \cap U_{P'} = \varnothing$ if $P \neq P'$.

Suppose now that $w, z \in F_A$ are such that $w\beta \mathscr{J} z\beta$ in B_A. Then there exist x, y, u, v in F_A^1 such that

$$(x\beta)(w\beta)(y\beta) = z\beta, \qquad (u\beta)(z\beta)(v\beta) = w\beta.$$

The first of these equalities implies that $C(w\beta) \subseteq C(z\beta)$ and the second that $C(z\beta) \subseteq C(w\beta)$; hence

$$w\beta \mathscr{J} z\beta \Rightarrow C(w\beta) = C(z\beta).$$

In fact, as remarked by McLean (1954), the converse of this is true also. For suppose that

$$C(w\beta) = C(z\beta) = \{a_1, a_2, \ldots, a_n\}.$$

Then since B_A is a union of groups, formulae (1.3) and (1.4) apply: that is, $a^2 \mathscr{J} a$ for all a in B_A, and $ab \mathscr{J} ba$ for all a, b in B_A. By repeated use of these results we can show that each of $w\beta$ and $z\beta$ is $\mathscr{J}$-equivalent to $(a_1 a_2 \ldots a_n)\beta$. Thus

$$w\beta \mathscr{J} z\beta \Leftrightarrow C(w\beta) = C(z\beta), \tag{4.4}$$

and we conclude that each U_P is a $\mathscr{J}$-class of B_A. The gross structure of B_A is thus fairly clear, the semilattice involved being effectively the set of finite subsets of A.

Each U_P is a rectangular band, by Theorem 3.1, and so we have the rather striking result that if $x\beta$, $y\beta$ and $z\beta$ are elements of B_A with the same content then

$$(x\beta)(y\beta)(z\beta) = (x\beta)(z\beta). \tag{4.5}$$

This in fact generalizes to the case where

$$C(y\beta) \subseteq C(x\beta) = C(z\beta),$$

for in such a case $C[(x\beta)(y\beta)(z\beta)] = C(x\beta) = C(z\beta)$, and so

$$(x\beta)(y\beta)(z\beta) = (x\beta)[(x\beta)(y\beta)(z\beta)](z\beta) = (x\beta)(z\beta).$$

It is not on the face of it obvious how one can tell whether or not two different words x and y in F_A determine the same element of B_A. The following analysis, based on the work of Green and Rees (1952) and Gerhard (1970), provides an algorithm for determining whether or not $x\beta = y\beta$.

If $x \in F_A$ and $|C(x)| = n \geqslant 1$, we define $\bar{x}(0)$ to be the letter in x that is *last to make its first appearance*, and $x(0)$ (if $n > 1$) to be the subword of x that precedes the first appearance of $\bar{x}(0)$. (Thus, for example, if $n = 4$ and $x = abacabcdbcbd$, then $\bar{x}(0) = d$ and $x(0) = abacabc$.) Dually, we define $\bar{x}(1)$ as the letter in x that is *first to make its last appearance*, and $x(1)$ (if $n > 1$) as the subword of x following the last appearance of $\bar{x}(1)$. (In our example, $\bar{x}(1) = a$ and $x(1) = bcdbcbd$.) If $|C(x)| = 1$ then the meaning of $x(0)$ and $x(1)$ is unclear: it is convenient to adjoin an identity element to F_A and to define $x(0) = x(1) = 1$ in this case. We can extend the congruence β from F_A to F_A^1 by the simple convention that $\{1\}$ is a β-class.

In the terminology of Green and Rees (1951), $\bar{x}(0)$ is the *initial mark*, $x(0)$ is the *initial*, $\bar{x}(1)$ is the *terminal mark* and $x(1)$ is the *terminal* of x. Since we shall require to iterate the process, it is more convenient here to adopt the more general notation of Gerhard (1970).

We now have the following result:

Lemma 4.6. If $x, y \in F_A$, then $x\beta = y\beta$ *if and only if*
(i) $C(x) = C(y)$,
(ii) $\bar{x}(0) = \bar{y}(0)$, $\bar{x}(1) = \bar{y}(1)$,
(iii) $(x(0), y(0)) \in \beta$, $(x(1), y(1)) \in \beta$.

Proof. If (i), (ii) and (iii) are given, then

$$C(x(0)\bar{x}(0)) = C(x) = C(y) = C(y(0)\bar{y}(0)),$$

$$C(\bar{x}(1)x(1)) = C(x) = C(y) = C(\bar{y}(1)y(1)).$$

Hence, modulo β, we have

$$\begin{aligned} x &\equiv x(0)\bar{x}(0)x\bar{x}(1)x(1) \\ &\equiv x(0)\bar{x}(0)\bar{x}(1)x(1), \quad \text{by (4.5)} \\ &\equiv y(0)\bar{y}(0)\bar{y}(1)y(1) \\ &\equiv y(0)\bar{y}(0)y\bar{y}(1)y(1), \quad \text{by (4.5)} \\ &\equiv y. \end{aligned}$$

Conversely, if $(x, y) \in \beta$, then x and y are connected by a finite sequence of elementary **B**-transitions. Now, if $x \to y$ by a single elementary **B**-transition, i.e. if (say) $x = puq$ and $y = pu^2q$, where $p, q \in F_A^1$, then it is evident that $\bar{x}(0) = \bar{y}(0)$. If the first occurrence of $\bar{x}(0)$ is in pu then we even have *equality* between $x(0)$ and $y(0)$; if $\bar{x}(0)$ occurs first in q then all we can say is that

$x(0) \to y(0)$ by an elementary **B**-transition. This, however, is enough, for we certainly conclude in both cases that

$$\bar{x}(0) = \bar{y}(0) \quad \text{and} \quad (x(0), y(0)) \in \beta.$$

Similarly,

$$\bar{x}(1) = \bar{y}(1) \quad \text{and} \quad (x(1), y(1)) \in \beta.$$

Clearly these results extend to the case where x and y are connected by a finite sequence of elementary **B**-transitions, and so the lemma is proved.■

If the β-equivalent elements x and y considered in the lemma are such that $|C(x)| = |C(y)| = k$, then the elements $x(0)$, $x(1)$, $y(0)$ and $y(1)$ figuring in part (iii) of the conditions all have content of cardinality $k - 1$. The lemma does therefore give useful information, particularly if we repeat the process. Before we do this, however, it is helpful to have some further notation. Let P_j be the set of words of length j in the alphabet $\{0, 1\}$, and let

$$\Sigma_k = \bigcup_{j=1}^{k} P_j.$$

Now for any x in F_A make the recursive definition that if $\alpha \in P_{j-1}$ then

$$\bar{x}(\alpha 0) = \overline{x(\alpha)}(0), \qquad x(\alpha 0) = x(\alpha)(0),$$

$$\bar{x}(\alpha 1) = \overline{x(\alpha)}(1), \qquad x(\alpha 1) = x(\alpha)(1).$$

(Thus, in the example $x = abacabcdbcbd$ already considered,

$$\bar{x}(00) = c, \quad x(00) = aba, \quad \bar{x}(01) = a, \quad x(01) = bc,$$

$$\bar{x}(10) = d, \quad x(10) = bc, \quad \bar{x}(11) = c, \quad x(11) = bd.)$$

If we now apply Lemma 4.6 repeatedly, and recall that by convention $w(0) = w(1) = 1$ for any word w such that $|C(w)| = 1$, we obtain

Theorem 4.7. *If x, y are elements of F_A such that $C(x) = C(y) = \{a_1, a_2, \ldots, a_k\}$, then $(x, y) \in \beta$ if and only if $\bar{x}(\alpha) = \bar{y}(\alpha)$ for every α in Σ_k.*■

One result of this analysis is that each U_P is finite. First, it is certainly the case that $|U_P|$ is determined solely by $|P|$; thus, if $|P| = k$ we shall denote $|U_P|$ by c_k. An element $x\beta$ in U_P by Lemma 4.6 uniquely determines a quadruple $(\bar{x}(0), \bar{x}(1), (x(0))\beta, (x(1))\beta)$; and any quadruple $(a_0, a_1, w_0\beta, w_1\beta)$ for which a_0 and a_1 belong to P and $w_0\beta \in U_{P\setminus\{a_0\}}$, $w_2\beta \in U_{P\setminus\{a_1\}}$ determines an element $(w_0 a_0 a_1 w_1)\beta$ of U_P. The number of elements of U_P is the number of quadruples of the kind described, and this leads to the recursion formula

$$c_k = k^2 c_{k-1}^2.$$

Hence, by iteration,

$$c_k = k^2(k-1)^4(k-2)^8\ldots$$

$$= \prod_{i=1}^{k-1} (k-i+1)^{2^i}, \tag{4.8}$$

since clearly $c_1 = 1$.

If A is finite—say $|A| = n$—we now readily obtain a formula for the number of elements in B_A:

Theorem 4.9. *The free band B_A on a finite set A is finite. Specifically, if $|A| = n$, then*

$$|B_A| = \sum_{k=1}^{n} \binom{n}{k} c_k,$$

where c_k is given by the formula (4.8).■

The number of elements in the free band on n generators increases with great rapidity as n increases. If we denote the number by b_n then by direct calculation we obtain

$c_1 = 1$	$b_1 = 1$
$c_2 = 4$	$b_2 = 6$
$c_3 = 144$	$b_3 = 159$
$c_4 = 331{,}776$	$b_4 = 332{,}380$
$c_5 = 2{,}751{,}882{,}854{,}400$	$b_5 = 2{,}751{,}884{,}514{,}765$,

while c_6 is approximately $2{\cdot}7 \times 10^{26}$.

Before we leave the topic of free bands it is of interest to record two further observations. First,

Proposition 4.10. *If $x\beta, y\beta \in B_A$, then $(x\beta, y\beta) \in \mathscr{R}$ if and only if $C(x) = C(y)$, $\bar{x}(0) = \bar{y}(0)$ and $(x(0), y(0)) \in \beta$.*

Proof. First, if $(x\beta, y\beta) \in \mathscr{R}$ then certainly $(x\beta, y\beta) \in \mathscr{J}$ and so x and y have the same content by (4.4). Moreover, there exist u, v in F_A such that (modulo β)

$$xu \equiv y, \qquad yv \equiv x.$$

Since by Lemma 4.6(i) we may deduce that $C(x) = C(y) = C(xu)$, it follows that $C(u) \subseteq C(x)$. Hence any letter that is going to appear at all in xu has already appeared in x, and so

$$\bar{y}(0) = \bar{x}\bar{u}(0) = \bar{x}(0), \qquad y(0) \equiv xu(0) = x(0)$$

as required.

Conversely, suppose that x, y in F_A are such that $\bar{x}(0) = \bar{y}(0) = a$ and

$$x(0) \equiv y(0) \equiv w \ (\beta).$$

Then (modulo β)

$$x \equiv x(0)\bar{x}(0)x \equiv wax,$$

where $C(wa) = C(x)$. Hence, by (4.5),

$$xwa \equiv waxwa \equiv (wa)^2 \equiv wa,$$

and so $x\beta \mathscr{R} (wa)\beta$ in B_A. Similarly $y\beta \mathscr{R} (wa)\beta$ in B_A and so $x\beta \mathscr{R} y\beta$ as required.■

An analogous argument gives

Proposition 4.11. *If $x\beta, y\beta \in B_A$, then $(x\beta, y\beta) \in \mathscr{L}$ if and only if $C(x) = C(y)$, $\bar{x}(1) = \bar{y}(1)$ and $(x(1), y(1)) \in \beta$.*■

5. VARIETIES OF BANDS

Let A be a countable set and let F_A be the free semigroup on A. Let S be a semigroup. If $p, q \in F_A$, then we shall say that *the identical relation $p = q$ is satisfied in S* if $p\phi = q\phi$ for *every* homomorphism $\phi : F_A \rightarrow S$. Informally, if we think of the elements of A as variables, we are requiring that the two words p and q give rise to equal elements of S for every possible assignment of values from S to the variables. Thus, for example, the identical relation $ab = ba$ is satisfied in S if S is a commutative semigroup.

The class of semigroups in which a finite or infinite collection

$$p_1 = q_1, p_2 = q_2, \ldots$$

of identical relations is satisfied is called the *variety* of semigroups determined by these identical relations. Thus, as we have already remarked, the class of *commutative* semigroups is a variety. The class of *all* semigroups is also a variety, determined, for example, by the identical relation $a = a$. Less trivially, the class $\mathscr{B}$ of bands is the variety determined by the single identical relation $a^2 = a$.

We shall find it convenient at times to denote the variety determined by the identical relations $p_1 = q_1$, $p_2 = q_2, \ldots$ by $[p_1 = q_1, p_2 = q_2, \ldots]$. The variety determined by a set $\mathbf{R}$ of identical relations will be written as $[\mathbf{R}]$.

Remark. Since $F_A \times F_A$ is countable, no more than $2^{\aleph_0}$ collections of identical relations are possible. Hence there are no more than $2^{\aleph_0}$ distinct varieties of semigroups. That there are in fact as many as $2^{\aleph_0}$ distinct varieties has been shown by Evans (1968). See also Isbell (1970).

The intersection of a non-empty collection of varieties $\mathscr{V}_i (i \in I)$ is a variety; for if $\mathscr{V}_i$ is determined by a set $\mathbf{R}_i$ of identical relations, then a semigroup S belongs to $\bigcap\{\mathscr{V}_i : i \in I\}$ if and only if the collection $\bigcup\{\mathbf{R}_i : i \in I\}$ of identical relations is satisfied in S. Thus, for example, the intersection $\mathscr{B} \cap \mathscr{C}$ of the variety $\mathscr{B}$ of bands and the variety $\mathscr{C}$ of commutative semigroups is the variety $[a^2 = a, ab = ba]$ of semilattices.

By contrast, the union of a collection of varieties need not be a variety. One does, however, have a well-defined *join* $\bigvee\{\mathscr{V}_i : i \in I\}$ of a collection $\{\mathscr{V}_i : i \in I\}$ of varieties, namely the intersection of the collection of all varieties containing every $\mathscr{V}_i$. (The collection is non-empty, since the variety of *all* semigroups necessarily contains every $\mathscr{V}_i$.) It can be hard to determine a reasonable set of identical relations characterizing $\bigvee\{\mathscr{V}_i : i \in I\}$, though in theory we can always obtain a suitable set by proceeding in the way now described.

First, we refer to a congruence ρ on a semigroup S as being *fully invariant* if $(\forall x, y \in S)$

$$(x, y) \in \rho \Rightarrow (x\alpha, y\alpha) \in \rho$$

for every endomorphism $\alpha: S \to S$. It is a routine matter to verify that the intersection of a non-empty family of fully invariant congruences is a fully invariant congruence.

Next, if $\mathbf{R}$ is a set of identical relations, which it is convenient now to regard as a subset of $F_A \times F_A$ in the obvious way, rather than as a set of equalities, then we define $\mathbf{R}^v$ to be the smallest fully invariant congruence on F_A containing $\mathbf{R}$. It is the intersection of the family (non-empty since the universal congruence is one such) of fully invariant congruences containing $\mathbf{R}$.

Now let S be a semigroup and let $\mathbf{I}(S)$ be the subset of $F_A \times F_A$ consisting of all pairs (w_1, w_2) for which the identical relation $w_1 = w_2$ is satisfied in S. Then

Proposition 5.1. *The relation* $\mathbf{I}(S)$ *is a fully invariant congruence on* F_A. *If* $\mathscr{V} = [\mathbf{R}]$ *is a variety, then* $S \in \mathscr{V}$ *if and only if* $\mathbf{I}(S) \supseteq \mathbf{R}^v$.

Proof. It is obvious that $\mathbf{I}(S)$ is an equivalence. Also, if $w_1 = w_2$ and $z_1 = z_2$ are satisfied identically in S, then $w_1\phi = w_2\phi$ and $z_1\phi = z_2\phi$ for every homomorphism $\phi : F_A \to S$. Hence

$$(w_1z_1)\phi = (w_1\phi)(z_1\phi) = (w_2\phi)(z_2\phi) = (w_2z_2)\phi$$

for every homomorphism $\phi : F_A \to S$ and so the identical relation $w_1z_1 = w_2z_2$ is satisfied in S. We have thus shown that

$$(w_1, w_2) \in \mathbf{I}(S), (z_1, z_2) \in \mathbf{I}(S) \Rightarrow (w_1z_1, w_2z_2) \in \mathbf{I}(S);$$

i.e. we have shown that $\mathbf{I}(S)$ is a congruence.

Now suppose that $(w_1, w_2) \in \mathbf{I}(S)$ and that $\gamma : F_A \to F_A$ is an endomorphism. Let $\phi : F_A \to S$ be a homomorphism. Then $\gamma\phi : F_A \to S$ is a homomorphism and so $w_1\gamma\phi = w_2\gamma\phi$. Since this last statement holds good for every homomorphism $\phi : F_A \to S$ we may interpret it as saying that $(w_1\gamma, w_2\gamma) \in \mathbf{I}(S)$. Thus $\mathbf{I}(S)$ is fully invariant.

If $S \in \mathscr{V} = [\mathbf{R}]$ then certainly $\mathbf{R} \subseteq \mathbf{I}(S)$. Since $\mathbf{I}(S)$ is then a fully invariant congruence on F_A containing $\mathbf{R}$, it must contain the smallest such congruence; that is, $\mathbf{R}^v \subseteq \mathbf{I}(S)$.

Conversely, if $\mathbf{R}^v \subseteq \mathbf{I}(S)$ then certainly $\mathbf{R} \subseteq \mathbf{I}(S)$. Thus the identical relations contained in $\mathbf{R}$ are all satisfied in S and so $S \in \mathscr{V}$.■

For a given S in a variety $\mathscr{V} = [\mathbf{R}]$ it may well be the case that $\mathbf{I}(S)$ *properly* contains $\mathbf{R}^v$, for we may have chosen an S that "accidentally" satisfies more identical relations than are implied by the original set $\mathbf{R}$. For example if $\mathbf{R} = \{(ab, ba)\}$ we may happen to choose an S in $[\mathbf{R}]$ that is a semilattice, and so contained in a variety strictly smaller than $[\mathbf{R}]$. However, we can reasonably regard $\mathbf{R}^v$ as providing us with the totality of identical relations associated with the variety $[\mathbf{R}]$ since there is always at least one semigroup S in $[\mathbf{R}]$ for which $\mathbf{I}(S) = \mathbf{R}^v$:

Proposition 5.2. *If* $\mathbf{R} \subseteq F_A \times F_A$ *and* $\mathbf{R}^v = \rho$, *then* $F_A/\rho \in [\mathbf{R}]$ *and* $\mathbf{I}(F_A/\rho) = \mathbf{R}^v$.

If $S \in [\mathbf{R}]$ *and* $\phi : A \to S$ *is an arbitrary mapping, then there exists a unique homomorphism* $\chi : F_A/\rho \to S$ *such that*

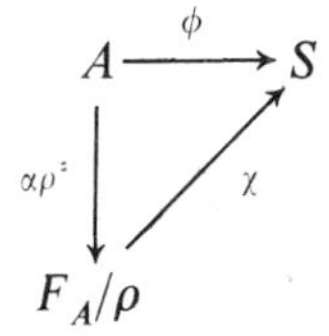

is a commutative diagram, where α *is the standard inclusion mapping of* A *into* F_A.

Proof. To show that $F_A/\rho \in [\mathbf{R}]$ we must show that for each (w_1, w_2) in $\mathbf{R}$ and for every homomorphism $\theta : F_A \to F_A/\rho$ it is the case that $w_1\theta = w_2\theta$. For each a in A we choose an element of $a\theta(\rho^\natural)^{-1}$ and call it $a\gamma$. This defines a mapping $\gamma : A \to F_A$ which we can extend to an endomorphism $\gamma : F_A \to F_A$ (since F_A is free) by the rule that

$$(a_1a_2 \dots a_n)\gamma = (a_1\gamma)(a_2\gamma)\dots(a_n\gamma) \qquad (a_1a_2 \dots a_n \in F_A).$$

Now certainly $a\gamma\rho^\natural = a\theta$ for every a in A. Indeed we can easily verify that $x\gamma\rho^\natural = x\theta$ for every x in F_A. Since $\rho = \mathbf{R}^v$ is fully invariant we have that $(w_1\gamma, w_2\gamma) \in \rho$. Hence

$$w_1\theta = w_1\gamma\rho^\natural = w_2\gamma\rho^\natural = w_2\theta.$$

Thus $F_A/\rho \in [\mathbf{R}]$.

It follows that $\mathbf{I}(F_A/\rho) \supseteq \mathbf{R}^v$. To show that equality holds, suppose that $(w_1, w_2) \in \mathbf{I}(F_A/\rho)$. Then $w_1\theta = w_2\theta$ for every homomorphism $\theta : F_A \to F_A/\rho$, and so in particular for the homomorphism $\rho^\natural$. Thus $w_1\rho^\natural = w_2\rho^\natural$ and so $(w_1, w_2) \in \rho = \mathbf{R}^v$. Hence $\mathbf{I}(F_A/\rho) = \mathbf{R}^v$ as required.

Now let $S \in [\mathbf{R}]$ and let ϕ be a mapping from A into S. Since F_A is a free semigroup there is a unique homomorphism $\psi : F_A \to S$ such that

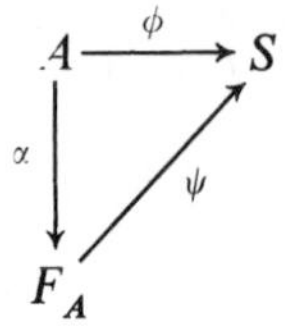

is a commutative diagram. Now $w_1\psi = w_2\psi$ for every (w_1, w_2) in $\mathbf{I}(S)$ and so certainly (by Proposition 5.1) for every (w_1, w_2) in $\mathbf{R}^v = \rho$. Hence $\psi \circ \psi^{-1} \supseteq \rho$ and so by Theorem I.5.4 there is a unique homomorphism $\chi : F_A/\rho \to S$ such that

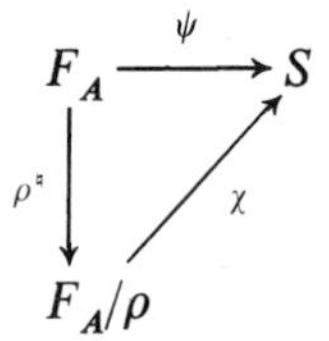

is a commutative diagram. Hence

$$\phi = \alpha\psi = \alpha\rho^\natural\chi$$

and so we obtain the commutative diagram described in the statement of the theorem.■

The property of F_A/ρ described by the commutative diagram justifies us in calling it the (*relatively*) *free* semigroup in the variety $[\mathbf{R}]$ on the set A of generators. We can obtain a relatively free semigroup in $[\mathbf{R}]$ for a more general infinite set A, and also for a finite set A provided A has enough symbols to spell the words appearing in $\mathbf{R}$. The free band $B_A = F_A/\beta$ discussed in Section 4 is an example of a relatively free semigroup, in which $\mathbf{R} = \{(a^2, a)\}$ and $\mathbf{R}^v = \beta$.

Proposition 5.3. *Let* $\mathbf{R}_1, \mathbf{R}_2$ *be subsets of* $F_A \times F_A$ *determining varieties* $[\mathbf{R}_1]$, $[\mathbf{R}_2]$ *respectively. Then*

$$[\mathbf{R}_1] \subseteq [\mathbf{R}_2] \text{ if and only if } \mathbf{R}_1^v \supseteq \mathbf{R}_2^v.$$

Proof. If $[\mathbf{R}_1] \subseteq [\mathbf{R}_2]$ then $F_A/\mathbf{R}_1^v \in [\mathbf{R}_2]$ and so $\mathbf{I}(F_A/\mathbf{R}_1^v) \supseteq \mathbf{R}_2^v$ (by Propositions 5.2 and 5.1). Hence, again by Proposition 5.2, $\mathbf{R}_1^v \supseteq \mathbf{R}_2^v$.

Conversely, if $\mathbf{R}_1^v \supseteq \mathbf{R}_2^v$ then

$$S \in [\mathbf{R}_1] \Rightarrow \mathbf{I}(S) \supseteq \mathbf{R}_1^v \Rightarrow \mathbf{I}(S) \supseteq \mathbf{R}_2^v \Rightarrow S \in [\mathbf{R}_2].$$

Thus $[\mathbf{R}_1] \subseteq [\mathbf{R}_2]$.∎

It is important to realise that two different sets $\mathbf{R}_1, \mathbf{R}_2$ of identical relations may determine the same variety. We have already encountered such a situation in Proposition 3.2, where the equivalence of (A), (C) and (D) can be interpreted as saying that the class of rectangular bands forms a variety which can be characterized either as $[aba = a]$ or as $[a^2 = a, abc = ac]$. It is, however, an immediate consequence of Proposition 5.3 that

Corollary 5.4. $[\mathbf{R}_1] = [\mathbf{R}_2]$ *if and only if* $\mathbf{R}_1^v = \mathbf{R}_2^v$.∎

Suppose now that we have a family $\{\mathscr{V}_i : i \in \mathbf{I}\}$ of varieties of semigroups, where $\mathscr{V}_i = [\mathbf{R}_i]$ $(i \in \mathbf{I})$. Let $\mathscr{V} = [\mathbf{R}]$, where

$$\mathbf{R} = \bigcap \{\mathbf{R}_i^v : i \in I\}.$$

Then $\mathbf{R}$ is a fully invariant congruence on F_A, and so $\mathbf{R} = \mathbf{R}^v$. Since $\mathbf{R}$ is the largest fully invariant congruence contained in every $\mathbf{R}_i^v$, it follows by the correspondence established in Proposition 5.3 that $\mathscr{V}$ is the smallest variety containing every $\mathscr{V}_i$, i.e. that

$$\mathscr{V} = \bigvee \{\mathscr{V}_i : i \in I\}.$$

In the case where I and all the sets $\mathbf{R}_i$ are finite, it is of interest, and is in some cases possible, to extract a finite generating set for $\mathbf{R}$; but there is no general routine for doing so, and there can be difficulty even in very simple cases. For example, if $\mathscr{B} = [a^2 = a]$ is the variety of bands and $\mathscr{C} = [ab = ba]$ is the variety of commutative semigroups, I know of no finite set of identical relations describing the variety $\mathscr{B} \vee \mathscr{C}$.

With respect to the operations $\cap$ and $\vee$, the set $\mathfrak{S}$ of varieties of semigroups becomes a complete lattice with greatest element the variety $[a = a]$ of all semigroups and least element the variety $[a = b]$ of trivial semigroups. The set of varieties of bands, i.e. the set of varieties $\mathscr{V}$ of semigroups such that $\mathscr{V} \subseteq \mathscr{B}$, is a sublattice $\mathfrak{B}$ of this lattice. Any logical difficulty involved here in talking of a set of classes is only apparent, for we can if we wish identify each variety with a fully invariant congruence on F_A, and talk instead of the lattice of fully invariant congruences on F_A, and of the sublattice consisting of all fully invariant congruences containing β.

By an *atom* in a lattice $(L, \wedge, \vee)$ with least element 0 we mean a minimal element a in the set $L\backslash\{0\}$. Any study of the lattice $\mathfrak{B}$ of varieties of bands must begin by identifying the atoms of $\mathfrak{B}$. Before we can do this, however, it is necessary to quote a result due to Birkhoff (1935). It applies to algebraic structures in general (see Cohn (1965)), but we shall quote it here without proof in the form in which we shall use it.

Theorem 5.5. *A non-empty class $\mathscr{V}$ of semigroups is a variety if and only if*
(*a*) *every subsemigroup of a semigroup in $\mathscr{V}$ is in $\mathscr{V}$;*
(*b*) *every homomorphic image of a semigroup in $\mathscr{V}$ is in $\mathscr{V}$;*
(*c*) *the direct product of a family of semigroups in $\mathscr{V}$ is in $\mathscr{V}$.* ∎

We can now identify the atoms of $\mathfrak{B}$. Let us write $\mathscr{SL}$ for the variety $[ab = ba, a^2 = a]$ of semilattices, $\mathscr{LZ}$ for the variety $[ab = a]$ of left zero semigroups, and $\mathscr{RZ}$ for the variety $[ab = b]$ of right zero semigroups.

Theorem 5.6. *The set of atoms in the lattice $\mathfrak{B}$ of varieties of bands is $\{\mathscr{SL}, \mathscr{LZ}, \mathscr{RZ}\}$.*

Proof. To show that $\mathscr{SL}$ is an atom in $\mathfrak{B}$, we show that the fully invariant congruence

$$\varepsilon = \{(ab, ba), (a^2, a)\}^v$$

corresponding to $\mathscr{SL}$ is a maximal non-universal fully invariant congruence on F_A. The congruence ε is the smallest semilattice congruence on F_A. Certainly $\varepsilon \supset \beta\,(= \{(a^2, a)\}^v)$, and the congruence ε/β on $B_A = F_A/\beta$ is the smallest semilattice congruence on B_A. Using Proposition 1.8 and formula (4.4) we now have that for all x, y in F_A,

$$\begin{aligned}(x, y) \in \varepsilon &\Leftrightarrow (x\beta, y\beta) \in \varepsilon/\beta\\ &\Leftrightarrow (x\beta, y\beta) \in \mathscr{J}\\ &\Leftrightarrow C(x) = C(y);\end{aligned}$$

that is, two elements of F_A are ε-equivalent if and only if they have the same content.

Now, if σ is a fully invarient congruence on F_A properly containing ε, there must exist y, z in F_A such that $(y, z) \in \sigma$ and $C(y) \neq C(z)$. Suppose, without loss of generality, that there is some a in A such that $a \in C(y)$, $a \notin C(z)$. Since $\varepsilon \subseteq \sigma$ we may assume that a appears exactly once in y and at the beginning.

Consider now an endomorphism of F_A mapping a to itself and all other elements of A to some element b ($\neq a$) of A. This endomorphism maps y to ab and z to b, and so, since σ is fully invariant, $(ab, b) \in \sigma$. Hence, again by the fully invariant property we have (modulo σ)

$$\begin{aligned} w_1 \equiv w_2 w_1 &\equiv w_1 w_2 \quad (\text{since } \varepsilon \subseteq \sigma) \\ &\equiv w_2 \end{aligned}$$

for every w_1, w_2 in F_A. Hence $\sigma = F_A \times F_A$ as required, and so $\mathscr{SL}$ is an atom in $\mathfrak{B}$.

The argument showing that $\mathscr{LZ}$ is an atom is based on the straightforward observation that if $\mathbf{Z} = \{(ab, a)\}$ then $(w_1, w_2) \in \mathbf{Z}^v$ if and only if w_1 and w_2 begin with the same letter. Thus any word w in F_A is $\mathbf{Z}^v$-equivalent to the one-letter word which is the first letter of w. Any fully invariant congruence σ properly containing $\mathbf{Z}^v$ must contain some (y, z) such that y and z have different first letters a, b (say) respectively. Then (modulo σ)

$$a \equiv y \equiv z \equiv b$$

and so by the fully invarient property it follows that $(w_1, w_2) \in \sigma$ for all w_1, w_2 in F_A.

A similar argument applies to $\mathscr{RZ}$.

To show that $\mathscr{SL}$, $\mathscr{LZ}$ and $\mathscr{RZ}$ are the only atoms, consider a variety $\mathscr{V} \notin \{\mathscr{SL}, \mathscr{LZ}, \mathscr{RZ}\}$ of bands containing a non-trivial band B, and suppose that $\mathscr{V}$ is an atom of $\mathfrak{B}$. Then $\mathscr{V}$ has trivial intersection with each of $\mathscr{SL}$, $\mathscr{LZ}$ and $\mathscr{RF}$. By Theorem 3.1 it follows that if B is not a rectangular band it has a non-trivial semilattice homomorphic image Y; hence (by Theorem 5.5) $Y \in \mathscr{SL} \cap \mathscr{V}$, a contradiction.

Hence B is a rectangular band, and so by Theorem I.1.9 is a direct product of a left zero semigroup L and a right zero semigroup R, at least one of which is non-trivial. Since we may in an obvious way regard L and R as subsemigroups of B it follows by Theorem 5.5 that $L \in \mathscr{LZ} \cap \mathscr{V}$, $R \in \mathscr{RZ} \cap \mathscr{V}$, and one way or the other we get a contradiction. ■

It is natural now to attempt to identify the varieties constituting the sublattice of $\mathfrak{B}$ generated by the atoms $\mathscr{SL}$, $\mathscr{LZ}$ and $\mathscr{RZ}$. It is possible to do this with informal, largely verbal arguments of the following sort. The fully invariant congruence ε corresponding to $\mathscr{SL}$ consists of all pairs (w_1, w_2) in $F_A \times F_A$ such that w_1 and w_2 have the same content. Equally, the fully invariant congruence ζ corresponding to $\mathscr{LZ}$ consists of all pairs (w_1, w_2) in $F_A \times F_A$ such that w_1 and w_2 have the same initial letter. Hence $\varepsilon \cap \zeta$, the fully invariant congruence corresponding to $\mathscr{SL} \vee \mathscr{LZ}$, consists of all pairs (w_1, w_2) in $F_A \times F_A$ such that w_1 and w_2 have the same content *and* the same initial letter. "Clearly" $\varepsilon \cap \zeta = \{(a^2, a), (abc, acb)\}^v$, and so $\mathscr{SL} \vee \mathscr{LZ} =$

$\mathscr{L}\mathscr{N}$, the variety $[a^2 = a, abc = acb]$ of *left normal* bands considered by Yamada and Kimura (1958).

There is nothing wrong with the argument, although the use of the word "clearly" involves an element of "handwaving". More formal arguments are available, and the results that we shall obtain on the way are sufficiently interesting to justify the more lengthy procedure.

First, we shall require an alternative version of Birkhoff's Theorem (5.5) involving the notion of a *subdirect product*. Again, this idea belongs to general algebra, but we shall here confine ourselves to semigroups.

There are several ways of approaching this construction, the simplest being to consider a subsemigroup T of the direct product P of a family $\{S_i : i \in I\}$ of semigroups. The semigroup P consists (as in Section I.1) of maps $p: I \to \cup \{S_i : i \in I\}$ such that $ip \in S_i$ for each i in I, and multiplication takes place in accordance with the "componentwise" formula

$$i(pq) = (ip)(iq) \qquad (i \in I).$$

For each i there is a natural "projection" homomorphism π_i from P onto S_i given by

$$p\pi_i = ip \qquad (p \in P).$$

The subsemigroup T will be called a *subdirect product* of the semigroups S_i $(i \in I)$ if $T\pi_i = S_i$ for each i in I.

Examples include P itself and (if each S_i has an identity element 1) the *weak* direct product W consisting of all w in P for which all but finitely many of the elements iw are equal to 1. But much smaller subsemigroups of P are possible: for example, suppose that the semigroups S_i are all equal—say $S_i = S$ for every i. Then for each s in S define $s^* \in P$ by

$$is^* = s \qquad (i \in I).$$

The *diagonal* $D = \{s^* : s \in S\}$ is then a subdirect product of $\{S_i : i \in I\}$, as may easily be verified.

An alternative approach to the idea of subdirect product is provided by the following theorem:

Theorem 5.7. *Let $\{S_i : i \in I\}$ be a family of semigroups and suppose that S is a semigroup with the property that for each i in I there is a homomorphism λ_i from S onto S_i. If, for all s, t in S,*

$$s\lambda_i = t\lambda_i \textit{ for all } i \textit{ in } I \Rightarrow s = t, \tag{5.8}$$

then S is isomorphic to a subdirect product of the semigroups $S_i (i \in I)$.

Conversely, if S is a subdirect product of the semigroups $S_i (i \in I)$ then for *each i in I there is a homomorphism λ_i from S onto S_i such that the implication* (5.8) *holds.*

Proof. To show that S is isomorphic to a subdirect product, let us denote, as before, the direct product to the semigroups $S_i(i \in I)$ by P. Now consider the map $\lambda: S \to P$ given by $s\lambda = \bar{s}$, where $\bar{s} \in P$ is defined by

$$i\bar{s} = s\lambda_i \qquad (i \in I).$$

Then λ is a homomorphism, since for every i in I

$$\begin{aligned} i(\bar{s}\bar{t}) &= (i\bar{s})(i\bar{t}) = (s\lambda_i)(t\lambda_i) \\ &= (st)\lambda_i = i\overline{st}. \end{aligned}$$

Also λ is one-one, since, for all s, t in S,

$$\begin{aligned} s\lambda = t\lambda &\Rightarrow i\bar{s} = i\bar{t} \text{ for all } i \text{ in } I \\ &\Rightarrow s\lambda_i = t\lambda_i \text{ for all } i \text{ in } I \\ &\Rightarrow s = t \text{ (by (5.8))}. \end{aligned}$$

Thus λ maps S isomorphically onto a subsemigroup $S\lambda$ of P. In fact $S\lambda$ is a subdirect product, for if $s_i \in S_i$ then since λ_i is given as mapping *onto* S_i there exists s in S such that $s\lambda_i = s_i$. Hence the element $\bar{s} = s\lambda$ of S has the property that

$$\bar{s}\pi_i = i\bar{s} = s\lambda_i = s_i,$$

and so $(S\lambda)\pi_i = S_i$ as required.

Conversely, if S is a subdirect product then the projection homomorphisms $\pi_i: S \to S_i$ are all onto. Also, if $s, t \in S$, then

$$\begin{aligned} (\forall i \in I)\, s\pi_i = t\pi_i &\Rightarrow (\forall i \in I)\, is = it \\ &\Rightarrow s = t \end{aligned}$$

by the very definition of equality in P. ■

Because of the correspondence (Theorem I.5.3) between homomorphisms and congruences we have the following alternative version of Theorem 5.7:

Theorem 5.9. *If S is a semigroup having a family of congruences* $\{\rho_i : i \in I\}$ *with the property that*

$$\bigcap \{\rho_i : i \in I\} = 1_S,$$

then S is isomorphic to a subdirect product of the semigroups S/ρ_i. ■

All this is leading to a statement of a theorem characterizing varieties. For the proof the reader is referred to (Cohn 1965; Theorem IV.3.5 and Corollary IV.3.6), where the result is attributed to P. Hall.

Theorem 5.10. *A non-empty class $\mathscr{V}$ of semigroups is a variety if and only if*
(a) every homomorphic image of a semigroup in $\mathscr{V}$ is in $\mathscr{V}$;
(b) every subdirect product of a family of semigroups in $\mathscr{V}$ is in $\mathscr{V}$. ■

A specialization of the subdirect product of semigroups has been considered by Kimura (1958). If S and T are two semigroups having a common homomorphic image H, and if $\phi: S \to H$ and $\psi: T \to H$ are homomorphisms onto H, then the *spined product of S and T with respect to H, ϕ and ψ is defined as*

$$Y = \{(s, t) \in S \times T : s\phi = t\psi\}.$$

This is the standard terminology in semigroup theory, but the idea occurs elsewhere in mathematics under different titles. The semigroup Y is the *pullback* or *fibre product* (Schubert 1972) of the diagram

$$\begin{array}{ccc} & & S \\ & & \big\downarrow{\scriptstyle\phi} \\ T & \xrightarrow[\psi]{} & H. \end{array}$$

It is easy to verify that Y is a subsemigroup of $S \times T$. Also, for any s in S there is at least one t in T such that $t\psi = s\phi$ and so there is at least one t in T for which $(s, t) \in Y$. Thus Y is a subdirect product. It can be thought of as a generalization of the diagonal and reduces to the diagonal when $S = T = H$ and $\phi = \psi = 1_S$. On the other hand, if H is the trivial semigroup then Y coincides with the direct product $S \times T$. If $\mathscr{V}$ is a variety of semigroups then certainly a spined product of two semigroups in $\mathscr{V}$ is contained in $\mathscr{V}$.

We now show how the strong semilattice construction discussed in the introduction and exemplified in Section 2 is relevant to varieties of semigroups. But first let us agree to denote by $^\circ S$ the semigroup obtained from S by adjoining an extra zero element 0 *whether or not S already has a zero*. This presupposes that if S already has a zero element we are denoting it by a symbol other than 0. In $^\circ S$ we always have the implication

$$xy = 0 \Rightarrow x = 0 \quad \text{or} \quad y = 0.$$

We now have

Lemma 5.11. *If $\mathscr{V}$ is a variety of semigroups that contains the variety of semilattices, and if S is a semigroup in $\mathscr{V}$, then $^\circ S \in \mathscr{V}$.*

Proof. We have that the multiplicative semigroup $\{0, 1\}$ is in $\mathscr{V}$ and so the direct product $T = S \times \{0, 1\}$ is in $\mathscr{V}$. The subset $I = \{(x, 0): x \in S\}$ is an ideal of T and so the Rees quotient T/I is in $\mathscr{V}$. But T/I is essentially $\{(s, 1): s \in S\} \cup \{0\}$ and so is isomorphic to $^\circ S$. ■

The following result is essentially in (Petrich 1973), though not there stated in terms of varieties:

Proposition 5.12. *Let $\mathscr{V}$ be a variety containing the variety of semilattices. If $S = \mathscr{S}(Y; S_\alpha; \phi_{\alpha,\beta})$, a strong semilattice of semigroups S_α, and if each S_α is in the variety $\mathscr{V}$, then $S \in \mathscr{V}$.*

Proof. Define $\psi_\alpha: S \to {}^\circ S_\alpha$ by the rule that

$$x\psi_\alpha = \begin{cases} x\phi_{\beta,\alpha} & \text{if } x \in S_\beta \quad \text{and} \quad \beta \geqslant \alpha \\ 0 \text{ otherwise.} \end{cases}$$

To see that this is a homomorphism, suppose that $x, y \in S$ are such that $x \in S_\beta$, $y \in S_\gamma$. Then either $\beta\gamma \geqslant \alpha$ or $\beta\gamma \ngeqslant \alpha$. If $\beta\gamma \geqslant \alpha$ then $\beta \geqslant \beta\gamma \geqslant \alpha$ and $\gamma \geqslant \beta\gamma \geqslant \alpha$ and so

$$\begin{aligned}(xy)\psi_\alpha &= [(x\phi_{\beta,\beta\gamma})(y\phi_{\gamma,\beta\gamma})]\,\phi_{\beta\gamma,\alpha} \\ &= (x\phi_{\beta,\alpha})(y\phi_{\gamma,\alpha}) = (x\psi_\alpha)(y\psi_\alpha).\end{aligned}$$

If $\beta\gamma \ngeqslant \alpha$ then $(xy)\psi_\alpha = 0$. Also, either $\beta \ngeqslant \alpha$ or $\gamma \ngeqslant \alpha$, for $\beta \geqslant \alpha$ and $\gamma \geqslant \alpha$ would imply that $\beta\gamma \geqslant \alpha$ contrary to assumption. Hence one or other of $x\psi_\alpha$ and $y\psi_\alpha$ is equal to 0 and so $(xy)\psi_\alpha = (x\psi_\alpha)(y\psi_\alpha)$ in this case also.

Next, notice that $S\psi_\alpha \supseteq S_\alpha\psi_\alpha = S_\alpha$. In fact unless $\beta \geqslant \alpha$ for every β in Y we have that $S\psi_\alpha = {}^\circ S_\alpha$. That is, $S\psi_\alpha = {}^\circ S_\alpha$ for every α in Y except for the minimum element ω of Y, if such an element exists. If ω exists, then $S\psi_\omega = S_\omega$.

Now suppose that $x \in S_\beta$ and $y \in S_\gamma$ are such that $x\psi_\alpha = y\psi_\alpha$ for every α in Y. Since $x \in S_\beta$ we have $x\psi_\beta = x\phi_{\beta,\beta} = x \neq 0$ and so $y\psi_\beta \neq 0$. Hence $\gamma \geqslant \beta$. But we may equally well show that $\beta \geqslant \gamma$. Hence $\beta = \gamma$ and it is now clear that

$$x = x\psi_\beta = y\psi_\beta = y.$$

It follows that S is isomorphic to a subdirect product of the semigroups ${}^\circ S_\alpha$ $(\alpha \in Y\backslash\{\omega\})$ and S_ω. Since $S_\omega \in \mathscr{V}$ and since by Lemma 5.11 each ${}^\circ S_\alpha$ is in $\mathscr{V}$, we deduce that $S \in \mathscr{V}$. ■

Let us now introduce notation for varieties of bands as follows:

$\mathscr{SL}$	semilattices	$ab = ba$	(5.13)
$\mathscr{LN}$	left normal bands	$abc = acb$	
$\mathscr{RN}$	right normal bands	$abc = bac$	
$\mathscr{N}$	normal bands	$abca = acba$	
$\mathscr{T}$	trivial bands	$a = b$	
$\mathscr{LZ}$	left zero bands	$ab = a$	
$\mathscr{RZ}$	right zero bands	$ab = b$	
$\mathscr{RB}$	rectangular bands	$aba = a$	

E

In each case the given identical relation characterizes the variety within the variety of bands. The characterizations within the variety of semigroups are obtained in each case by adding the identical relation $a^2 = a$. (Of course it is not always necessary to do so: e.g. in the case of rectangular bands we have seen that the identical relation $aba = a$ *implies* the identical relation $a^2 = a$.)

As a first step towards identifying the sublattice $\mathfrak{A}$ of $\mathfrak{B}$ generated by the atoms of $\mathfrak{B}$, notice that $\mathscr{LZ} \subseteq \mathscr{RB}$, $\mathscr{RZ} \subseteq \mathscr{RB}$, so that certainly

$$\mathscr{LZ} \vee \mathscr{RZ} \subseteq \mathscr{RB}.$$

But any rectangular band may be thought of as the direct product of a left zero semigroup L and a right zero semigroup R (Proposition I.1.9). Since

$$L \in \mathscr{LZ} \subseteq \mathscr{LZ} \vee \mathscr{RZ} \quad \text{and} \quad R \in \mathscr{RZ} \subseteq \mathscr{LZ} \vee \mathscr{RZ},$$

it follows that $L \times R \in \mathscr{LZ} \vee \mathscr{RZ}$. We deduce that

$$\mathscr{LZ} \vee \mathscr{RZ} = \mathscr{RB}.$$

Further information about $\mathfrak{A}$ is obtained from the following result, due to Yamada and Kimura (1958):

Proposition 5.14. *A band B is normal if and only if it is a strong semilattice of rectangular bands.*

Proof. If $B = \mathscr{S}(Y; E_\alpha; \phi_{\alpha,\beta})$, a strong semilattice of rectangular bands, and if a, b, c are arbitrary elements of B, with $a \in E_\alpha$, $b \in E_\beta$ and $c \in E_\gamma$, let us write δ for $\alpha\beta\gamma$. Then

$$\begin{aligned} abca &= (a\phi_{\alpha,\delta})(b\phi_{\beta,\delta})(c\phi_{\gamma,\delta})(a\phi_{\alpha,\delta}) \\ &= a\phi_{\alpha,\delta} \quad \text{(since } E_\delta \text{ is a rectangular band)} \end{aligned}$$

and similarly $acba = a\phi_{\alpha,\delta}$. Hence $abca = acba$ for all a, b, c *in* B and so B is normal.

Conversely, if the band B is normal, we consider the structure given to B by Theorem 3.16. Recall that if $a \in E_\alpha$ and $(x, \xi) \in E_\beta = I_\beta \times \Lambda_\beta$, where $\beta \leqslant \alpha$, then formula (3.11) gives

$$a(x, \xi) = (\phi_\beta^a x, \xi), \qquad (x, \xi)\, a = (x, \xi\psi_\beta^a).$$

Hence if $a \in E_\alpha$, and if $b = (x, \xi)$ and $c = (y, \eta)$ are elements of $E_\beta (\beta \leqslant \alpha)$, then

$$abca = (\phi_\beta^a x, \xi)(y, \eta\psi_\beta^a) = (\phi_\beta^a x, \eta\psi_\beta^a),$$

while

$$acba = (\phi_\beta^a y, \eta)(x, \xi\psi_\beta^a) = (\phi_\beta^a y, \xi\psi_\beta^a).$$

Thus if the band is normal we conclude that ϕ_β^a and ψ_β^a are *constant* mappings for every a in E_α and for every pair (α, β) of elements of Y for which

$\alpha \geqslant \beta$. The mapping $\Phi_{\alpha\beta}: E_\alpha \to \mathscr{T}^*(I_\beta) \times \mathscr{T}(\Lambda_\beta)$ sending each a in E_α to the ordered pair $(\phi^a_\alpha, \psi^a_\beta)$ may thus be identified with the mapping $\phi_{\alpha,\beta}: E_\alpha \to I_\beta \times \Lambda_\beta$ sending a to the ordered pair $(\langle\phi^a_\beta\rangle, \langle\psi^a_\beta\rangle)$. Moreover, if $a, a' \in E_\alpha$, then

$$(aa')\,\phi_{\alpha,\beta} = (\langle\phi^{aa'}_\beta\rangle, \langle\psi^{aa'}_\beta\rangle)$$

$$= (\langle\phi^a_\beta\phi^{a'}_\beta\rangle, \langle\psi^a_\beta\psi^{a'}_\beta\rangle) \quad \text{(by (3.14) and (3.15))}$$

$$= (\langle\phi^a_\beta\rangle, \langle\psi^{a'}_\beta\rangle) \quad \text{(by the rules for composing constant left and right mappings)}$$

$$= (\langle\phi^a_\beta\rangle, \langle\psi^a_\beta\rangle)(\langle\phi^{a'}_\beta\rangle, \langle\psi^{a'}_\beta\rangle)$$

$$= (a\phi_{\alpha,\beta})(b\phi_{\alpha,\beta}),$$

and so $\phi_{\alpha,\beta}: E_\alpha \to E_\beta$ is a homomorphism.

Next, notice that if $\beta \leqslant \alpha$ then for any a in E_α and any $b = (x, \xi)$ in E_β,

$$aba = (\phi^a_\beta x, \xi\psi^a_\beta) = (\langle\phi^a_\beta\rangle, \langle\psi^a_\beta\rangle) = a\phi_{\alpha,\beta}. \tag{5.15}$$

If $\beta = \alpha$ it follows that

$$a\phi_{\alpha,\alpha} = aaa = a$$

for every a in E_α. Thus $\phi_{\alpha,\alpha}$ is the identical automorphism of E_α.

If $\alpha \geqslant \beta \geqslant \gamma$ and $a \in E_\alpha$ then it follows from (5.15) that for any b in E_β and c in E_γ,

$$(a\phi_{\alpha,\beta})\phi_{\beta,\gamma} = abacaba.$$

Since $bacab \in E_\gamma$ it follows, again by (5.15), that

$$abacaba = a\phi_{\alpha,\gamma}.$$

Hence $\phi_{\alpha,\beta}\phi_{\beta,\gamma} = \phi_{\alpha,\gamma}$.

Finally, if $a \in E_\alpha$, $b \in E_\beta$ and $\alpha\beta = \gamma$, then the multiplication rule (3.13) gives

$$ab = (\langle\phi^a_\gamma\phi^b_\gamma\rangle, \langle\psi^a_\gamma\psi^b_\gamma\rangle)$$

$$= (\langle\phi^a_\gamma\rangle, \langle\psi^b_\gamma\rangle) \quad \text{(by the rules for composing constant left and right hand mappings)}$$

$$= (\langle\phi^a_\gamma\rangle, \langle\psi^a_\gamma\rangle)(\langle\phi^b_\gamma\rangle, \langle\psi^b_\gamma\rangle)$$

$$= (a\phi_{\alpha,\gamma})(b\phi_{\beta,\gamma}).$$

Hence B is a strong semilattice $\mathscr{S}(Y; E_\alpha; \phi_{\alpha,\beta})$ of rectangular bands. ■

Before considering the consequences of this result in the context of varieties, we mentioned two obvious corollaries. First, if the band is left normal then so is every sub-band and hence so in particular is each of the rectangular bands E_α. Now a band E_α that has both the identical relations

$$abc = acb \quad \text{and} \quad aba = a$$

is necessarily a left zero semigroup, since for all a, b in E_α,

$$ab = aab = aba = a.$$

Hence a left normal band is a strong semilattice of left zero semigroups.

Conversely, if $B = \mathscr{S}(Y; L_\alpha; \phi_{\alpha,\beta})$, a strong semilattice of left zero semigroups L_α, then for arbitrary elements $a \in L_\alpha$, $b \in L_\beta$ and $c \in L_\gamma$ in B,

$$abc = (a\phi_{\alpha,\delta})(b\phi_{\beta,\delta})(c\phi_{\gamma,\delta}) = a\phi_{\alpha,\delta},$$

where $\delta = \alpha\beta\gamma$. and similarly $acb = a\phi_{\alpha,\beta}$. Thus we have

Corollary 5.16. *A band is left normal if and only if it is a strong semilattice $\mathscr{S}(Y; L_\alpha; \phi_{\alpha,\beta})$ of left zero semigroups.* ■

Notice that any mapping whatever between left zero semigroups is a homomorphism. Notice too that the multiplication in

$\mathscr{S}(Y; L_\alpha; \phi_{\alpha,\beta})$ is given by

$$l_\alpha l_\beta = l_\alpha \phi_{\alpha,\alpha\beta}. \tag{5.17}$$

A closely analogous argument leads to

Corollary 5.18. *A band is right normal if and only if it is a strong semilattice $\mathscr{S}(Y; R_\alpha; \phi_{\alpha,\beta})$ of right zero semigroups.* ■

Once again, any mapping whatever between right zero semigroups is a homomorphism. Also, multiplication in $\mathscr{S}(Y; R_\alpha; \phi_{\alpha,\beta})$ is given by

$$r_\alpha r_\beta = r_\beta \phi_{\beta,\alpha\beta}. \tag{5.19}$$

We now use Propositions 5.12 and 5.14 as follows. First since $\mathscr{SL} \subseteq \mathscr{N}$ and $\mathscr{RB} \subseteq \mathscr{N}$, we have that $\mathscr{SL} \vee \mathscr{RB} \subseteq \mathscr{N}$. Conversely, if $B \in \mathscr{N}$ then B is a strong semilattice of semigroups in $\mathscr{SL} \vee \mathscr{RB}$ and so is itself in $\mathscr{SL} \vee \mathscr{RB}$. Hence

$$\mathscr{SL} \vee \mathscr{RB} = \mathscr{N}. \tag{5.20}$$

Similar arguments based on Corollaries 5.16 and 5.18 give that

$$\mathscr{SL} \vee \mathscr{LZ} = \mathscr{LN}, \mathscr{SL} \vee \mathscr{RZ} = \mathscr{RN}. \tag{5.21}$$

Yamada and Kimura (1958) proved the following result:

Proposition 5.22. *A band is normal if and only if it is isomorphic to a spined product of a left normal and a right normal band.*

Proof. First, since $\mathcal{N}$ is a variety and since $\mathcal{L}\mathcal{N} \subseteq \mathcal{N}$ and $\mathcal{R}\mathcal{N} \subseteq \mathcal{N}$, a spined product (indeed any subdirect product whatever) of a left normal and a right normal band must necessarily be normal.

Conversely, if B is normal then by Proposition 5.14 B is a strong semilattice $\mathcal{S}(Y; E_\alpha; \phi_{\alpha,\beta})$ of rectangular bands E_α. Each E_α is the direct product $L_\alpha \times R_\alpha$ of a left zero semigroup L_α and a right zero semigroup R_α. Moreover, as shown in Corollary 3.6 the homomorphism $\phi_{\alpha,\beta}: E_\alpha \to E_\beta$ determines homomorphisms $\phi^l_{\alpha,\beta}: L_\alpha \to L_\beta$ and $\phi^r_{\alpha,\beta}: R_\alpha \to E_\beta$ such that

$$(l_\alpha, r_\alpha)\phi_{\alpha,\beta} = (l_\alpha\phi^l_{\alpha,\beta}, r_\alpha\phi^r_{\alpha,\beta})$$

for every (l_α, r_α) in E_α.

Now $L = \bigcup \{L_\alpha : \alpha \in Y\}$ becomes a strong semilattice of left zero bands if we define

$$l_\alpha \circ l_\beta = l_\alpha\phi^l_{\alpha,\alpha\beta},$$

and similarly $R = \bigcup \{R_\alpha : \alpha \in Y\}$ becomes a strong semilattice of right zero bands if we define

$$r_\alpha * r_\beta = r_\beta\phi^r_{\beta,\alpha\beta}.$$

(Compare these formulae with (5.17) and (5.19).) Thus L is a left normal band and R is a right normal band.

Certainly the semigroups L and R have an obvious common homomorphic image, namely the semilattice Y. If $\phi: L \to Y$ and $\psi: R \to Y$ are the obvious homomorphisms, defined respectively by

$$l_\alpha\phi = \alpha, \qquad r_\alpha\psi = \alpha \qquad (l_\alpha \in L_\alpha, r_\alpha \in R_\alpha),$$

then the spined product of L and R with respect to Y, ϕ and ψ consists of those pairs (l, r) in $L \times R$ for which $l\phi = r\psi$, i.e. consists of

$$\bigcup \{L_\alpha \times R_\alpha : \alpha \in Y\}.$$

Moreover, the multiplication in the spined product is given by

$$\begin{aligned}(l_\alpha, r_\alpha)(l_\beta, r_\beta) &= (l_\alpha \circ l_\beta, r_\alpha * r_\beta)\\ &= (l_\alpha\phi^l_{\alpha,\alpha\beta}, r_\beta\phi^r_{\beta,\alpha\beta})\\ &= (l_\alpha\phi^l_{\alpha,\alpha\beta}, r_\alpha\phi^r_{\alpha,\alpha\beta})(l_\beta\phi^l_{\beta,\alpha\beta}, r_\beta\phi^r_{\beta,\alpha\beta})\\ &= [(l_\alpha, r_\alpha)\phi_{\alpha,\alpha\beta}][(l_\beta, r_\beta)\phi_{\beta,\alpha\beta}],\end{aligned}$$

and so coincides with the multiplication in the normal band B. ■

It is an easy consequence of this result that

$$\mathscr{L}\mathscr{N} \vee \mathscr{R}\mathscr{N} = \mathscr{N}.$$

We can summarize the observations we have made about the sublattice $\mathfrak{A}$ generated by the atoms of $\mathfrak{B}$ by means of the diagram:

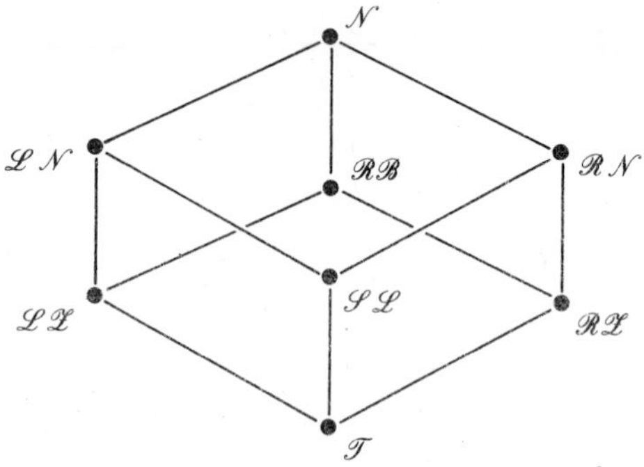

Not all the statements implied by the diagram have been proved, but it is easy to fill in the gaps. For example, to show that $\mathscr{L}\mathscr{N} \vee \mathscr{R}\mathscr{B} = \mathscr{N}$, note that since $\mathscr{R}\mathscr{B} \subseteq \mathscr{N}$ and $\mathscr{S}\mathscr{L} \subseteq \mathscr{L}\mathscr{N} \subseteq \mathscr{N}$, we have

$$\mathscr{N} = \mathscr{S}\mathscr{L} \vee \mathscr{R}\mathscr{B} \subset \mathscr{L}\mathscr{N} \vee \mathscr{R}\mathscr{B} \subseteq \mathscr{N}$$

by (5.20), and so $\mathscr{L}\mathscr{N} \vee \mathscr{R}\mathscr{B} = \mathscr{N}$ as required. The various statements about intersections implied by the diagram are all all obvious: for example, since $\mathscr{S}\mathscr{L} \subseteq \mathscr{L}\mathscr{N}$ and $\mathscr{S}\mathscr{L} \subseteq \mathscr{R}\mathscr{N}$, we have $\mathscr{S}\mathscr{L} \subseteq \mathscr{L}\mathscr{N} \cap \mathscr{R}\mathscr{N}$; but conversely, if $B \in \mathscr{L}\mathscr{N} \cap \mathscr{R}\mathscr{N}$ then, for all a, b in B,

$$\begin{aligned} ab = aab &= aba \text{ (by left normality)} \\ &= baa \text{ (by right normality)} \\ &= ba \end{aligned}$$

and so $B \in \mathscr{S}\mathscr{L}$.

It is reasonable to ask whether identities other than those listed in (5.13) will determine interesting varieties of bands. This is not in general an easy question and we shall not pursue it. The reader is referred to Kimura (1957; 1958*a*, *b*, *c*), Yamada and Kimura (1958), Yamada (1958) and to Petrich (1971). Petrich classifies all identities in up to three variables, and relates the identities to properties of the mappings ϕ^a_β, ψ^a_β considered in Section 3. Complete descriptions of the lattice $\mathfrak{B}$ of varieties of bands have been given (independently) by Birjukov (1970), Gerhard (1970) and Fennemore (1971a, 1971b). A most useful survey article on varieties of semigroups has been written by Evans (1971).

EXERCISES

1. (Tamura and Kimura 1954). If S is a commutative semigroup and $a, b \in S$, write $a|b$ (*a divides b*) if there exists x in S such that $ax = b$. Define η as the set of all pairs (a, b) in $S \times S$ for which

$$(\exists m \in \mathbf{N})\, a|b^m \quad \text{and} \quad (\exists n \in \mathbf{N})\, b|a^n.$$

(i) Show that η is a congruence on S.
(ii) Show that S/η is a semilattice.
(iii) Show that if ρ is a congruence on S such that S/ρ is a semilattice, then $\eta \subseteq \rho$.
A commutative semigroup T is called *archimedean* if

$$(\forall a, b \in T)(\exists m, n \in \mathbf{N}) \quad a|b^m \text{ and } b|a^n.$$

Show that each η-class of a commutative semigroup S is an archimedean subsemigroup of S. Deduce that any commutative semigroup can be expressed as a semilattice of archimedean semigroups.

2. If S is a regular semigroup and E is the set of idempotents of S, show that the following statements are equivalent:

(A) S is a Clifford semigroup;
(B) $(\forall a \in S)(\forall a' \in V(a))\, aa' = a'a$;
(C) $\mathscr{L} = \mathscr{R}$.

3. (i) If the semigroup S is a semilattice Y of completely simple semigroups S_α and if L is a left ideal of S_α, show that $L \cup [\bigcup\{S_\beta : \beta < \alpha\}]$ is a left ideal of S.
(ii) If $S = \mathscr{S}(Y; G_\alpha; \phi_{\alpha,\beta})$ is a strong semilattice of groups and L is a left ideal of S, show that

$$L \cap G_\alpha \neq \varnothing \Rightarrow L \subseteq \bigcup\{G_\beta : \beta \leqslant \alpha\}.$$

Deduce that L is a two-sided ideal of S.
(iii) Show that S is a Clifford semigroup if and only if it is a union of groups and every one-sided ideal is a two-sided ideal.
For other characterizations of Clifford semigroups, see Lajos (1969; 1970a, b, c; 1971a, b; 1972a, b).

4. A regular semigroup is called *E-unitary* if its idempotents form a unitary subsemigroup. (For the definition of *unitary*, see Exercise II.12). Show that a strong semilattice $\mathscr{S}(Y; G_\alpha; \phi_{\alpha,\beta})$ of groups is E-unitary if and only if $\phi_{\alpha,\beta}$ is one–one for every $\alpha \geqslant \beta$ in Y.

5. Show that if λ is a left translation of a group G, then $\lambda = \lambda_{\lambda e}$, where e is the identity element of G. Deduce that all left translations (and similarly all right translations) of G are inner, and hence that G is isomorphic to its translational hull $\Omega(G)$.

6. Show that the homomorphism $a \rightarrow (\lambda_a, \rho_a)$ from S into $\Omega(S)$ is not one–one if S is the two-element null semigroup.

7. Show that a band B is

$$\begin{array}{llll} & \text{rectangular} & \text{if and only if } \mathscr{D} = B \times B, \\ & \text{left zero} & \text{if and only if } \mathscr{L} = B \times B \\ & \text{right zero} & \text{if and only if } \mathscr{R} = B \times B, \\ \text{and trivial} & & \text{if and only if } \mathscr{H} = B \times B. \end{array}$$

8. If B is a band, show that, for all a, b in B,

$$ab \,\mathscr{L}\, bab, \qquad ab \,\mathscr{R}\, aba.$$

9. In the semigroup $\mathscr{T}(X)$, where $X = \{0, 1, \ldots, 6\}$, let

$$p = \begin{pmatrix} 0 & 1 & 2 & 3 & 4 & 5 & 6 \\ 1 & 2 & 2 & 5 & 5 & 5 & 2 \end{pmatrix}, \qquad q = \begin{pmatrix} 0 & 1 & 2 & 3 & 4 & 5 & 6 \\ 3 & 6 & 6 & 4 & 4 & 4 & 6 \end{pmatrix}.$$

Show that the elements p and q generate a six-element subsemigroup $S = \{p, p^2, q, q^2, pq, qp\}$ of $\mathscr{T}(X)$ with Cayley table

	p	p^2	q	q^2	pq	qp
p	p^2	p^2	pq	pq	pq	p^2
p^2	p^2	p^2	pq	pq	pq	p^2
q	qp	qp	q^2	q^2	q^2	qp
q^2	qp	qp	q^2	q^2	q^2	qp
pq	p^2	p^2	pq	pq	pq	p^2
qp	qp	qp	q^2	q^2	q^2	qp.

Show that the identical relation $abc = ac$ is satisfied in S, but that S is not a band.

10. The free band B_A is expressible in the usual way as a semilattice of rectangular bands. The underlying semilattice is the semilattice of finite non-empty subsets of A, and the "product" of two subsets P and Q in the semilattice is their *union*. Thus, if $\leqslant$ is defined in the usual way in the semilattice,

$$Q \leqslant P \text{ if and only if } Q \supseteq P.$$

The rectangular bands are the $\mathscr{J}$-classes of B_A. There is one such $\mathscr{J}$-class U_P corresponding to each finite non-empty subset P of A, and U_P is the set of all elements of B_A with content P.

If $|P| > 1$, let $I_P = \bigcup\{U_{P\setminus\{a\}}: a \in P\}$, and let $\alpha: U_P \to I_P \times I_P$ be defined by

$$x\alpha = (x(0), x(1)) \qquad (x \in U_P).$$

Show that α is an isomorphism from U_P onto the rectangular band $I_P \times I_P$. If $|P| = 1$, let $I_P = \{1\}$.

If $P \subseteq Q$, so that $Q \leqslant P$ in the semilattice, show that for each x in U_P the maps $\phi_Q^x: I_Q \to I_Q$ and $\psi_Q^x: I_Q \to I_Q$ (see (3.11)) are given by

$$\phi_Q^x y(0) = (xy)(0), \qquad y(1)\psi_Q^x = (yx)(1) \qquad (y \in U_Q).$$

11. Let $A = \{a, b, c\}$. Then the $\mathscr{J}$-classes of the free band B_A are

$$U_{\{a\}}, U_{\{b\}}, U_{\{c\}}, U_{\{a,b\}}, U_{\{a,c\}}, U_{\{b,c\}}, U_A.$$

For simplicity of notation, denote a β-class by any chosen representative of it in F_A. Then

$$U_{\{a\}} = \{a\}, \qquad U_{\{a,b\}} = \{ab, aba, ba, bab\},$$

while U_A may be identified with the 12×12 rectangular band $I_A \times I_A$, where

$$I_A = \{ab, aba, ba, bab, ac, aca, ca, cac, bc, bcb, cb, cbc\}.$$

(i) Determine the isomorphism α from $U_{\{a,b\}}$ onto the rectangular band $\{a, b\} \times \{a, b\}$. (See Exercise 10 above for the definition of α).

(ii) Show that, in the notation of (3.11),

$$\phi^a_{\{a,b\}} = \begin{pmatrix} a & b \\ a & a \end{pmatrix}, \qquad \phi^b_{\{a,b\}} = \begin{pmatrix} a & b \\ b & b \end{pmatrix},$$

and similarly compute $\psi^a_{\{a,b\}}, \psi^b_{\{a,b\}}$.

(iii) Show that

$$\phi^a_A = \begin{pmatrix} ab & aba & ba & bab & ac & aca & ca & cac & bc & bcb & cb & cbc \\ ab & aba & aba & ab & ac & aca & aca & ac & ab & ab & ac & ac \end{pmatrix}$$

and similarly compute $\phi^b_A, \psi^a_A, \psi^b_A, \phi^{ab}_A, \psi^{ab}_A$.

(iv) Verify that ϕ^{abc}_A and ψ^{abc}_A are constant maps and that

$$\langle \phi^{abc}_A \rangle = ab, \qquad \langle \psi^{abc}_A \rangle = bc.$$

(v) Verify that $\phi^{ab}_A = \phi^a_A \phi^b_A$ (composed as left mappings), and that $\psi^{ab}_A = \psi^a_A \psi^b_A$ (composed as right mappings).

12. If B is a band, the relation $\leqslant$ defined by

$$e \leqslant f \text{ if and only if } ef = fe = e$$

is a partial order on B. Show that if B is normal then $\leqslant$ is compatible; i.e. $(\forall e, f, g, h \in B)$

$$e \leqslant f \text{ and } g \leqslant h \;\Rightarrow\; eg \leqslant fh. \qquad (1)$$

Conversely, show that if B is a band and if the order relation $\leqslant$ is compatible in the sense of (1) then B is normal. [Hint: observe that $xyx \leqslant x$ for all x, y in B; then use Theorem 3.1 to show that for all a, b, c in B the elements $abca$ and $acba$ lie in the same rectangular band E_α; hence show that

$$abca = abca \,.\, acba \,.\, abca = (abca)cb(abca) \leqslant acba,$$

and show similarly that $acba \leqslant abca$.]

13. Show that the variety $\mathscr{N}$ of normal bands can be described alternatively as

$$\mathscr{N} = [a^2 = a, abcd = acbd].$$

14. Let $N_A = F_A/\nu$ be the (relatively) free normal band on the set A of generators ($|A| \geqslant 3$). Show that if $w, z \in F_A$ then $(w, z) \in \nu$ if and only if w and z have the same initial letter, the same final letter and the same content. Deduce that if $|A| = n$ then

$$|N_A| = \sum_{k=1}^{n} \binom{n}{k} k^2 = 2^{n-2} n(n+1).$$

Chapter V

Inverse Semigroups

INTRODUCTION

In many ways this was the hardest chapter to write, because of the difficulty in selecting material from the vast and rapidly growing corpus of published work on inverse semigroups. In 1961 Clifford and Preston offered the opinion that inverse semigroups were the most promising class of semigroups for future study, and the intervening years have amply justified their forecast. It will be possible here to cover only a small part of the recent work on the subject. The interested reader may wish to consult some of the references I shall give to topics not covered in the text.

Inverse semigroups were studied first by Vagner (1952, 1953) and independently by Preston (1954a, b, c). Vagner called them "generalized groups", and this nomenclature is still standard in the Russian literature. The origin of the idea in both cases was the study of semigroups of partial one–one mappings of a set, and one of the earliest results was a representation theorem (analogous to Cayley's Theorem in group theory) to the effect that every inverse semigroup has a faithful representation as an inverse semigroup of partial one–one mappings. This important result has become known as the Vagner–Preston Representation Theorem and is proved as Theorem 1.9 below.

The theory of inverse semigroups has many features in common with the theory of groups, but there are some important differences. Among the new features is the non-trivial natural partial order that exists in each inverse semigroup. This order relation is introduced and discussed in Section 2.

In Section 3 two special congruences on an inverse semigroup are described, namely the minimum group congruence σ and the maximum idempotent-separating congruence μ. The existence of *fundamental* inverse semigroups (i.e. inverse semigroups having no non-trivial idempotent-separating congruences) is established.

The set of idempotents of an inverse semigroup forms a semilattice, and at a certain stage it becomes natural to ask to what extent the structure of an inverse semigroup is determined by the structure of its semilattice of idempotents. If the semilattice is what is called "anti-uniform" it is possible to assert that the inverse semigroup must be a Clifford semigroup. This is Theorem 5.2. In other cases the answer is less complete, but in all cases the crucial notion is that of the *Munn semigroup* T_E of a semilattice E. The fundamental fact (Theorem 4.8) concerning this semigroup is that if S is any inverse semigroup whatever with semilattice of idempotents E, then there exists a homomorphism $\phi: S \to T_E$ whose kernel is μ, the maximum idempotent-separating congruence on S.

In Sections 6 and 7 this apparatus is used to give descriptions respectively of fundamental *bisimple* inverse semigroups (Theorem 6.4) and fundamental *simple* inverse semigroups (Theorem 7.4). If the restriction "fundamental" is removed then the problem becomes vastly more complicated, and both in Section 6 and Section 7 only the case of ω-semigroups is considered (i.e. the case of semigroups for which $E = \{e_0, e_1, e_2, \ldots\}$ with $e_0 > e_1 > e_2 > \ldots$). A satisfactory structure theorem is obtained in each case (Theorems 6.11 and 7.8), and reference is made to more general work along the same lines.

In Section 8 the study of representations of inverse semigroups by partial one-one mappings is resumed. Schein's (1962) elegant theory of such representations is of interest in its own right, but a further justification for its inclusion in this chapter is that we shall in Section VII.4 make an interesting application of the theory to prove a result of T. E. Hall (1975) on amalgams of inverse semigroups.

It is appropriate in this introduction to make brief references to certain topics that will not be covered in this chapter. McAlister's (1973, 1974) approach to the structure theory of inverse semigroups *via* "proper" inverse semigroups is referred to briefly in Exercise 12. The approach to the structure of bisimple inverse semigroups *via* "RP-systems" by Reilly (1968) and Reilly and Clifford (1968) deriving from Clifford's (1953) paper on bisimple inverse semigroups with an identity is not covered at all.

In recent years several authors have studied *free* inverse semigroups. See Scheiblich (1972), Reilly (1972), Preston (1973), Schein (1975) and Munn (1974).

In Section III.6 a brief study was made of finite congruence-free semigroups. For some interesting work on (infinite) congruence-free inverse semigroups see the recent papers by Munn (1972, 1975).

1. PRELIMINARIES

A semigroup S is called an *inverse semigroup* if every a in S possesses a unique inverse, i.e. if there exists a unique element a^{-1} in S such that

$$aa^{-1}a = a, \qquad a^{-1}aa^{-1} = a^{-1}. \tag{1.1}$$

Such a semigroup is certainly regular, but not every regular semigroup is an inverse semigroup: a rectangular band, in which every element is an inverse of every other element, is an obvious example.

The definition we have given, while the most natural one, is not in accord with our general theme of investigating the effect of properties of the idempotents on the structure of a semigroup. The equivalence of (a) and (b) in the next theorem shows, however, that a definition in terms of idempotents is possible.

Theorem 1.2. *The following statements about a semigroup S are equivalent:*
(a) *S is an inverse semigroup;*
(b) *S is regular and idempotent elements commute;*
(c) *each $\mathscr{L}$-class and each $\mathscr{R}$-class of S contains a unique idempotent;*
(d) *each principal left ideal and each principal right ideal of S contains a unique idempotent generator.*

Proof. It is clear by the definitions of $\mathscr{L}$ and $\mathscr{R}$ that (c) and (d) are equivalent. Notice the connection between (c) and Proposition II.3.2.

To show that (a) $\Rightarrow$ (b), let e, f be idempotents and let $x = (ef)^{-1}$. Then

$$efxef = ef, \qquad xefx = x.$$

The element fxe is idempotent, since

$$(fxe)^2 = f(xefx)e = fxe;$$

also,

$$(ef)(fxe)(ef) = efxef = ef,$$
$$(fxe)(ef)(fxe) = f(xefx)e = fxe,$$

and so ef is an inverse of fxe. But fxe, being idempotent, is its own unique inverse, and so

$$fxe = ef.$$

It follows that ef is idempotent, and similarly we obtain that fe is idempotent. Hence

$$(ef)(fe)(ef) = (ef)^2 = ef, \qquad (fe)(ef)(fe) = (fe)^2 = fe$$

and so fe is an inverse of ef. But ef, being idempotent, is its own unique inverse, and so we finally obtain that $ef = fe$.

To show that (b) $\Rightarrow$ (c), note first that since S is regular every $\mathscr{L}$-class contains at least one idempotent (Proposition II.3.2). If e, f are $\mathscr{L}$-equivalent idempotents, then by Proposition II.3.3

$$ef = e, \qquad fe = f.$$

Since, by hypothesis, $ef = fe$, it follows that $e = f$. Similar remarks apply to $\mathscr{R}$-classes. We can express the property (c) of inverse semigroups as follows:

$$\mathscr{L} \cap (E \times E) = \mathscr{R} \cap (E \times E) = 1_E, \tag{1.3}$$

where E is the set of idempotents of S.

To show that (c) $\Rightarrow$ (a), notice first that a semigroup with the property (c) is necessarily regular, since assuredly every $\mathscr{D}$-class contains an idempotent. If a', a'' are inverses of a then aa' and aa'' are idempotents in S that are $\mathscr{R}$-equivalent to a and hence to each other. By property (c) we thus have that $aa' = aa''$. Equally, $a'a = a''a$, and so

$$a' = a'aa' = a''aa' = a''aa'' = a''. \blacksquare$$

We shall consistently write $E(S)$, or simply E, for the set of idempotents of the inverse semigroup S. It is a subsemigroup of S, since if $e, f \in E$ then $(ef)^2 = e^2f^2 = ef$. Indeed it is a commutative semigroup of idempotents and so we are justified in referring to the *semilattice* of idempotents of S. Notice that a semilattice is itself an example of an inverse semigroup.

It is convenient to list certain elementary properties of inverse semigroups in a proposition as follows:

Proposition 1.4. *Let S be an inverse semigroup with semilattice of idempotents E.*

(a) $(a^{-1})^{-1} = a$ *for every a in S.*

(b) $e^{-1} = e$ *for every e in E.*

(c) $(ab)^{-1} = b^{-1}a^{-1}$ *for every a, b in S.*

(d) $aea^{-1} \in E$, $a^{-1}ea \in E$ *for every a in S and every e in E.*

(e) $a \mathscr{R} b$ *if and only if* $aa^{-1} = bb^{-1}$; $a \mathscr{L} b$ *if and only if* $a^{-1}a = b^{-1}b$.

(f) *If $e, f \in E$, then $l \mathscr{D} f$ in S if and only if there exists a in S such that* $aa^{-1} = e$, $a^{-1}a = f$.

Proof. Part (a) follows by the mutuality of the inverse property, and (b) is immediate. To prove (c), notice that, since bb^{-1} and $a^{-1}a$ are idempotents,

$$(ab)(b^{-1}a^{-1})(ab) = a(bb^{-1})(a^{-1}a)b = aa^{-1}abb^{-1}b = ab,$$

$$(b^{-1}a^{-1})(ab)(b^{-1}a^{-1}) = b^{-1}(a^{-1}a)(bb^{-1})a^{-1} = b^{-1}bb^{-1}a^{-1}aa^{-1}$$
$$= b^{-1}a^{-1}.$$

Thus $b^{-1}a^{-1}$ is an inverse, and hence *the* inverse, of ab. That is, $(ab)^{-1} = b^{-1}a^{-1}$.

To prove (d), note that

$$(aea^{-1})^2 = ae(a^{-1}a)ea^{-1} = aa^{-1}ae^2a^{-1} = aea^{-1},$$

and similarly $(a^{-1}ea)^2 = a^{-1}ea$.

Part (e) is simply a specialization of Proposition II.4.1 to the case where inverses are unique, and (f) is likewise a specialization of Proposition II.3.6.■

Part (c) of the proposition generalizes in the obvious way as follows

Corollary 1.5. *If $a_1, a_2, \ldots, a_n$ are elements of an inverse semigroup S, then*

$$(a_1a_2 \ldots a_n)^{-1} = a_n^{-1} \ldots a_2^{-1}a_1^{-1}.■$$

In particular, $(a^n)^{-1} = (a^{-1})^n$ for every a in S, and so we may unambiguously use the notation a^{-n} for any positive integer n.

We have already mentioned in passing that semilattices are examples of inverse semigroups. So of course are groups. Less trivially, the Clifford semigroups discussed in Section IV.2 are examples of inverse semigroups. We shall very soon see, however, that not every inverse semigroup is of this type.

The elementary results proved so far, and especially Proposition 1.4(c), show that the theory of inverse semigroups has some features in common with group theory. The analogy can be carried further, and the description of a congruence on a group in terms of a normal subgroup (the class containing the identity) can be generalized to a description of a congruence on an inverse semigroup in terms of a "kernel normal system", which is the collection of classes containing the idempotents. We shall not pursue this aspect of the theory at all, however, and refer the reader to Vagner (1953), Preston (1954a) and Clifford and Preston (1967).

Inverse semigroups do not constitute a variety (see Section IV.5) of semigroups—no more do groups—but the class is closed under the taking of homomorphic images:

Proposition 1.6. *Let S be an inverse semigroup, let T be a semigroup and let $\phi : S \to T$ be a homomorphism. Then $S\phi$ is an inverse semigroup.*

Proof. It is immediate that $S\phi$ is regular. If g, h are idempotents in $S\phi$ then by Lallement's Lemma (II.4.6) there exist idempotents e, f in S such that $e\phi = g$ and $f\phi = h$. Hence

$$gh = (e\phi)(f\phi) = (ef)\phi = (fe)\phi = (f\phi)(e\phi) = hg,$$

and so $S\phi$ is an inverse semigroup.■

Notice also in this connection that for each s in S the element $s^{-1}\phi$ is an inverse in $S\phi$ of $s\phi$, since

$$(s\phi)(s^{-1}\phi)(s\phi) = (ss^{-1}s)\phi = s\phi$$

$$\text{and } (s^{-1}\phi)(s\phi)(s^{-1}\phi) = (s^{-1}ss^{-1})\phi = s^{-1}\phi.$$

Since inverses in $S\phi$ are unique, we can express this as

$$(s\phi)^{-1} = s^{-1}\phi \qquad (s \in S), \tag{1.7}$$

another result that echoes group theory.

The way in which inverse semigroups arise most naturally is as sets of one–one partial mappings of a set. A *partial one–one mapping* of a set X is a mapping whose domain is a (possibly empty) subset of X and which is one–one. Such mappings are of course elements of the set $\mathscr{B}(X)$ of binary relations on X and so can be "multiplied" by means of the composition law $\circ$ defined on $\mathscr{B}(X)$. In fact the set $\mathscr{I}(X)$ of partial one–one mappings is closed under $\circ$; for if $\alpha, \beta \in \mathscr{I}(X)$ then $(x, y) \in \alpha \circ \beta$ if and only if there exists z in X

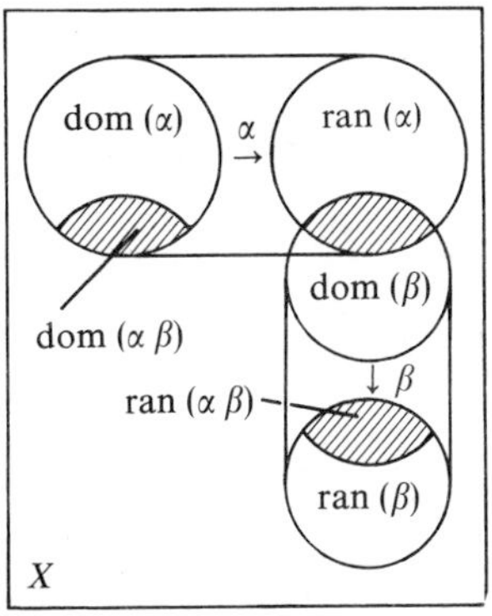

such that $(x, z) \in \alpha$ and $(z, y) \in \beta$, i.e. if and only if there exists z in $\operatorname{ran}(\alpha) \cap \operatorname{dom}(\beta)$ such that $x\alpha = z$ and $z\beta = y$, i.e. if and only if $y = (x\alpha)\beta$, where $x \in [\operatorname{ran}(\alpha) \cap \operatorname{dom}(\beta)]\alpha^{-1}$. Thus $\alpha \circ \beta$ is a partial one–one mapping with domain $[\operatorname{ran}(\alpha) \cap \operatorname{dom}(\beta)]\alpha^{-1}$ and range $[\operatorname{ran}(\alpha) \cap \operatorname{dom}(\beta)]\beta$. It is often useful to visualize $\alpha \circ \beta$ (which, following our usual practice, we shall write simply as $\alpha\beta$) as in the diagram. Notice that $\mathscr{I}(X)$ includes the empty relation with domain and range $\varnothing$. Since this relation is the empty subset of $X \times X$ we denote it too by $\varnothing$ and remark that in the semigroup $\mathscr{I}(X)$ we have $\alpha \circ \beta = \varnothing$ if and only if $\operatorname{ran}(\alpha) \cap \operatorname{dom}(\beta) = \varnothing$.

Proposition 1.8. *$\mathscr{I}(X)$ is an inverse semigroup.*

Proof. We have seen that $\mathscr{I}(X)$ is closed under the associative operation of composition. An element $\alpha : \text{dom}(\alpha) \to \text{ran}(\alpha)$ of $\mathscr{I}(X)$ has an obvious inverse in $\mathscr{I}(X)$, namely $\alpha^{-1} : \text{ran}(\alpha) \to \text{dom}(\alpha)$, the ordinary set-theoretic inverse of the bijection α. It is obvious that

$$\alpha\alpha^{-1} = 1_{\text{dom}(\alpha)}, \qquad \alpha^{-1}\alpha = 1_{\text{ran}(\alpha)}, \qquad \alpha\alpha^{-1}\alpha = \alpha, \qquad \alpha^{-1}\alpha\alpha^{-1} = \alpha^{-1}.$$

Thus certainly $\mathscr{I}(X)$ is regular.

The most satisfactory way of showing that $\mathscr{I}(X)$ is an inverse semigroup is to show that its idempotents commute. This requires that we be able to recognize idempotents. If $\alpha \in \mathscr{I}(X)$ then

$$\text{dom}(\alpha^2) = [\text{dom}(\alpha) \cap \text{ran}(\alpha)]\alpha^{-1}, \qquad \text{ran}(\alpha^2) = [\text{dom}(\alpha) \cap \text{ran}(\alpha)]\alpha.$$

If $\alpha^2 = \alpha$ then in particular $\text{dom}(\alpha^2) = \text{dom}(\alpha)$ and so

$$[\text{dom}(\alpha) \cap \text{ran}(\alpha)]\alpha^{-1} = \text{dom}(\alpha) = [\text{ran}(\alpha)]\alpha^{-1}.$$

Since α^{-1} is one–one it follows that $\text{dom}(\alpha) \cap \text{ran}(\alpha) = \text{ran}(\alpha)$, i.e. that $\text{ran}(\alpha) \subseteq \text{dom}(\alpha)$. Similarly, by considering ranges, we may show that $\text{dom}(\alpha) \subseteq \text{ran}(\alpha)$. Hence $\text{dom}(\alpha) = \text{ran}(\alpha)$. Thus α can be idempotent only if it is a one–one mapping of a subset A of X onto itself. This alone is not enough to make α idempotent, for we must in addition have that $x\alpha^2 = x\alpha$ for every x in A. Since α is one–one this implies that $x\alpha = x$ for every x in A. The following lemma is now obvious:

Lemma 1.9. *An element α of $\mathscr{I}(X)$ is idempotent if and only if $\alpha = 1_A$ for some subset A of X.*■

If now we consider the product of two idempotents 1_A and 1_B in $\mathscr{I}(X)$, we find that

$$\text{dom}(1_A 1_B) = (A \cap B)1_A^{-1} = A \cap B, \qquad \text{ran}(1_A 1_B) = (A \cap B)1_B = A \cap B.$$

Indeed, since $1_A 1_B$ clearly maps $A \cap B$ identically, we have

$$1_A 1_B = 1_{A \cap B}.$$

It now follows immediately that idempotents commute in $\mathscr{I}(X)$.■

We call $\mathscr{I}(X)$ the *symmetric inverse semigroup* on X. Notice that the semilattice of idempotents of $\mathscr{I}(X)$ is isomorphic to the semilattice of subsets of X under intersection. The idempotent $1_X = 1$ is the identity element of $\mathscr{I}(X)$, while the idempotent $1_\varnothing = \varnothing$ is the zero element. The set of elements of $\mathscr{I}(X)$ having domain and range equal to X is the subgroup H_1 of elements α in $\mathscr{I}(X)$ for which $\alpha\alpha^{-1} = \alpha^{-1}\alpha = 1$. It coincides with $\mathscr{G}(X)$, the *symmetric group* on the set X.

Just as every group can be embedded up to isomorphism in a symmetric group (Cayley's Theorem) and every semigroup can be embedded up to isomorphism in a full transformation semigroup (Theorem I.1.9), so every inverse semigroup can be embedded up to isomorphism in a symmetric inverse semigroup. The proof of this result, while necessarily more complicated than in the group case, is basically very similar to it. It is due to Vagner (1952) and Preston (1954c).

Theorem 1.10. (The Vagner–Preston Representation Theorem). *If S is an inverse semigroup then there exists a set X and a monomorphism $\phi : S \to \mathscr{I}(X)$.*

Proof. Let $X = S$, and for each a in S consider the mapping $\alpha_a : Saa^{-1} \to Sa^{-1}a$ given by

$$x\alpha_a = xa \quad (x \in Saa^{-1}).$$

First, notice that $xa = xaa^{-1}a \in Sa^{-1}a$ for *any* x in S, and so the codomain of α_a is $Sa^{-1}a$ as stated. In fact α_a maps Saa^{-1} *onto* $Sa^{-1}a$, for if $sa^{-1}a \in Sa^{-1}a$ then

$$sa^{-1}a = sa^{-1}aa^{-1}a = [sa^{-1}(aa^{-1})]\alpha_a \in (Saa^{-1})\alpha_a.$$

In addition, α_a is one–one, for if xaa^{-1} and yaa^{-1} in Saa^{-1} are such that $(xaa^{-1})\alpha_a = (yaa^{-1})\alpha_a$, then

$$\begin{aligned} xaa^{-1} &= xaa^{-1}aa^{-1} = (xaa^{-1})\alpha_a . a^{-1} = (yaa^{-1})\alpha_a . a^{-1} \\ &= yaa^{-1}aa^{-1} = yaa^{-1}. \end{aligned}$$

Thus $\alpha_a \in \mathscr{I}(S)$ and so we have a mapping $\phi : S \to \mathscr{I}(S)$ defined by

$$a\phi = \alpha_a \qquad (a \in S).$$

In fact ϕ is one–one, for if $\alpha_a = \alpha_b$ $(a, b \in S)$ then $\mathrm{dom}(\alpha_a) = \mathrm{dom}(\alpha_b)$, $\mathrm{ran}(\alpha_a) = \mathrm{ran}(\alpha_b)$. That is,

$$Saa^{-1} = Sbb^{-1}, \qquad Sa^{-1}a = Sb^{-1}b$$

and so by Theorem 1.2 (d) we have

$$aa^{-1} = bb^{-1}, \qquad a^{-1}a = b^{-1}b.$$

Moreover, since $\alpha_a = \alpha_b$ we have that a^{-1} $(=a^{-1}aa^{-1} \in Saa^{-1})$ has the same image under α_a and α_b. That is, $a^{-1}a = a^{-1}b$, and it now easily follows that

$$a = aa^{-1}a = aa^{-1}b = bb^{-1}b = b.$$

It remains to prove that ϕ is homomorphic. The following lemma is useful:

Lemma 1.11. *If e and f are idempotents in an inverse semigroup S, then $Se \cap Sf = Sef$.*

Proof. First, if $z = xef \in Sef$ then $z \in Sf$ and $z = xfe \in Se$; hence $z \in Se \cap Sf$. Conversely, if $z = xe = yf \in Se \cap Sf$, then

$$z = yf = (yf)f = zf = xef$$

and so $z \in Sef$. ■

Returning now to the proof of Theorem 1.10, notice first that for each a in S the element $\alpha_{a^{-1}}$ of $\mathscr{I}(S)$ maps $Sa^{-1}(a^{-1})^{-1}$ onto $S(a^{-1})^{-1}a^{-1}$, i.e. maps $Sa^{-1}a$ onto Saa^{-1}. Moreover, if $xaa^{-1} \in Saa^{-1}$, then

$$(xaa^{-1})\alpha_a\alpha_{a^{-1}} = xaa^{-1}aa^{-1} = xaa^{-1}.$$

Thus $\alpha_a\alpha_{a^{-1}}$ is the identity mapping of Saa^{-1}, and a similar argument shows that $\alpha_{a^{-1}}\alpha_a$ is the identity mapping of $Sa^{-1}a$. We conclude that

$$\alpha_{a^{-1}} = \alpha_a^{-1} \qquad (a \in S).$$

Next, if a and b are elements of S, then

$$\mathrm{dom}(\alpha_a\alpha_b) = [Sa^{-1}a \cap Sbb^{-1}]\alpha_a^{-1} = Sa^{-1}abb^{-1}a^{-1}$$

and

$$\mathrm{ran}(\alpha_a\alpha_b) = [Sa^{-1}a \cap Sbb^{-1}]\alpha_b = Sa^{-1}ab.$$

Since $Sa \subseteq S$ and $Sa^{-1} \subseteq S$, we have

$$Sa^{-1}abb^{-1}a^{-1} \subseteq Sabb^{-1}a^{-1} = Saa^{-1}abb^{-1}a^{-1} \subseteq Sa^{-1}abb^{-1}a^{-1},$$

and similarly

$$Sa^{-1}ab \subseteq Sab = Sabb^{-1}a^{-1}ab \subseteq Sb^{-1}a^{-1}ab \subseteq Sa^{-1}ab.$$

Hence

$$\mathrm{dom}(\alpha_a\alpha_b) = Sa^{-1}abb^{-1}a^{-1} = Sabb^{-1}a^{-1} = S(ab)(ab)^{-1} = \mathrm{dom}(\alpha_{ab}),$$

$$\mathrm{ran}(\alpha_a\alpha_b) = Sa^{-1}ab = Sb^{-1}a^{-1}ab = S(ab)^{-1}(ab) = \mathrm{ran}\,(\alpha_{ab}).$$

Moreover, if $x \in \mathrm{dom}(\alpha_a\alpha_b)$, then

$$x\alpha_a\alpha_b = (x\alpha_a)\alpha_{bab} = (xa)b = x(ab) = x\alpha$$

and so $\alpha_a\alpha_b = \alpha_{ab}$. Thus ϕ is homomorphic. ■

Notice that the representation ϕ reduces to the well-known right regular representation for groups when S is a group, for in this case $Saa^{-1} = Sa^{-1}a = S$ for every a in S.

We shall return to the question of representations of inverse semigroups by partial one–one mappings in Section 8.

2. THE NATURAL ORDER RELATION ON AN INVERSE SEMIGROUP

If a, b are elements of an inverse semigroup S, let us write $a \leqslant b$ if there exists an idempotent e in S such that $a = eb$.

Lemma 2.1. *The relation $\leqslant$ defined above is a partial order relation on the inverse semigroup S.*

Proof. To show that $\leqslant$ is reflexive we need only note that for any a in S we have $a = ea$, where $e = aa^{-1}$. If $a = eb$ and $b = fa$, where $e, f \in E$, then

$$ea = e(eb) = eb = a$$

and so $a = eb = efa = fea = fa = b$. Thus $\leqslant$ is anti-symmetric. If $a = eb$ and $b = fc$ $(e, f \in E)$ then $a = (ef)c$; thus $\leqslant$ is transitive.■

Several alternative characterizations of $\leqslant$ are available. The following list is by no means exhaustive:

Proposition 2.2. *If a, b are elements of an inverse semigroup S then the following statements are equivalent:*

(i) $a \leqslant b$;	(ii) $(\exists e \in E)\ a = be$;
(iii) $aa^{-1} = ba^{-1}$;	(iv) $aa^{-1} = ab^{-1}$;
(v) $a^{-1}a = b^{-1}a$;	(vi) $a^{-1}a = a^{-1}b$;
(vii) $a = ab^{-1}a$;	(viii) $a = aa^{-1}b$.

Proof. Only sample proofs are necessary; we shall show that (i) ≡ (iii) ≡ (vii). If $a = eb$ with $e \in E$, then

$$aa^{-1} = ebb^{-1}e = bb^{-1}e^2 = bb^{-1}e = ba^{-1}.$$

Thus (i) ⇒ (iii).

Now assume that (iii) holds. Then $a^{-1} = a^{-1}aa^{-1} = a^{-1}ba^{-1}$, and taking inverses of both sides we obtain $a = ab^{-1}a$. Thus (iii) ⇒ (vii).

If (vii) holds, then $a^{-1} = a^{-1}ba^{-1}$ and so

$$a = aa^{-1}a = aa^{-1}ba^{-1}a = aa^{-1}b(b^{-1}b)(a^{-1}a) = aa^{-1}ba^{-1}ab^{-1}b = eb,$$

where $e = (aa^{-1})(ba^{-1}ab^{-1}) \in E$. Thus (vii) ⇒ (i).■

The restriction of $\leqslant$ to E is easily seen to be the natural semilattice ordering on E:

$$e \leqslant f \text{ if and only if } ef = e.$$

It is of interest to examine the underlying set-theoretic meaning of the relation $\leqslant$ when applied to the symmetric inverse semigroup $\mathscr{I}(X)$:

Proposition 2.3. *In the inverse semigroup* $\mathscr{I}(X)$, $\alpha \leqslant \beta$ *if and only if* $\alpha \subseteq \beta$ *(considered as subsets of* $X \times X$*).*

Proof. If $\alpha \leqslant \beta$ in $\mathscr{I}(X)$ then there exists a subset C of X such that $\alpha = 1_C\beta$. If $(x, y) \in \alpha$ then there exists z in X such $(x, z) \in 1_C$, $(z, y) \in \beta$. Hence $x = z \in C$ and so $(x, y) \in \beta$. Thus $\alpha \subseteq \beta$.

Conversely, if $\alpha \subseteq \beta$ consider the element $1_A\beta$, where $A = \mathrm{dom}(\alpha)$. Then

$$(x, y) \in \alpha \Rightarrow (x, x) \in 1_A,\ (x, y) \in \beta$$
$$\Rightarrow (x, y) \in 1_A\beta,$$

and so $\alpha \subseteq 1_A\beta$. To see the reverse inclusion, suppose that $(x, y) \in 1_A\beta$. Then $x \in A = \mathrm{dom}(\alpha)$ and $(x, y) \in \beta$. Hence there exists $x\alpha$ in X such that $(x, x\alpha) \in \alpha \subseteq \beta$. Since β is a partial mapping we must have $x\alpha = y$. Hence $(x, y) \in \alpha$.

We have shown that $\alpha = 1_A\beta$ and hence that $\alpha \leqslant \beta$ in $\mathscr{I}(X)$.■

In the language of mappings, to say that $\alpha \subseteq \beta$ is to say that α is a *restriction* of β, or that β is an *extension* of α, i.e. to say that $\mathrm{dom}(\alpha) \subseteq \mathrm{dom}(\beta)$ and that $x\beta = x\alpha$ for all x in $\mathrm{dom}(\alpha)$.

The next result gives that the order relation on an arbitrary inverse semigroup is compatible with the operations of multiplication and inversion:

Proposition 2.4. *If* $a \leqslant b$ *in an inverse semigroup* S, *then*

$$ca \leqslant cb \quad \text{and} \quad ac \leqslant bc$$

for every c *in* S. *Also* $a^{-1} \leqslant b^{-1}$.

Proof. If $a = eb$, with $e \in E$, then it is immediate that $ac = ebc$, so that $ac \leqslant bc$. Also,

$$ca = ceb = c(c^{-1}c)eb = (cec^{-1})cb$$

and so $ca \leqslant cb$. Also $a^{-1} = b^{-1}e$ and so $a^{-1} \leqslant b^{-1}$ by Proposition 2.2 (ii).■

If H is an arbitrary subset of an inverse semigroup S, then, following Schein (1962) and Clifford and Preston (1967), we shall define the *closure* $H\omega$ of H by

$$H\omega = \{x \in S : (\exists h \in H)\ h \leqslant x\}.$$

The term "closure" is justified by the following easily verified facts (where H, K are subsets of S):

$$H \subseteq H\omega, \tag{2.5}$$

$$H \subseteq K \Rightarrow H\omega \subseteq K\omega, \tag{2.6}$$

$$(H\omega)\omega = H\omega. \tag{2.7}$$

We shall say that H is *closed* if $H\omega = H$.

Not every subsemigroup of an inverse semigroup is an inverse semigroup. A subsemigroup H of an inverse semigroup S is an inverse semigroup if and only if

$$(\forall x \in S)\ x \in H \Rightarrow x^{-1} \in H.$$

In such a case we all H an *inverse subsemigroup* of S.

Proposition 2.8. *If H is an inverse subsemigroup of an inverse semigroup S, then $H\omega$ is a closed inverse subsemigroup of S.*

Proof. That $H\omega$ is closed follows immediately from (2.7). To see that $H\omega$ is a subsemigroup, consider x, y in $H\omega$. By definition there exist h, k in H such that $x \geqslant h$, $y \geqslant k$. By Proposition 2.4 it follows that $xy \geqslant hk \in H$. Hence $xy \in H\omega$. To see that $H\omega$ is an inverse subsemigroup, observe that if $x \in H\omega$ then there exists h in H such that $x \geqslant h$. By Proposition 2.4 we may deduce that $x^{-1} \geqslant h^{-1} \in H$, and so $x^{-1} \in H\omega$ as required.■

3. CONGRUENCES ON INVERSE SEMIGROUPS

A congruence ρ on an inverse semigroup S (or indeed on any semigroup) will be called a *group congruence* if S/ρ is a group. It is a consequence of Proposition 1.6 that S/ρ is an inverse semigroup for any congruence ρ whatever. Since, as may easily be shown, a group can be characterized as an inverse semigroup having only one idempotent, and since any ρ-class that is an idempotent in S/ρ must contain at least one idempotent of S (Lemma II.4.6), it follows that ρ is a group congruence if and only if $(e, f) \in \rho$ for all e, f in E.

Using this characterization of group congruences, we may now easily verify that if $\{\rho_i : i \in I\}$ is a non-empty family of group congruences then their intersection $\bigcap\{\rho_i : i \in I\}$ is again a group congruence. This implies the existence of a minimum group congruence on an inverse semigroup, characterized as the intersection of *all* the group congruences on S. (There is always at least one group congruence on a semigroup S, namely the universal congruence). A more useful characterization of the minimum group congruence on an inverse semigroup was given by Munn (1961) as follows:

Theorem 3.1. *If S is an inverse semigroup with semilattice of idempotents E, then the relation*

$$\sigma = \{(x, y) \in S \times S; (\exists e \in E)\ ex = ey\}$$

is the minimum group congruence on S.

Proof. We show first that σ is an equivalence. Certainly $x\ \sigma\ x$ for every x in S, since $(xx^{-1})x = (xx^{-1})x$. Symmetry is equally immediate. To prove transitivity, note that if $ex = ey$ and $fy = fz$, where $e, f \in E$, then

$$(ef)x = fex = fey = efy = (ef)z,$$

where $ef \in E$, and so $x\ \sigma\ z$.

In fact σ is a congruence, for if $x\ \sigma\ y$ and $z \in S$ it is immediate that $xz\ \sigma\ yz$. If $ex = ey$ then $zex = zey$ and so

$$(zez^{-1})zx = z.e(z^{-1}z)x = zz^{-1}zex = zz^{-1}zey$$
$$= (zez^{-1})zy.$$

Hence $zx\ \sigma\ zy$.

If $e, f \in E$ then

$$ef.e = ef.f \quad (=ef).$$

Since $ef \in E$, this implies that $e\ \sigma\ f$ for every e, f in E. Hence S/σ is a group.

Finally, if ρ is a group congruence and if $x\ \sigma\ y$, then $ex = ey$ for some e in E, and so certainly

$$(e\rho)(x\rho) = (e\rho)(y\rho)$$

in S/ρ. Now $e\rho$, being an idempotent in the group S/ρ, is necessarily the identity of the group. Hence it follows that $x\rho = y\rho$, i.e. that $x\ \rho\ y$. We have thus shown that $\sigma \subseteq \rho$. Hence σ is the *minimum* group congruence on S.■

We remark that if S has a zero then the minimum group congruence is the universal congruence $S \times S$; for clearly $0x = 0y$ for every pair x, y of elements of S.

It is important to realize that it is not always possible to find a minimum group congruence on a semigroup that is *not* an inverse semigroup. For an example where no minimum group congruence exists, see Exercise 26 below.

We have seen that group congruences are such that all idempotents fall in a single congruence class. Opposite in a sense are the *idempotent-separating* congruences ρ, with the property that

$$\rho \cap (E \times E) = 1_E,$$

i.e. such that if e, f are idempotents

$$(e, f) \in \rho \Rightarrow e = f.$$

We have encountered such congruences before, in Section II.4, and we now need to make use of Proposition II.4.8, which states that on a regular semigroup (and so certainly on an inverse semigroup) a congruence is idempotent-separating if and only if it is contained in $\mathscr{H}$.

It then follows that there exists a maximum idempotent-separating congruence on an inverse semigroup S, namely $\mathscr{H}^{\flat}$. An alternative characterization of this congruence, due to Howie (1964), is frequently useful:

Theorem 3.2. *If S is an inverse semigroup with semilattice of idempotents E, then the relation*

$$\mu = \{(a, b) \in S \times S : (\forall e \in E)\ a^{-1}ea = b^{-1}eb\}$$

is the maximum idempotent-separating congruence on S.

Proof. First, it is clear that μ is an equivalence. If $(a, b) \in \mu$ and $c \in S$ then $c^{-1}(a^{-1}ea)c = c^{-1}(b^{-1}eb)c$ for every e in E, and so $(ac, bc) \in \mu$. Also, since $c^{-1}ec \in E$ for every e in E, we have $a^{-1}(c^{-1}ec)a = b^{-1}(c^{-1}ec)b$ for every e in E, and so $(ca, cb) \in \mu$. Thus μ is a congruence.

If $(e, f) \in \mu \cap (E \times E)$ then

$$e = e^{-1}ee = f^{-1}ef = ef$$

and

$$f = f^{-1}ff = e^{-1}fe = ef,$$

and so $e = f$. Thus μ is idempotent-separating.

Finally, if ρ is any idempotent-separating congruence and if $(a, b) \in \rho$, then $(a^{-1}, b^{-1}) \in \rho$ by (1.7) and so $(a^{-1}ea, b^{-1}eb) \in \rho$ for every e in E. Since $a^{-1}ea$ and $b^{-1}eb$ are idempotents, it follows that $a^{-1}ea = b^{-1}eb$ for every e in E, i.e. that $(a, b) \in \mu$. Thus $\rho \subseteq \mu$ and so μ is the maximum idempotent-separating congruence on S. ■

Let us call an inverse semigroup S *fundamental* if $\mu = 1_S$, i.e. if

$$[(\forall e \in E)\ a^{-1}ea = b^{-1}eb] \Rightarrow a = b. \tag{3.3}$$

Theorem 3.4. *If S is an inverse semigroup with semilattice of idempotents E, and if μ is the maximum idempotent-separating congruence on S, then S/μ is fundamental and has semilattice of idempotents isomorphic to E.*

Proof. Every idempotent in S/μ has the form $e\mu$, where $e \in E$. If, for every e in E,

$$(a\mu)^{-1}(e\mu)(a\mu) = (b\mu)^{-1}(e\mu)(b\mu),$$

then $(a^{-1}ea, b^{-1}eb) \in \mu$. Since μ is idempotent-separating it follows that $a^{-1}ea = b^{-1}eb$, i.e. that $a\mu = b\mu$. Hence the implication (3.3) holds in S/μ

and so S/μ is fundamental. The homorphism $\mu^\natural : S \to S/\mu$ is an isomorphism when restricted to E, since for e, f in E,

$$e\mu^\natural = f\mu^\natural \Rightarrow (e,f) \in \mu \Rightarrow e = f.$$

Hence $E(S/\mu) \simeq E(S)$.■

4. FUNDAMENTAL INVERSE SEMIGROUPS

Theorem 3.4 assures us that fundamental inverse semigroups do exist, since S/μ is a fundamental inverse semigroup for any inverse semigroup S whatever. More directly, it is obvious that every semilattice is a fundamental inverse semigroup, and it is not hard to show (see Exercises 18, 19) that every symmetric inverse semigroup $\mathscr{I}(X)$ is fundamental.

Munn (1966) showed how an important fundamental inverse semigroup could be constructed from any semilattice E. The crucial observation, which was the basis of a passing remark in (Howie 1964), was that every element a in an inverse semigroup S determines an isomorphism α_a from the principal ideal Eaa^{-1} of E onto the principal ideal $Ea^{-1}a$. The isomorphism α_a is defined by

$$e\alpha_a = a^{-1}ea \qquad (e \in Eaa^{-1}). \tag{4.1}$$

It is clear that α_a maps into $Ea^{-1}a$. To see that it is one–one, notice that if $e(= eaa^{-1} = aa^{-1}e)$ and $f(= faa^{-1} = aa^{-1}f)$ are elements of Eaa^{-1}, then

$$\begin{aligned} e\alpha_a = f\alpha_a &\Rightarrow a^{-1}ea = a^{-1}fa \Rightarrow aa^{-1}eaa^{-1} = aa^{-1}faa^{-1} \\ &\Rightarrow e = f. \end{aligned}$$

To see that α_a is onto, observe that if $e \in Ea^{-1}a$ then $e = a^{-1}aea^{-1}a = (aea^{-1})\alpha_a$, where $aea^{-1} = aea^{-1}aa^{-1} \in Eaa^{-1}$. To see that α_a is an isomorphism, observe that if $e, f \in Eaa^{-1}$, then

$$(e\alpha_a)(f\alpha_a) = a^{-1}eaa^{-1}fa = a^{-1}efa = (ef)\alpha_a.$$

The structure of S is clearly dependent on these partial isomorphisms of E, and it is natural to hope that the structure of S is actually determined by E and by the partial isomorphisms α_a. This turns out to be over-optimistic, but nevertheless some interesting results will emerge from this speculation.

If we are given a semilattice E, the following construction is now natural. Let $\mathscr{U}$ be the equivalence relation on E given by

$$\mathscr{U} = \{(e,f) \in E \times E : Ee \simeq Ef\}. \tag{4.2}$$

If $(e,f) \in \mathscr{U}$ let $T_{e,f}$ be the set of all isomorphisms from Ee onto Ef. Let

$$T_E = \bigcup_{(e,f)\in\mathscr{U}} T_{e,f}. \tag{4.3}$$

We shall call T_E the *Munn semigroup* of the semilattice E.

This terminology presupposes that we can define a multiplication on T_E. We show now that this is indeed possible. Clearly T_E is a subset of $\mathscr{I}(E)$, the symmetric inverse semigroup on the set E, since all the elements of T_E are partial one–one mappings of E. We may therefore multiply elements of T_E as elements of $\mathscr{I}(E)$. If $\alpha: Ee \to Ef$ and $\beta: Eg \to Eh$ are elements of T_E then the product of α and β in $\mathscr{I}(E)$ maps $(Ef \cap Eg)\alpha^{-1}$ onto $(Ef \cap Eg)\beta$, i.e. maps $(Efg)\alpha^{-1}$ onto $(Efg)\beta$. Write $(fg)\alpha^{-1} = i$ and $(fg)\beta = j$. Then

$$\begin{aligned} & x \in (Efg)\alpha^{-1} \\ & \Leftrightarrow x\alpha \in Efg, \\ \text{i.e.}\quad & \Leftrightarrow x\alpha \leqslant fg, \\ \text{i.e.}\quad & \Leftrightarrow x \leqslant (fg)\alpha^{-1}, \\ \text{i.e.}\quad & \Leftrightarrow x \in Ei, \end{aligned}$$

and similarly $x \in (Efg)\beta \Leftrightarrow x \in Ej$. Thus $\alpha\beta$ maps the principal ideal Ei onto the principal ideal Ej. Since it is clearly an isomorphism, we have that $\alpha\beta \in T_E$. Thus T_E is a subsemigroup of $\mathscr{I}(E)$.

It is even an *inverse* subsemigroup of $\mathscr{I}(E)$, since if $\alpha: Ee \to Ef$ is in T_E, it is clear that $\alpha^{-1}: Ef \to Ee$ is also in T_E.

We have thus established part of the following result:

Theorem 4.4. *If E is a semilattice, then the Munn semigroup T_E of E is an inverse semigroup whose semilattice of idempotents is isomorphic to E.*

Proof. It remains to consider the idempotents of the inverse semigroup T_E. A typical idempotent of T_E is the identical mapping 1_{Ee} of Ee onto itself, and so we do have a one–one mapping $e \mapsto 1_{Ee}$ from E onto $E(T_E)$. Now $1_{Ee}1_{Ef}$ has domain

$$(Ee \cap Ef)1_{Ee}^{-1} = (Eef)1_{Ee}^{-1} = Eef$$

and range

$$(Ee \cap Ef)1_{Ef} = (Eef)1_{Ef} = Eef;$$

hence $1_{Ee}1_{Ef} = 1_{Eef}$ and so the mapping $e \mapsto 1_{Ee}$ is an isomorphism. ■

We shall very often want to identify 1_{Ee} with e and to think of T_E as an inverse semigroup having E itself as semilattice of idempotents.

Before investigating the general consequences of this construction we pause to consider two examples.

Example 4.5. Let $E = \{0, 1, \ldots\}$, with the natural order, i.e. with $0 < 1 < 2 < \ldots$. Then

$$En = \{0, 1, \ldots, n\}$$

and so $Em \simeq En$ if and only if $m = n$. In this case $\mathscr{U} = 1_E$ and we say that E is *anti-uniform*. The only isomorphism in $T_{n,n}$ is 1_{En} and so

$$T_E = \{1_{E0}, 1_{E1}, 1_{E2}, \ldots\} \simeq E.$$

Example 4.6. Let $E = C_\omega = \{e_0, e_1, e_2, \ldots\}$, with

$$e_0 > e_1 > e_2 > \ldots$$

Then

$$Ee_n = \{e_n, e_{n+1}, e_{n+2}, \ldots\},$$

and $Ee_m \simeq Ee_n$ for every m, n in $N = \{0, 1, 2, \ldots\}$. Here $\mathscr{U} = E \times E$ and we say that E is *uniform*. The only isomorphism from Ee_m onto Ee_n is $\alpha_{m,n}$, given by

$$e_k\alpha_{m,n} = e_{k-m+n} \qquad (k \geqslant m),$$

the inverse being $\alpha_{n,m}\colon Ee_n \to Ee_m$, defined by

$$e_l\alpha_{n,m} = e_{l-n+m} \qquad (l \geqslant n).$$

If $\alpha_{m,n}$ and $\alpha_{p,q}$ are elements of T_E then their product maps $(Ee_n \cap Ee_p)\alpha_{m,n}^{-1}$ onto $(Ee_n \cap Ee_p)\alpha_{p,q}$. If we write $t = \max(n, p)$, the greater of n and p, we can say that $\alpha_{m,n}\alpha_{p,q}$ maps Ee_{t-n+m} onto Ee_{t-p+q}. That is,

$$\alpha_{m,n}\alpha_{p,q} = \alpha_{m-n+t,\, q-p+t}. \tag{4.7}$$

We can thus identify the Munn semigroup of the semilattice $C_\omega = \{e_0, e_1, e_2, \ldots\}$ with the semigroup $(N \times N, .)$ in which multiplication is given by

$$(m, n)(p, q) = (m - n + \max(n, p), q - p + \max(n, p)). \tag{4.8}$$

This is the so-called *bicyclic* semigroup, which plays a very important role in the theory of inverse semigroups. By Theorem 4.4 it is an inverse semigroup with semilattice of idempotents isomorphic to C_ω.

The importance of the Munn semigroup T_E lies in the following result:

Theorem 4.9. *If S is an inverse semigroup with semilattice of idempotents E, then there is a homomorphism $\phi\colon S \to T_E$ whose kernel is μ, the maximum idempotent-separating congruence on S.*

Proof. At the beginning of this section we saw that for each a in S there is an isomorphism $\alpha_a: Eaa^{-1} \to Ea^{-1}a$ given by

$$e\alpha_a = a^{-1}ea \qquad (e \in Eaa^{-1})$$

and with inverse $\alpha_a^{-1} = \alpha_{a^{-1}}$ given by

$$f\alpha_a^{-1} = afa^{-1} \qquad (f \in Ea^{-1}a).$$

Thus $\alpha_a \in T_E$ and it is natural to define the mapping $\phi: S \to T_E$ by

$$a\phi = \alpha_a \qquad (a \in S).$$

To see that ϕ is homomorphic, notice that if $a, b \in S$ then the product $\alpha_a\alpha_b$ in T_E maps Ei onto Ej, where

$$i = (a^{-1}abb^{-1})\alpha_a^{-1} = aa^{-1}abb^{-1}a^{-1} = abb^{-1}a^{-1} = (ab)(ab)^{-1},$$

and

$$j = (a^{-1}abb^{-1})\alpha_b = b^{-1}a^{-1}abb^{-1}b = b^{-1}a^{-1}ab = (ab)^{-1}(ab).$$

Also, for all e in Ei,

$$e\alpha_a\alpha_b = (a^{-1}ea)\alpha_b = b^{-1}a^{-1}eab = e\alpha_{ab}.$$

Hence $\alpha_a\alpha_b = \alpha_{ab}$ and so ϕ is a homomorphism.

If $(a, b) \in \mu$ then certainly $(a, b) \in \mathscr{H}$ and so $aa^{-1} = bb^{-1}$, $a^{-1}a = b^{-1}b$. Hence α_a and α_b have the same domain and range. Also, $a^{-1}ea = b^{-1}eb$ for every e in E and so certainly for every e in Eaa^{-1}. Hence $\alpha_a = \alpha_b$ and so $(a, b) \in \ker \phi$. Conversely, if $(a, b) \in \ker \phi$ then certainly α_a and α_b have the same domain and range and so $aa^{-1} = bb^{-1}$, $a^{-1}a = b^{-1}b$. Also, for every e in E,

$$\begin{aligned} a^{-1}ea &= a^{-1}(eaa^{-1})a = (eaa^{-1})\alpha_a = (eaa^{-1})\alpha_b \\ &= (ebb^{-1})\alpha_b = b^{-1}ebb^{-1}b = b^{-1}eb, \end{aligned}$$

and so $(a, b) \in \mu$. ∎

If we could assert that the homomorphism $\phi: S \to T_E$ was onto, then the theorem would imply that every fundamental inverse semigroup was isomorphic to the Munn semigroup of its semilattice of idempotents. Unfortunately this is not so, and we must be content with a weaker result. We call an inverse subsemigroup T of an inverse semigroup S *full* if T contains all the idempotents of S. Then

Theorem 4.10. *An inverse semigroup S with semilattice of idempotents E is fundamental if and only if it is isomorphic to a full inverse subsemigroup of T_E.*

Proof. If S is fundamental then in Theorem 4.9 we have $\ker \phi = \mu = 1_S$. Hence ϕ embeds S in T_E and we must show that $S\phi\,(\simeq S)$ is a full inverse subsemigroup of T_E. By Proposition 1.6, $S\phi$ is an inverse subsemigroup of T_E. Also, if $e \in E$, then $e\phi = \alpha_e\colon Eee^{-1} \to Ee^{-1}e$, where

$$x\alpha_e = e^{-1}xe \qquad (x \in Eee^{-1}).$$

That is, α_e maps Ee onto Ee, and $x\alpha_e = xe = x$ for each x in Ee. That is, $\alpha_e = 1_{Ee}$. Thus ϕ maps the semilattice E of idempotents of S onto the isomorphic semilattice $\{1_{Ee}\colon e \in E\}$ of idempotents of T_E. Thus $S\phi$ is full.

Conversely, if S is isomorphic to a full inverse subsemigroup S' of T_E, then S' contains 1_{Ee} for every e in E. Let α, β in S' be such that $(\alpha, \beta) \in \mu$, the maximum idempotent-separating congruence on S'. Then α and β are $\mathscr{H}$-equivalent in S' and so $\alpha\alpha^{-1} = \beta\beta^{-1}$, $\alpha^{-1}\alpha = \beta^{-1}\beta$. If we suppose that $\operatorname{dom}(\alpha) = Ee$, $\operatorname{ran}(\alpha) = Ef$, $\operatorname{dom}(\beta) = Eg$ and $\operatorname{ran}(\beta) = Eh$, then

$$\alpha\alpha^{-1} = 1_{Ee}, \qquad \alpha^{-1}\alpha = 1_{Ef}, \qquad \beta\beta^{-1} = 1_{Eg}, \qquad \beta^{-1}\beta = 1_{Eh}.$$

Hence from the $\mathscr{H}$-equivalence of α and β we can deduce that $g = e$ and $h = f$, i.e. that

$$\operatorname{dom}(\alpha) = \operatorname{dom}(\beta) = Ee, \qquad \operatorname{ran}(\alpha) = \operatorname{ran}(\beta) = Ef.$$

By Theorem 3.2 we have that

$$\alpha^{-1}1_{Ex}\alpha = \beta^{-1}1_{Ex}\beta$$

for every x in E. In particular, then, these two elements of T_E have the same range; i.e. $E((xe)\alpha) = E((xe)\beta)$. It follows that $(xe)\alpha = (xe)\beta$ for every x in E, i.e. that α and β are equal mappings of Ee onto Ef. We have shown that $(\alpha, \beta) \in \mu \Rightarrow \alpha = \beta$; hence S', and so also S, is fundamental. ■

As a consequence, we have

Corollary 4.11. *The inverse semigroup T_E is fundamental for any semilattice E.* ■

5. ANTI-UNIFORM SEMILATTICES

We recall that a semilattice E is *uniform* if $\mathscr{U} = E \times E$, i.e. if

$$(\forall e, f \in E) \qquad Ee \simeq Ef,$$

and is *anti-uniform* if $\mathscr{U} = 1_E$, i.e. if

$$Ee \simeq Ef \Rightarrow e = f.$$

We have seen (Proposition 1.4 (f)) that two idempotents e, f in an inverse semigroup S are $\mathscr{D}$-equivalent if and only if there exists an element a in S

such that $aa^{-1} = e$ and $a^{-1}a = f$. In such a case it is easy to verify that the mapping $x \mapsto a^{-1}xa$ $(x \in Ee)$ is an isomorphism from Ee onto Ef, and so we have $(e, f) \in \mathscr{U}$. We have shown that

$$\mathscr{D} \cap (E \times E) \subseteq \mathscr{U}. \tag{5.1}$$

If E is anti-uniform we deduce from this that

$$\mathscr{D} \cap (E \times E) = 1_E,$$

and from Theorem IV.2.1 we now immediately deduce half of the following result:

Theorem 5.2 (Howie and Schein 1969). *A semilattice E has the property that EVERY inverse semigroup having E as semilattice of idempotents is a Clifford semigroup if and only if E is anti-uniform.*

Proof. To show the remaining half, suppose that E is a semilattice which is *not* anti-uniform; i.e. suppose that there exist distinct elements e, f in E for which $Ee \simeq Ef$. Then there exists α in T_E such that $\operatorname{dom}(\alpha) = Ee$, $\operatorname{ran}(\alpha) = Ef$. That is, $\alpha\alpha^{-1} = 1_{Ee}$ and $\alpha^{-1}\alpha = 1_{Ef}$, and so the idempotent 1_{Ee} and 1_{Ef} of T_E (which, following our usual practice, we identify with e and f) are $\mathscr{D}$-equivalent in T_E. Thus, by Proposition IV.2.1, T_E is an inverse semigroup having E (effectively) as semilattice of idempotents and which is *not* a Clifford semigroup. ■

In view of Theorem 5.2 it is of interest to identify anti-uniform semilattices. We have seen (Example 4.5) that a chain order-isomorphic to the non-negative integers is anti-uniform. So is a finite chain; indeed we have:

Proposition 5.3. *Any well-ordered chain is anti-uniform.*

Proof. Let E be a well-ordered chain and let $\phi: Ee \to Ef$ be an isomorphism, where we may suppose without loss of generality that $e \leqslant f$, i.e. that $Ee \subseteq Ef$. Certainly $e\phi = f$, since the maximum element of Ee must map under ϕ to the maximum element of Ef. Suppose now that the set

$$Y = \{y \in Ee \colon y\phi \neq y\}$$

is non-empty. By the well-ordering property there is a least element x in Y, i.e. an element for which $x\phi \neq x$ and

$$z < x \Rightarrow z\phi = z.$$

Now either $x\phi < x$ or $x\phi > x$. If $x\phi < x$ then $x\phi \in Ee$ and $(x\phi)\phi = x\phi$.

This contradicts the hypothesis that ϕ is one–one. If $x\phi > x$ then for each y in Ee,

$$y \geqslant x \Rightarrow y\phi \geqslant x\phi > x,$$

while

$$y < x \Rightarrow y\phi = y < x.$$

It follows that there is no element y of Ee such that $y\phi = x$, and this contradicts the hypothesis that ϕ is onto.

The assumption that $Y \neq \varnothing$ has led us to a contradiction. Hence $Y = \varnothing$; that is, $x\phi = x$ for every x in Ee. In particular, $f = e\phi = e$. Thus E is anti-uniform as required.■

Schein (1964) effectively conjectured that the *only* anti-uniform semilattices are well-ordered chains. This proved over-optimistic, as we shall see, but in 1969 Howie and Schein proved

Proposition 5.4. *If E is a semilattice with the minimal condition and E is anti-uniform, then E is a well-ordered chain.*

Proof. To say that E has the minimal condition is to say that every non-empty subset of E contains a minimal element. We shall show that in such a semilattice, if E is not totally ordered then E is not anti-uniform. Following the usual terminology for ordered sets, let us say that a, b in E are *comparable* if either $a \leqslant b$ or $b \leqslant a$, and *incomparable* if neither of these holds. A partially ordered set is totally ordered if and only if there are no pairs of incomparable elements.

We define a subset K of E by the rule that $x \in K$ if and only if there exist elements of E that are incomparable with x. We are assuming that E is not totally ordered and so we have $K \neq \varnothing$. By the minimal condition there is at least one minimal member e of K. Then $K_e \neq \varnothing$, where K_e is the set of elements of E that are incomparable with e; let f be a minimal element of K_e. Certainly e and f are incomparable, and so $ef < e$, $ef < f$.

In fact ef *covers* e, in the sense (see also Sections I.8 and III.4) that there is no g in E for which $ef < g < e$. To see this, suppose that $ef \leqslant g < e$; then $fef \leqslant fg \leqslant ef$ and so $fg = ef$. But since $g < e$ it is comparable with every element of E, and with f in particular. If $f \leqslant g$ then $f \leqslant g < e$ and so f and e are comparable—a contradiction; hence $g < f$. But then we have $g < e$ and $g < f$ and so $g \leqslant ef$. Hence $g = ef$ and so there is no g such that $ef < g < e$.

A closely similar argument shows that f covers ef, and we deduce that

$$Ee = Eef \cup \{e\}, \qquad Ef = Eef \cup \{f\},$$

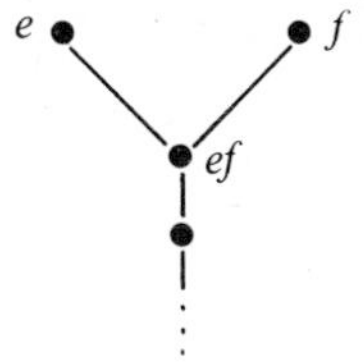

where Eef is a well-ordered chain. Now clearly $(e, f) \in \mathscr{U}$, an obvious isomorphism between Ee and Ef being that which sends e to f and every other element to itself. Thus E is not anti-uniform. ■

It is not possible to remove all mention of the minimum condition in this proposition. Consider for example the following semilattice, described by Howie and Schein (1969). The set $\mathbf{Q}$ of rational numbers is known to be countable: let ε be a one–one map from $\mathbf{Q}$ onto the set $N = \{0, 1, 2, \ldots\}$ of non-negative integers and let

$$E = \bigcup_{q \in \mathbf{Q}} \{(q, 0), (q, 1), \ldots, (q, q\varepsilon)\},$$

with the *lexicographic* order: $(q, m) \leqslant (r, n)$ if and only if either $q < r$, or $q = r$ and $m \leqslant n$.

Notice that an element $(q, 0)$ $(q \in \mathbf{Q})$ has no immediate predecessor in the totally ordered set E—i.e. $\{x \in E\colon x < (q, 0)\}$ has no maximum element. Indeed it is easy to see that the set of elements of E having no immediate predecessors is precisely $\{(q, 0)\colon q \in \mathbf{Q}\}$. Similar, the set of elements of E having no immediate successors is $\{(q, q\varepsilon)\colon q \in \mathbf{Q}\}$.

Let us suppose that E is not anti-uniform, i.e. that there exist distinct elements (q, m), (r, n) in E such that there is an isomorphism $\phi\colon E(q, m) \to E(r, n)$. We distinguish two cases:

$$\text{(i) } m \neq n, \qquad \text{(ii) } m = n \quad \text{and} \quad q \neq r.$$

In case (i) we may without loss of generality suppose that $m > n$. Then certainly

$$(q, m)\phi = (r, n), \qquad (q, m - 1)\phi = (r, n - 1), \ldots,$$

and finally $(q, m - n) = (r, 0)$. Now an isomorphism must preserve the property of having or not having an immediate predecessor. Hence we have arrived at a contradiction, since $(q, m - n)$ has an immediate predecessor and $(r, 0)$ does not.

In case (ii) we may suppose that $q > r$. Let $s \in \mathbf{Q}$ be such that $q > s > r$. Then $(s, 0) \in E(q, m)$ and has no immediate predecessor. Hence $(s, 0)\phi \in E(r, n)$ and has no immediate predecessor; thus $(s, 0)\phi = (t, 0)$, where $t \leqslant r$. Certainly $t < s$ and so $t\varepsilon \neq s\varepsilon$. If $s\varepsilon > t\varepsilon$ we have

$$(s, 1)\phi = (t, 1), \qquad (s, 2)\phi = (t, 2), \ldots, (s, t\varepsilon)\phi = (t, t\varepsilon),$$

a contradiction, since $(s, t\varepsilon)$ has an immediate successor in $E(q, m)$ whereas $(t, t\varepsilon)$ has no immediate successor in $E(r, n)$. If $s\varepsilon < t\varepsilon$, we similarly conclude that $(s, s\varepsilon)\phi = (t, s\varepsilon)$, which again gives a contradiction. Hence E is anti-uniform. E is certainly not well-ordered, since, for example, the set $\{x \in E: x > (q, q\varepsilon)\}$ has no least element.

Anti-uniform semilattices need not be chains, as is shown by an example due to Munn and quoted in (Howie and Schein 1969). A verbal description of this example is tedious and perhaps it is best to allow the diagram to speak for itself.

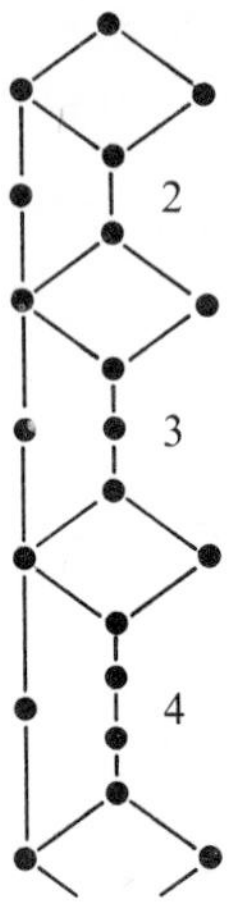

6. BISIMPLE INVERSE SEMIGROUPS

If S is a bisimple inverse semigroup with semilattice of idempotents E, then in particular all the idempotents are mutually $\mathscr{D}$-equivalent:

$$\mathscr{D} \cap (E \times E) = E \times E.$$

Hence it follows by (5.1) that $\mathscr{U} = E \times E$, i.e. that E is a *uniform* semilattice.

If conversely we start with a uniform semilattice E then we cannot expect that every inverse semigroup having E as semilattice of idempotents will be bisimple: E itself is one such inverse semigroup and is assuredly not bisimple. We do, however, have a method of constructing at least one bisimple inverse semigroup whose semilattice of idempotents is isomorphic to E.

More precisely, we have

Proposition 6.1. *If E is a uniform semilattice, then T_E is a bisimple inverse semigroup.*

Proof. We shall in fact prove more generally that if E is any semilattice whatever then, in T_E,

$$\mathscr{D} \cap (E \times E) = \mathscr{U}. \tag{6.2}$$

Since T_E is an inverse semigroup whose semilattice of idempotents is (effectively) E, one half of this result is obvious from (5.1). Suppose now that $(e,f) \in \mathscr{U}$. Then $Ee \simeq Ef$ and so there exists at least one α in T_E such that $\text{dom}(\alpha) = Ee$ and $\text{ran}(\alpha) = Ef$. That is, $\alpha\alpha^{-1} = 1_{Ee} (= e)$ and $\alpha^{-1}\alpha = 1_{Ef} (= f)$, and so e, f are $\mathscr{D}$-equivalent in T_E.

Applying this to the uniform case, we find that *all* idempotents in T_E are $\mathscr{D}$-equivalent. Hence, since every element of a regular semigroup is $\mathscr{D}$-equivalent (indeed even $\mathscr{R}$- or $\mathscr{L}$-equivalent) to some idempotent, T_E is bisimple. ∎

This gives a useful recipe for constructing bisimple inverse semigroups. It is not of course a universal recipe, for T_E is fundamental, and we can easily produce a non-fundamental bisimple inverse semigroup simply by taking the direct product of a T_E (with E uniform) and an arbitrary non-trivial group. Recall, however, that we can embed any fundamental inverse semigroup as a full inverse subsemigroup S' of the approropriate T_E (Theorem 4.10). In the bisimple case we can be somewhat more precise, for if the subsemigroup S' of T_E is itself to be bisimple, there must exist for each e, f in E at least one α *contained in* S' for which $\text{dom}(\alpha) = Ee$ and $\text{ran}(\alpha) = Ef$; that is, for every (e,f) in $\mathscr{U} = E \times E$,

$$S' \cap T_{e,f} \neq \varnothing. \tag{6.3}$$

If we decide to apply the term *transitive* to subsemigroups of T_E with this property, it is easy to see that that transitivity is not only a necessary but also a sufficient condition for bisimplicity. Noting finally that a transitive inverse subsemigroup of T_E is necessarily full, we deduce:

Theorem 6.4. *An inverse semigroup S with semilattice of idempotents E is fundamental and bisimple if and only if E is uniform and S is isomorphic to a transitive inverse subsemigroup of T_E.* ∎

If we apply this result to the uniform semilattice C_ω (Example 4.6) we notice immediately that in this case $|T_{e,f}| = 1$ for every e, f. Hence the condition (6.3) can be satisfied only if

$$S' \cap T_{e,f} = T_{e,f}$$

for every e, f, i.e. only if $S' = T_E$. We thus have

Corollary 6.5. *Up to isomorphism, the only fundamental bisimple inverse semigroup S having C_ω as semilattice of idempotents is the bicyclic semigroup.* ■

A further fairly easy observation at this stage will later prove very useful.

Proposition 6.6. *Let S be an inverse semigroup with semilattice of idempotents E and let E have the property that $|T_{e,f}| = 1$ for all $(e,f) \in \mathscr{U}$. Then the semigroup S has the property that $\mu = \mathscr{H}$, and so $\mathscr{H}$ is a congruence.*

Proof. Since $\mu = \mathscr{H}^\flat$ by Proposition II.4.7, we certainly have $\mu \subseteq \mathscr{H}$. Conversely, if as in the proof of Theorem 4.9 we denote the image of the element a in S under the standard homomorphism $\phi: S \to T_E$ by α_a, then, for all a, b in S,

$$\begin{aligned}(a, b) \in \mathscr{H} &\Rightarrow aa^{-1} = bb^{-1}, a^{-1}a = b^{-1}b \\ &\Rightarrow \operatorname{dom}(\alpha_a) = \operatorname{dom}(\alpha_b), \operatorname{ran}(\alpha_a) = \operatorname{ran}(\alpha_b) \\ &\Rightarrow \alpha_a = \alpha_b \text{ (by the given property of } E) \\ &\Rightarrow (a, b) \in \ker \phi = \mu.\end{aligned}$$

Thus we also have $\mathscr{H} \subseteq \mu$ and so $\mathscr{H} = \mu$ as required. ■

Corollary 6.7. *If S is an inverse semigroup with semilattice of idempotents C_ω, then $\mathscr{H} = \mu$.* ■

We should not expect the simple situation indicated by Corollaries 6.5 and 6.7 to obtain for more general uniform semilattices. Indeed it does not, as is illustrated by Exercise 20 below.

We can, however, think of a general bisimple inverse semigroup as in some sense an "inflation" of a fundamental bisimple inverse semigroup, and Munn's (1970b) structure theory for bisimple inverse semigroups effectively makes this vague notion precise. Another attempt at a general description (from a different point of view) has been made by Reilly and Clifford (1968), and recently McAlister (1974) has given an extremely general unified account showing the connection between the two approaches. All of these structure theories suffer from an excess of complication, the fault being not with the authors, but with bisimple inverse semigroups.

In the special case of *ω-semigroups* i.e. of semigroups whose semilattice of idempotents is isomorphic to C_ω, the situation is much more encouraging, partly because of the tidy situation described in Corollaries 6.5 and 6.7 and the existence of an ingenious special "inflation" method first considered in special cases by Bruck (1958) and Reilly (1966) and in general form by Munn (1970c).

Let T be a semigroup with identity 1 and let θ be a homomorphism from T into H_1, the $\mathscr{H}$-class containing the identity of T (what is often called the *group of units* of T). Let $N = \{0, 1, 2, \ldots\}$. We can make $N \times T \times N$ into a semigroup by defining

$$(m, a, n)(p, b, q) = (m - n + t, a\theta^{t-n}b\theta^{t-p}, q - p + t), \tag{6.8}$$

where $t = \max(n, p)$, and where θ^0 is interpreted as the identity map of T. We must of course check that the given composition is associative. Observe that

$$[(m, a, n)(p, b, q)](r, c, s) = (m - n - q + p + u, a\theta^{u-n-q+p}b\theta^{u-q}c\theta^{u-r}, s - r + u),$$

while

$$(m, a, n)[(p, b, q)(r, c, s)] = (m - n + w, a\theta^{w-n}b\theta^{w-p}c\theta^{w-r-p+q}, s - r - p + q + w),$$

where

$$\left.\begin{aligned} u &= \max(q - p + \max(n, p), r), \\ w &= \max(n, p - q + \max(q, r)). \end{aligned}\right\} \tag{6.9}$$

Now the outer coordinates in the multiplication (6.8) combine exactly as in the bicyclic semigroup, which we know to be associative since it is isomorphic to T_{C_ω}. Hence, equating first coordinates (or equivalently third coordinates) we obtain

$$w = u + p - q,$$

a result that it is possible to obtain alternatively by a tedious direct verification from the definitions (6.9) of u and w. It is clear that this result implies also the equality of the two middle coordinates, and so the composition (6.8) is indeed associative. We shall denote the semigroup obtained in this way by $S = BR(T, \theta)$, and refer to it as the *Bruck–Reilly extension of T determined by θ*. Bruck's original construction (1958) corresponds to the case where θ maps the whole of S to the element 1. Reilly's construction (1966) corresponds to the case where T is a group and so coincides with its group of units.

Several properties of the Bruck–Reilly extension can now be described:

Proposition 6.10. *If T is a semigroup with identity 1 and $S = BR(T, \theta)$, then*
(a) *S is a simple semigroup with identity $(0, 1, 0)$;*
(b) *two elements (m, a, n) and (p, b, q) of S are $\mathscr{D}$-equivalent in S if and only if a and b are $\mathscr{D}$-equivalent in T.*
(c) *the element (m, a, n) of S is idempotent if and only if $m = n$ and $a^2 = a$.*
(d) *S is an inverse semigroup if and only if T is an inverse semigroup.*

Proof. (a) We show that if (m, a, n) and (p, b, q) are arbitrary elements of S then there exist (r, x, s) and (t, y, u) in S such that

$$(r, x, s)(m, a, n)(t, y, u) = (p, b, q);$$

by Corollary III.1.3 this is just what we require. If we take

$$(r, x, s) = (p, (a\theta)^{-1}, m + 1) \quad \text{and} \quad (t, y, u) = (n + 1, b, q),$$

where $(a\theta)^{-1}$ is the inverse of $a\theta$ in the group H_1, then it is easy to check that the desired equality holds. That $(0, 1, 0)$ is the identity of S is a matter of routine verification.

(b) Since there is some scope for confusion here, we shall use superscripts S and T to distinguish between the Green equivalences on S and those on T.

If $(m, a, n)\,\mathscr{R}^S(p, b, q)$ then

$$(m, a, n)(r, x, s) = (p, b, q) \tag{6.11}$$

for some (r, x, s) in S. Hence

$$p = m - n + \max(n, r) \geqslant m.$$

But equally we may show that $m \geqslant p$, and so in fact $m = p$. It follows that

$$m - n + \max(n, r) = m$$

and hence that $n \geqslant r$. Hence, equating the middle coordinates in (6.11), we have

$$a(x\theta^{n-r}) = b$$

and so $R_b \leqslant R_a$ in T. But we may similarly show that $R_a \leqslant R_b$, and so $a\,\mathscr{R}^T\,b$. Conversely if $a\,\mathscr{R}^T\,b$ then $ax = b$, $bx' = a$ for some x, x' in $T\,(= T^1)$. Hence

$$(m, a, n)(n, x, q) = (m, b, q),$$

$$(m, b, q)(q, x', n) = (m, a, n)$$

in S, and so $(m, a, n)\,\mathscr{R}^S\,(m, b, q)$.

We have shown so far that

$$(m, a, n)\,\mathscr{R}^S\,(p, b, q) \Leftrightarrow m = p \quad \text{and} \quad a\,\mathscr{R}^T\,b. \tag{6.12}$$

A dual argument establishes that

$$(m, a, n)\,\mathscr{L}^S\,(p, b, q) \Leftrightarrow n = q \quad \text{and} \quad a\,\mathscr{L}^T\,b. \tag{6.13}$$

Now suppose that (m, a, n) and (p, b, q) in S are such that $(m, a, n)\,\mathscr{D}^S\,(p, b, q)$. Then there exists (r, c, s) in S for which

$$(m, a, n)\,\mathscr{R}^S\,(r, c, s), \qquad (r, c, s)\,\mathscr{L}^S(p, b, q).$$

By (6.12) and (6.13) it follows that $a\mathscr{R}^T c$ and $c\mathscr{L}^T b$ (and $r = m$, $s = q$); hence $a\,\mathscr{D}^T\,b$.

Conversely, if $a\,\mathscr{D}^T\,b$ then for some c in T we have $a\,\mathscr{R}^T\,c$, $c\,\mathscr{L}^T\,b$. Hence, for every m, n, p, q in N,

$$(m, a, n)\,\mathscr{R}^S\,(m, c, q), \quad (m, c, q)\,\mathscr{L}^S\,(p, b, q),$$

and so $(m, a, n)\,\mathscr{D}^S\,(p, b, q)$.

(c) $(m, a, n)^2 = (m - n + t, a\theta^{t-n}b\theta^{t-m}, n - m + t)$, where $t = \max(m, n)$. Hence (m, a, n) can be idempotent only if $m = n$. Since $(m, a, m)^2 = (m, a^2, m)$, the element (m, a, m) is idempotent if and only if $a^2 = a$.

(d) If T is an inverse semigroup, then each element (m, a, n) of S has an inverse (n, a^{-1}, m). Thus S is regular. To show that it is an inverse semigroup, let (m, e, m), (n, f, n) be idempotents in S (with $m \geqslant n$, say). Then

$$\left.\begin{aligned}(m, e, m)(n, f, n) &= (m, e(f\theta^{m-n}), m)\\ (n, f, n)(m, e, m) &= (m, (f\theta)^{m-n}e, m).\end{aligned}\right\} \qquad (6.14)$$

Now $f\theta^{m-n}$ is an idempotent in T. (Indeed, if $m \neq n$ we must have $f\theta^{m-n} = 1$, the only idempotent in H_1.) Hence $e(f\theta^{m-n}) = (f\theta^{m-n})e$ and so idempotents commute in S.

Conversely, if S is an inverse semigroup and if (p, b, q) is the inverse of (m, a, n), then

$$(m, a, n)(p, b, q) = (m - n + t, a\theta^{t-n}b\theta^{t-p}, q - p + t)$$

(with $t = \max(n, p)$) is an idempotent $\mathscr{R}^S$-equivalent to (m, a, n) and $\mathscr{L}^S$-equivalent to (p, b, q). Hence

$$m = m - n + t = q - p + t = q,$$

and so $n = p(= t)$, $m = q$. The inverse property now gives

$$(m, a, n) = (m, a, n)(n, b, m)(m, a, n) = (m, aba, n),$$

$$(n, b, m) = (n, b, m)(m, a, n)(n, b, m) = (n, bab, n);$$

thus $aba = a$, $bab = a$, and so b is an inverse for a in T. Thus T is regular. If e, f are idempotents in T, then the commuting of the idempotents $(0, e, 0)$, $(0, f, 0)$ of S immediately implies that $ef = fe$ in T. ■

As a consequence of this theorem, if T is a group with identity e (so that θ is simply an endomorphism of T) then $BR(T, \theta)$ becomes a bisimple inverse semigroup with idempotents (m, e, m) $(m = 0, 1, 2, \ldots)$. From (6.14) it is easy to see moreover that

$$(0, e, 0) > (1, e, 1) > (2, e, 2) > \ldots;$$

thus $BR(T, \theta)$ is a bisimple inverse ω-semigroup.

Reilly (1966) showed that the converse of this result also holds. See also Warne (1966). We summarize the situation in a theorem as follows:

Theorem 6.15. *Let G be a group and let θ be an endomorphism of G. Let $S = BR(G, \theta)$ be the Bruck–Reilly extension of G determined by θ; i.e., let $S = N \times G \times N$, with multiplication given by*

$$(m, a, n)(p, b, q) = (m - n + t, a\theta^{t-n}b\theta^{t-p}, q - p + t),$$

where $t = \max(n, p)$. Then S is a bisimple inverse ω-semigroup. Conversely, any bisimple inverse ω-semigroup is isomorphic to some $BR(G, \theta)$.

Proof. It remains to prove the second (converse) half. Let S be a bisimple inverse semigroup with

$$E = C_\omega = \{e_0, e_1, e_2, \ldots\}$$

as semilattice of idempotents. We know that T_E is isomorphic to the bicyclic semigroup:

$$T_E = \{\alpha_{m,n}: m, n \in N\},$$

where $\alpha_{m,n}$ is the unique isomorphism from Ee_m onto Ee_n and where

$$\alpha_{m,n}\alpha_{p,q} = \alpha_{m-n+t,\,q-p+t},$$

with $t = \max(n, p)$. By Theorem 4.8 we have a homomorphism $\phi: S \to T_E$ whose kernel is μ. In fact, by Corollary 6.7, $\ker \phi = \mathscr{H}$.

A typical $\mathscr{H}$-class of S is

$$H_{m,n} = \{a \in S: aa^{-1} = e_m, a^{-1}a = e_n\},$$

and each element a of $H_{m,n}$ maps under ϕ to an element α_a of T_E whose domain is $Eaa^{-1} = Ee_m$ and whose range is $Ea^{-1}a = Ee_n$. That is,

$$H_{m,n}\phi = \alpha_{m,n}.$$

Hence

$$H_{m,n}H_{p,q} \subseteq H_{m-n+t,\,q-p+t}. \tag{6.16}$$

The $\mathscr{H}$-class $H_{0,0}$ is a group with identity e_0; let us write $H_{0,0} = G$. Choose an element a in $H_{0,1}$ and fix it. Then $a^{-1} \in H_{1,0}$ by the standard eggbox argument (Section II.3). Also $a^2 \in H_{0,2}$ by (6.16); and more generally by induction we find that

$$a^n \in H_{0,n}, \qquad a^{-n} \in H_{n,0} \qquad (n = 0, 1, 2, \ldots),$$

where a^0 is defined as e_0. Notice next that

$$a^n a^{-n} = e_0, \qquad a^{-n}a^n = e_n.$$

	L_0	L_1	- - -	L_n	
R_0	e_0	a		a^n	
R_1	a^{-1}	e_1			
⋮					
R_m	a^{-m}			$H_{m,n}$	

Lemma 6.17. *The map*

$$g \mapsto a^{-m}ga^n \qquad (g \in G)$$

is a bijection from $G = H_{0,0}$ *onto* $H_{m,n}$ *with inverse given by*

$$x \mapsto a^m x a^{-n} \qquad (x \in H_{m,n}).$$

Proof. This is a simple application of Green's Lemmas (II.2.1 and II.2.2). We have $e_0 a^n = a^n$ and $a^n a^{-n} = e_0$, and so the map $g \mapsto ga^n$ is a bijection from $G = H_{0,0}$ onto $H_{0,n}$, with inverse $y \mapsto ya^{-n}$. Also $a^{-m}e_0 = a^{-m}$ and $a^m a^{-m} = e_0$, and so the map $y \mapsto a^{-m}y$ is a bijection from $H_{0,n}$ onto $H_{m,n}$, with inverse $x \mapsto a^m x$. Combining these two bijections gives the required result. ■

Returning now to the proof of Theorem 6.15, we see that, once a is chosen, we have a bijection $\psi: S \to N \times G \times N$ given by

$$(a^{-m}ga^n)\psi = (m, g, n). \tag{6.18}$$

If $g \in G$ then

$$ag \in H_{0,1}H_{0,0} = H_{0,1}$$

(by (6.16)) and so by Lemma 6.17 there is a unique expression $a^0 g' a^1$ for eg. That is, for each g in G there is a unique g' in G such that $ag = g'a$. We can use this fact to define a mapping $\theta: G \to G$; that is, we define $\theta: G \to G$ by the rule

$$ag = (g\theta)a \qquad (g \in G). \tag{6.19}$$

Notice that if $g_1, g_2 \in G$ then

$$[(g_1g_2)\theta]a = a(g_1g_2) = (ag_1)g_2 = [(g_1\theta)a]g_2$$
$$= (g_1\theta)(ag_2) = (g_1\theta)[(g_2\theta)a] = [(g_1\theta)(g_2\theta)]a;$$

hence, postmultiplying by a^{-1}, we find that

$$(g_1g_2)\theta = (g_1\theta)(g_2\theta),$$

i.e. that θ is an endomorphism of G.

Notice next that

$$a^2g = a(ag) = a(g\theta)a = (g\theta^2)a^2;$$

in fact, by induction,

$$a^n g = (g\theta^n)a^n \qquad (n = 0, 1, 2, \ldots),$$

where θ^0 is interpreted as the identity map of G. Also, from (6.19) we can deduce that

$$g^{-1}a^{-1} = a^{-1}(g\theta)^{-1} = a^{-1}(g^{-1}\theta).$$

for every g in G. Hence, changing the notation, we have

$$ga^{-1} = a^{-1}(g\theta) \qquad (g \in G).$$

More generally, we can prove by induction that

$$ga^{-n} = a^{-n}(g\theta^n) \qquad (n = 0, 1, 2, \ldots).$$

Lemma 6.17 yields a unique expression of the type $a^{-m}ga^n$ for every element of S. We are now in a position to describe the manner in which two such expressions multiply. If $n \geqslant p$, then

$$(a^{-m}ga^n)(a^{-p}ha^q) = a^{-m}g(a^{n-p}h)a^q = a^{-m}g(h\theta^{n-p})a^{q-p+n},$$

while if $n \leqslant p$,

$$(a^{-m}ga^n)(a^{-p}ha^q) = a^{-m}(ga^{-(p-n)})ha^q = a^{-(m-n+p)}(g\theta^{p-n})ha^q.$$

We can summarize these two formulae as follows:

$$(a^{-m}ga^n)(a^{-p}ha^q) = a^{-(m-n+t)}(g\theta^{t-n})(h\theta^{t-p})a^{q-p+t},$$

where $t = \max(n, p)$. Thus the bijection $\psi : S \to N \times G \times N$ given by (6.18) is an isomorphism from S onto the Bruck–Reilly extension $BR(G, \theta)$. ∎

This important theorem has served as a model for several later and more complicated results. See for example the papers of Warne (1966a, b, c; 1967a, b; 1968, 1969, 1970, 1971a, b), and the Ph.D. theses of Hickey (1970) and Patricia McLean (1973). As long as the semilattice E has the crucial property of Proposition 6.6, the difficulties of classifying the bisimple inverse semigroups seem to be manageable; but the structure necessarily becomes much more complicated when E does not have this property.

The Reilly Theorem (6.15) passes the "congruence test" mentioned at the beginning of Section III.4. For a description of the congruences and for various results about congruences on a bisimple inverse ω-semigroup see Munn and Reilly (1966) and Munn (1966).

7. SIMPLE INVERSE SEMIGROUPS

We now briefly describe a theory for simple inverse semigroups that parallels in many respects the theory we have developed for the bisimple case. In the bisimple case we were able to show that a semilattice E can be the semilattice of idempotents of a bisimple inverse semigroup if and only if it is uniform. The first stage in generalizing the theory must be to obtain the appropriate analogue of the notion of uniformity. The key is the following lemma:

Lemma 7.1. *An inverse semigroup S with semilattice of idempotents E is simple if and only if*

$$(\forall e, f \in E)(\exists g \in E)[g \leqslant f \quad \text{and} \quad e \mathscr{D} g].$$

Proof. Let S be simple. If $e, f \in E$ then $e \mathscr{J} f$ and so there exist x, y in S such that $e = xfy$. If we define g to be $fyex$, then

$$g^2 = fye(xfy)ex = fye^3x = fyex = g$$

and so $g \in E$. Also $fg = g$ and so $g \leqslant f$. If $z = x^{-1}e$ then

$$xz = xx^{-1}e = xx^{-1}xfy = xfy = e,$$

and so $e \mathscr{L} z$. Also,

$$zx = x^{-1}ex = x^{-1}e^2x = x^{-1}xfyex = x^{-1}xg = gx^{-1}x$$
$$= fyexx^{-1}x = fyex = g,$$

$$gx^{-1} = gx^{-1}xx^{-1} = x^{-1}xgx^{-1} = x^{-1}xfyexx^{-1} = x^{-1}e^2xx^{-1}$$
$$= x^{-1}xx^{-1}e = x^{-1}e = z,$$

and so $z \mathscr{R} g$. Thus $e \mathscr{D} g$ as required.

Conversely, if S has the property described in the statement of the lemma, consider any two idempotents e, f in S. Then there exists $g \in E$ such that $g \leqslant f$ and $e \mathscr{D} g$, and so

$$J_e = J_g \leqslant J_f.$$

Equally, there exists $h \in E$ such that $h \leqslant e$ and $f \mathscr{D} h$, and so

$$J_f = J_h \leqslant J_e.$$

Hence $J_e = J_f$, and so all the idempotents of S fall in a single $\mathscr{J}$-class. But every element of S is $\mathscr{J}$-equivalent (indeed even $\mathscr{R}$- or $\mathscr{L}$-equivalent) to some idempotent and so it follows that S is simple. ∎

As a consequence of this lemma and of (5.1), if S is a simple inverse semigroup with semilattice of idempotents E, then E has the property

$$(\forall e, f \in E)(\exists g \in E)[g \leqslant f \quad \text{and} \quad Ee \simeq Eg]. \tag{7.2}$$

This condition is reminiscent of the condition of uniformity encountered in the bisimple case. Following Munn (1970a) we call a semilattice E *subuniform* if (7.2) holds. We have proved half of

Proposition 7.3. *A semilattice E is the semilattice of idempotents of a simple inverse semigroup if and only if it is subuniform.*

Proof. To establish the other half, let us consider T_E, where E is a subuniform semilattice. If $e, f \in E$ then $Ee \simeq Eg$ for some $g \leqslant f$ and so, by (6.2), $e \mathscr{D} g$ in T_E. By Lemma 7.1 it follows that T_E is simple. ■

If we have a simple inverse semigroup which is also *fundamental* then by Theorem 4.9 we can effectively regard it as a full inverse subsemigroup of T_E (where E is the semilattice of idempotents of S). As in the bisimple case we can be a bit more precise than this. Recall that an inverse subsemigroup S of T_E is called *transitive* if for every pair e, f of idempotents of E there exists an of E there exists α in S such that

$$\mathrm{dom}(\alpha) = Ee, \qquad \mathrm{ran}(\alpha) = Ef.$$

Let us now decide to call S *subtransitive* if for every pair e, f of idempotents of E there exists α in S such that

$$\mathrm{dom}(\alpha) = Ee, \qquad \mathrm{ran}(\alpha) \subseteq Ef.$$

We now have

Theorem 7.4 (Munn 1970a). *If E is a subuniform semilattice, then every subtransitive inverse subsemigroup of T_E is a fundamental simple inverse semigroup with semilattice of idempotents isomorphic to E. Conversely, if S is a fundamental simple inverse semigroup with (necessarily subuniform) semilattice of idempotents E, then S is isomorphic to a subtransitive inverse subsemigroup of T_E.*

Proof. Let S be a subtransitive inverse subsemigroup of T_E, where E is subuniform. If $e \in E$ then by subtransitivity there exists $\alpha \in S$ with $\mathrm{dom}(\alpha) = Ee$, $\mathrm{ran}(\alpha) \subseteq Ee$, and so $(e =)\, 1_{Ee} = \alpha\alpha^{-1} \in S$; hence S is a full inverse subsemigroup of T_E and so by Theorem 4.10 is a fundamental inverse semigroup with semilattice of idempotents isomorphic to E. To show that S is simple, note that for all e, f in E there exists α in S such that

$$\mathrm{dom}(\alpha) = Ee, \qquad \mathrm{ran}(\alpha) \subseteq Ef,$$

i.e. such that

$$\mathrm{dom}(\alpha) = Ee, \qquad \mathrm{ran}(\alpha) = Eg,$$

where $g \leqslant f$. Moreover, $1_{Eg} = \alpha^{-1}\alpha \in S$ and $\alpha\alpha^{-1} = 1_{Ee}$. Thus the idempotents e, g (strictly $1_{Ee}, 1_{Eg}$) are $\mathscr{D}$-equivalent in S, and so by Lemma 7.1 we deduce that S is simple.

Conversely, if S is a fundamental simple inverse semigroup, we begin by embedding S in T_E in the usual way, where E is the semilattice of idempotents of S. Thus $S \simeq S'$, a full inverse subsemigroup of T_E. Since S' is simple we have by Lemma 7.1 that for all e, f in E there exists g that $g \leqslant f$ and $1_{Ee}\,\mathscr{D}\,1_{Eg}$ in S'. Hence there exists α in S' such that $\alpha\alpha^{-1} = 1_{Ee}$, $\alpha^{-1}\alpha = 1_{Eg}$, i.e. such that

$$\mathrm{dom}(\alpha) = Ee, \qquad \mathrm{ran}(\alpha) = Eg \subseteq Ef.$$

Thus S' is subtransitive. ■

It is possible to find an analogue for simple semigroups of Corollary 6.5, but the answer is necessarily more complicated. Let d be a positive integer. If (m, n) and (p, q) are elements of the bicyclic semigroup such that

$$m \equiv n \pmod d, \qquad p \equiv q \pmod d,$$

then

$$m - n + \max(n, p) \equiv q - p + \max(n, p) \pmod d;$$

hence, for $d = 1, 2, 3, \ldots$, the subset

$$B_d = \{(m, n) \in N \times N : m \equiv n \pmod d\}$$

is a subsemigroup of the bicyclic semigroup. It is even an inverse subsemigroup, since evidently the inverse (n, m) of an element (m, n) in B_d also belongs to B_d.

Lemma 7.5. *In* B_d,

$$(m, n)\,\mathscr{R}\,(p, q) \quad \textit{if and only if} \quad m = p;$$

$$(m, n)\,\mathscr{L}\,(p, q) \quad \textit{if and only if} \quad n = q.$$

Proof. If $(m, n)\,\mathscr{R}\,(p, q)$ then there exist (x, y), (z, t) in B_d such that

$$(m, n)(x, y) = (p, q), \qquad (p, q)(z, t) = (m, n).$$

Hence

$$p = m - n + \max(n, x) \geqslant m, \qquad m = p - q + \max(q, z) \geqslant p,$$

and so $m = p$. Conversely, if $(m, n), (m, q) \in B_d$, then (n, q) and (q, n) are in B_d, since $m \equiv n \equiv q \pmod d$. Hence, since

$$(m, n)(n, q) = (m, q) \quad \text{and} \quad (m, q)(q, n) = (m, n)$$

we have that $(m, n) \mathscr{R} (m, q)$ in B_d. The other half of the lemma is proved similarly. ■

Lemma 7.6. *In* B_d,

$$(m, n) \mathscr{D} (p, q) \quad \textit{if and only if} \quad m \equiv n \equiv p \equiv q (\text{mod } d).$$

Proof. By the previous lemma

$$(m, n) \mathscr{D} (p, q) \Leftrightarrow (\exists (x, y) \in B_d) \ \ (m, n) \mathscr{R} (x, y), (x, y) \mathscr{L} (p, q),$$

$$\text{i.e.} \quad \Leftrightarrow (\exists (x, y) \in B_d)\, m = x, y = q$$

$$\text{i.e.} \quad \Leftrightarrow (m, q) \in B_d$$

$$\text{i.e.} \quad \Leftrightarrow m \equiv q \ (\text{mod } d)$$

$$\text{i.e.} \quad \Leftrightarrow m \equiv n \equiv p \equiv q \ (\text{mod } d). \ ■$$

From this lemma it follows that in B_d there are precisely d $\mathscr{D}$-classes, namely

$$D_{(0, d)}, \ D_{(1, d+1)}, \ \ldots, \ D_{(d-1, 2d-1)}.$$

We can now prove the analogue of Corollary 6.5:

Proposition 7.7. *Up to isomorphism, the only fundamental simple inverse* ω*-semigroups are the semigroups* B_d $(d = 1, 2, 3 \ldots)$.

Proof. By virtue of Theorem 7.4, what we have to show is that the semigroups

$$B'_d = \{\alpha_{m,n} : m \equiv n \ (\text{mod } d)\}$$

are the only subtransitive inverse subsemigroups of

$$T_{C_\omega} = \{\alpha_{m,n} : m, n \in N\}.$$

First, to show that B'_d is subtransitive, notice that if m, n are any two non-negative integers then there exists $p \geqslant n$ such that $m \equiv p \ (\text{mod } d)$. Hence for any e_m, e_n in C_ω there exists $\alpha_{m,p}$ in B'_d such that

$$\text{dom}(\alpha_{m,p}) = Ee_m, \qquad \text{ran}(\alpha_{m,p}) = Ee_p \subseteq Ee_n.$$

If S is a subtransitive inverse subsemigroup of T_{C_ω} then there exists α in S such that $\text{dom}(\alpha) = Ee_0$, $\text{ran}(\alpha) \subseteq Ee_1$, i.e. such that $\text{dom}(\alpha) = Ee_0$, $\text{ran}(\alpha) = Ee_d$ $(d \geqslant 1)$. That is, for some $d \geqslant 1$, we have $\alpha_{0,d} \in S$. Let us suppose that d is the least positive integer for which this is true: i.e. $\alpha_{0,r} \in S$ with $0 \leqslant r < d$ implies $r = 0$.

The standard multiplication rule (4.7) in T_{C_ω} gives that $\alpha_{0,d}^2 = \alpha_{0,2d}$; indeed more generally we have $\alpha_{0,d}^k = \alpha_{0,kd}$ $(k = 1, 2\ldots)$. As remarked in the proof of Theorem 7.4, a subtransitive inverse subsemigroup of T_{C_ω} is necessarily full, i.e. necessarily contains all the idempotents of T_{C_ω}. Hence, for every $m \geqslant 0$,

$$\alpha_{m,m}\alpha_{0,kd} \in S;$$

that is, $\alpha_{m,m+kd} \in S$. Hence also $\alpha_{m+kd,m} \in S$ for every $m, k \geqslant 0$. Thus $B'_d \subseteq S$. To show that equality holds, let us suppose by way of contradiction that there exists $\alpha_{m,n}$ in S for which $m \not\equiv n \pmod d$. We may then write $n = m + kd + r$, where $0 < r < d$. It follows that S contains

$$\alpha_{m,m+kd+r}\alpha_{m+kd,m} = \alpha_{m,m+r}.$$

Hence, if $m > 0$ the subsemigroup S contains

$$\alpha_{m-1,m-1+d}\alpha_{m,m+r}\alpha_{m-1+d,m-1} = \alpha_{m-1,m-1+r}.$$

We can continue this argument until we obtain that $\alpha_{0,r} \in S$ in contradiction to the choice of d. Hence $S = B'_d$. ■

We observed in the last section that the Bruck–Reilly extension $BR(T, \theta)$ of a *group* T is a bisimple ω-semigroup. More generally, we now take T to be a semilattice of groups—a Clifford semigroup—of a special form. Specifically, let T be the union of disjoint groups $G_0, \ldots, G_{d-1}$, and for $i = 0, \ldots, d-2$ let $\gamma_i: G_i \to G_{i+1}$ be a homomorphism. Then for all i, j in $\{0, \ldots, d-1\}$ such that $i < j$ we have a homomorphism $\alpha_{i,j}: G_i \to G_j$ given by

$$\alpha_{i,j} = \gamma_i\gamma_{i+1}\cdots\gamma_{j-1}.$$

For each i in $\{0, \ldots, d-1\}$ we define $\alpha_{i,i}$ to be the identical automorphism of G_i. Then, if $i \geqslant j \geqslant k$,

$$\alpha_{i,j}\alpha_{j,k} = \alpha_{i,k}.$$

Thus, if Y denotes the semilattice $\{0, \ldots, d-1\}$ in which $0 > 1 > 2 > \ldots > d-1$, we have a Clifford semigroup $T = \mathscr{S}(Y; G_i; \alpha_{i,j})$. We denote the idempotents of T by $e_0, e_1, \ldots, e_{d-1}$ (the identity elements respectively of the groups $G_0, G_1, \ldots, G_{d-1}$) and remark that in T

$$e_0 > e_1 > \ldots > e_{d-1}.$$

The element e_0 is the identity of T. We shall refer to T as a *finite chain of groups* (of *length* d).

If T is such a semigroup then it follows by Theorem 6.9 that $S = BR(T, \theta)$ (for any homomorphism $\theta: T \to G_0$) is a simple inverse semigroup in which the $\mathscr{D}$-classes are the subsets $N \times G_i \times N$ $(i = 0, \ldots, d-1)$. Thus S has precisely d $\mathscr{D}$-classes. Now consider two idempotents (m, e_i, m) and (n, e_j, n).

If $m = n$, then it is clear that $(m, e_i, m) < (m, e_j, m)$ if and only if $i > j$. If $m > n$ then

$$(m, e_i, m)(n, e_j, n) = (m, e_i(e_j\theta^{m-n}), m) = (m, e_i, m)$$

(since $e_j\theta^{m-n} = e_0$, the identity of T); hence $(m, e_i, m) < (n, e_j, n)$ whatever i and j may be. In summary, the idempotents of S form a chain as follows:

$$\begin{aligned} &(0, e_0, 0) > (0, e_1, 0) > \ldots > (0, e_{d-1}, 0) \\ &> (1, e_0, 1) > (1, e_1, 1) > \ldots > (1, e_{d-1}, 1) \\ &> (2, e_0, 2) > (2, e_1, 2) > \ldots . > (2, e_{d-1}, 2) \\ &> \ldots . \end{aligned}$$

Thus $S = BR(T, \theta)$ is a simple inverse ω-semigroup.

More remarkably, Munn (1968), and independently Kočin (1968), were able to show in essence that every simple inverse ω-semigroup is of this form:

Theorem 7.8. *Let T be a finite chain of groups of length d. If θ is a homomorphism from T into the group of units of T then the Bruck-Reilly extension $BR(T, \theta)$ of T determined by θ is a simple inverse ω-semigroup with d $\mathscr{D}$-classes. Conversely, every simple inverse ω-semigroup is isomorphic to one of this type.*

Proof. We have already proved the direct half. To prove the converse half, let us consider a simple inverse semigroup S whose semilattice of idempotents is

$$E = C_\omega = \{f_0, f_1, f_2, \ldots\}.$$

(The change in notation from e to f is so as to avoid confusion with the notation used above in describing the chain of groups $\mathscr{S}(Y; G_i; \alpha_{i,j})$.)

By Theorem 4.9 there is a homomorphism $\phi: S \to T_E$ mapping each a in S to the element $\alpha_a: Eaa^{-1} \to Ea^{-1}a$ of T_E, where

$$x\alpha_a = a^{-1}xa \qquad (x \in Eaa^{-1}).$$

The kernel of ϕ is μ, which in this instance coincides with $\mathscr{H}$, by Corollary 6.7. The subsemigroup S of T_E is a fundamental simple inverse ω-semigroup and so by Proposition 7.7 is isomorphic to some B_d $(d \geqslant 1)$. We write

$$S\phi = \{\alpha_{m,n}: m, n \in N, m \equiv n \pmod{d}\},$$

where $\alpha_{m,n}: Ef_m \to Ef_n$ is given as usual by

$$f_k\alpha_{m,n} = f_{k-m+n} \qquad (k \geqslant m),$$

and where

$$\alpha_{m,n}\alpha_{p,q} = \alpha_{m-n+t,\,q-p+t} \quad (t = \max(n, p)).$$

For all $m, n \in N$ we define

$$H_{m,n} = \{a \in S: aa^{-1} = f_m, a^{-1}a = f_n\};$$

then either $H_{m,n} = \varnothing$ or $H_{m,n}$ is an $\mathscr{H}$-class of S. The essential difference between this situation and the corresponding situation in the proof of Reilly's Theorem (6.15) now appears, for the former possibility can arise here. Indeed we can say exactly when it does arise, for if $H_{m,n} \neq \varnothing$ then $H_{m,n}\phi = \alpha_{m,n} \in S\phi$; hence $H_{m,n} \neq \varnothing$ if and only if $m \equiv n \pmod{d}$. We easily deduce that the semigroup S has precisely d $\mathscr{D}$-classes

$$D^0, D^1, \ldots, D^{d-1},$$

where

$$D^i = \bigcup\{H_{m,n}: m \equiv n \equiv i \pmod{d}\}$$
$$= \bigcup\{H_{pd+i,\, qd+i}: p, q \in N\}.$$

Lemma 7.9. *For $i = 0, 1, \ldots, d-1$, the $\mathscr{D}$-class D^i is a bisimple inverse ω-semigroup with identity element f_i.*

Proof. Since $H_{m,n}\phi = \alpha_{m,n}$ (whenever $m \equiv n \pmod{d}$) we can deduce from the multiplication formula for the elements $\alpha_{m,n}$ that

$$H_{m,n}H_{p,q} \subseteq H_{m-n+t,\, q-p+t} \qquad (t = \max(n, p)). \tag{7.10}$$

If $H_{m,n}, H_{p,q} \subseteq D^i$, that is, if $m \equiv n \equiv i \pmod{d}$ and $p \equiv q \equiv i \pmod{d}$, then

$$m - n + t \equiv q - p + t \equiv t \equiv i \pmod{d}$$

and so $H_{m,n}H_{p,q} \subseteq D^i$. Thus D^i is a subsemigroup of S.

Next, from the definition of $H_{m,n}$ it easily follows that the set $H_{m,n}^{-1}$ (defined as $\{a^{-1}: a \in H_{m,n}\}$) coincides with the set $H_{n,m}$. Thus the inverse of an element of D^i lies in D^i, and so D^i is an inverse subsemigroup of S.

Of the $\mathscr{H}$-classes

$$H_{pd+i,\, qd+i} = \{a \in S: aa^{-1} = f_{pd+i}, a^{-1}a = f_{qd+i}\}$$

that make up D^i, the ones that contain idempotents are precisely the ones in which $p = q$. More precisely, $H_{pd+i,\, pd+i}$ contains the idempotent f_{pd+i}. Thus the idempotents of the inverse subsemigroup D^i of S form an infinite chain

$$f_i > f_{d+i} > f_{2d+i} > \cdots.$$

Thus D^i is an inverse ω-semigroup.

Finally, all the idempotents in D^i are $\mathscr{D}$-equivalent in D^i, since for any p, q in N there exists an element a in D^i (in fact in $H_{pd+i,\, qd+i}$) such that $aa^{-1} = f_{pd+i}$, $a^{-1}a = f_{qd+i}$. It follows that D^i is a bisimple inverse ω-semigroup as required. ■

Returning now to the main proof, we observe that the group of units of D^i is $H_{i,i}$. For simplicity let us write G_i for $H_{i,i}$. The identity of G_i is f_i. Let

$$T = \bigcup_{i=0}^{d-1} G_i.$$

If $i, j \in \{0, \ldots, d-1\}$ with $i \leqslant j$, then by formula (7.10) and by the fact that $H_{m,n}^{-1} = H_{n,m}$ we have that

$$G_iG_j = H_{i,i}H_{j,j} \subseteq H_{j,j}, \qquad G_jG_i = H_{j,j}H_{i,i} \subseteq H_{j,j},$$

$$G_i^{-1} = H_{i,i}^{-1} = H_{i,i} = G_i;$$

hence T is an inverse subsemigroup of S. Since it is evidently a union of disjoint groups, we have that T is a Clifford semigroup in which the idempotents form the finite chain

$$f_0 > f_1 > \ldots > f_{d-1}.$$

That is, T is a finite chain of groups.

The result stated in Lemma 7.9 suggests that we might apply Reilly's Theorem (6.15) to each D^i, obtaining that $D^i \simeq BR(G_i, \theta_i)$, where θ_i is an endomorphism of G_i. This in fact is almost what we do. Reilly's method of proof suggests that we ought for each i to choose $a_i \in H_{i,d+i}$ and express each element of D^i as $a_i^{-m}g_ia_i^n (\in H_{md+i,nd+i})$ with $m, n \in N$ and $g_i \in G_i$. It pays, however, not to choose the elements a_i independently but to make all the elements a_i $(i = 1, \ldots, d-1)$ depend on the chosen element a_0 in $H_{0,d}$.

Simplifying the notation, let us choose $a \in H_{0,d}$. Then $a^{-1} \in H_{d,0}$ and by definition of $H_{0,d}$ we have

$$aa^{-1} = f_0, \qquad a^{-1}a = f_d.$$

Also, for $i = 0, \ldots, d-1$,

$$f_ia \in H_{i,i}H_{0,d} = H_{i,d+i}.$$

Since f_i is the identity element of D^i it follows that

$$(f_ia)f_i = f_ia,$$

from which we easily deduce that $(f_ia)^2 = f_ia^2$. Indeed it is easy to show by induction that

$$(f_ia)^n = f_ia^n \qquad (n = 1, 2, \ldots).$$

Taking inverses, we have that

$$(f_ia)^{-n} = a^{-n}f_i \qquad (n = 1, 2, \ldots).$$

Thus, using $f_i a$ for a_i in the Reilly method as indicated above, we express each element of D^i uniquely as

$$(f_i a)^{-m} g_i (f_i a)^n = a^{-m} f_i g_i f_i a^n = a^{-m} g_i a^n.$$

(since f_i is the identity element of G_i), where $m, n \in N$ and $f_i \in G_i$. The element $a^{-m} g_i a^n$ belongs to the $\mathscr{H}$-class $H_{md+i,\, nd+i}$. There is no difficulty about zero values of m and n if we interpret $a^{-0} g_i a^n$ as $g_i a^n$ and $a^{-m} g_i a^0$ as $a^{-m} g_i$ in the obvious way.

We thus have a bijection Ψ from S onto $N \times T \times N$ defined by

$$(a^{-m} g_i a^n)\Psi = (m, g_i, n).$$

It remains to find a homomorphism $\theta : T \to G_0$ such that Ψ becomes an isomorphism from S onto $BR(T, \theta)$.

Again following the lead given by Reilly's proof, we observe that for $i = 0, \ldots, d-1$,

$$ag_i \in H_{0,\, d} H_{i,\, i} \subseteq H_{0,\, d}.$$

Hence ag_i is uniquely expressible as $g'_0 a$, with $g'_0 \in G_0$. Since g'_0 is determined by g_i we can regard the resulting formula $ag_i = g'_0 a$ as defining a mapping $\phi_i : G_i \to G_0$:

$$ag_i = (g_i \phi_i)a. \tag{7.11}$$

Since the groups $G_0, \ldots, G_{d-1}$ are disjoint, we can use the maps $\phi_0, \ldots, \phi_{d-1}$ thus obtained to piece together a map $\theta : T \to G_0$:

$$t\theta = t\phi_i \qquad (t \in G_i).$$

This θ is the homomorphism we are looking for. To show the homomorphic property, consider x, y in T, with $x \in G_i$, $y \in G_j$ and $i \leqslant j$. Then $xy \in G_j$ and so, by (7.11),

$$a(xy) = [(xy)\phi_j]a = [(xy)\theta]a.$$

On the other hand

$$a(xy) = (ax)y = (x\phi_i)ay = (x\phi_i)(y\phi_j)a = [(x\theta)(y\theta)]a.$$

Hence $[(xy)\theta]a = [(x\theta)(y\theta)]a$ and so, since $aa^{-1} = f_0$,

$$(xy)\theta = (x\theta)(y\theta).$$

Exactly as in the proof of Theorem 6.15 we have that for every x in I and for $k = 0, 1, 2, \ldots,$

$$a^k x = (x\theta^k)a^k, \quad xa^{-k} = a^{-k}(x\theta^k),$$

where a^0 is interpreted as f_0 and $x\theta^0$ as x.

Now let $x, y \in T$ and let $m, n, p, q \in N$. If $n \geqslant p$, then

$$(a^{-m}xa^n)(a^{-p}ya^q) = a^{-m}x(y\theta^{n-p})a^{q-p+n},$$

while if $n \leqslant p$,

$$(a^{-m}xa^n)(a^{-p}ya^q) = a^{-(m-n+p)}(x\theta^{p-n})ya^q.$$

If we write $t = \max(n, p)$ we can combine these results in the single formula

$$(a^{-m}xa^n)(a^{-p}ya^q) = a^{-(m-n+t)}(x\theta^{t-n})(y\theta^{t-p})a^{q-p+t},$$

and it is now clear that Ψ is an isomorphism from S onto $BR(T, \theta)$. ∎

Remark. In Exercise 25 below it is indicated that the homomorphism $\theta: T \to G_0$ is effectively determined by a homomorphism $\gamma: G_{d-1} \to G_0$. This observation makes it easy to recover Munn's theorem (1968) in its original form.

Munn's result has been used by G. R. Baird (1972) to study congruences on simple ω-semigroups.

8. REPRESENTATIONS OF INVERSE SEMIGROUPS

In this section we describe an analysis of representations of inverse semigroups by partial one–one mappings. This is due to Schein (1962); see also (Clifford and Preston 1967, Sections 7.2 and 7.3). The results obtained are of considerable interest in their own right, but the most compelling reason for including them in this rather crowded chapter is that we shall require to make use of them in Section VII.4. The idea is essentially the same as in group theory, where one shows that every transitive permutational representation of a group is equivalent to a representation by permutations of the cosets of a certain subgroup of the group. (See, for example, M. Hall (1959).)

Let us begin by reminding ourselves that a *representation* of an inverse semigroup S is a homomorphism ϕ of S into some symmetric inverse semigroup $\mathscr{I}(X)$. If ϕ is one–one we call the representation *faithful*. The particular representation $\phi: S \to \mathscr{I}(S)$ described in Theorem 1.10 we shall call the *Vagner–Preston representation of S*.

If $\phi: S \to T$ is a homomorphism from an inverse semigroup S into an inverse semigroup T, then Proposition 1.6 assures us that $S\phi$ is an inverse subsemigroup of T. In particular, then, if $\phi: S \to \mathscr{I}(X)$ is a representation, $S\phi$ is an inverse subsemigroup of $\mathscr{I}(X)$.

For any inverse subsemigroup H of $\mathscr{I}(X)$ we have a relation τ_H on X called the *transitivity relation* of H, defined by the rule that $(a, b) \in \tau_H$ if and only if there exists κ in H for which $a \in \operatorname{dom}(\kappa)$ and $a\kappa = b$.

Lemma 8.1. *If H is an inverse subsemigroup of a symmetric inverse semigroup $\mathscr{I}(X)$, then τ_H, the transitivity relation of H, is a symmetric and transitive relation on X.*

Proof. If $(x, y) \in \tau_H$ then $x\kappa = y$ for some κ in H. Hence $y \in \text{ran}(\kappa) = \text{dom}(\kappa^{-1})$ and $y\kappa^{-1} = x$. Since $\kappa^{-1} \in H$, it follows that $(y, x) \in \tau_H$. If (x, y) $(y, z) \in \tau_H$, then $x\kappa = y$, $y\lambda = z$ for some κ, λ in H. It follows that $x \in \text{dom}(\kappa\lambda)$ and that $z = x(\kappa\lambda)$. Since $\kappa\lambda \in H$, we deduce that $(x, z) \in \tau_H$. ■

We cannot in general assert that τ_H is an equivalence relation, since there may exist elements x in X that are not in the domain of any κ in H. What this amounts to is that there exist elements x in X such that $(x, x) \notin \tau_H$, for if we had $(x, y) \in \tau_H$ for some y then we would have $(y, x) \in \tau_H$ by the symmetry of τ_H and hence $(x, x) \in \tau_H$ by transitivity. We shall refer to the set

$$X\tau_H = \{x \in X : (x, x) \in \tau_H\}$$

as the *domain* of τ_H. Then τ_H is an equivalence relation on its domain $X\tau_H$.

The inverse subsemigroup H of $\mathscr{I}(X)$ will be called *effective* if $X\tau_H = X$. In such a case τ_H is an equivalence relation on X.

The τ_H-classes in $X\tau_H$ are called the *transitivity classes of* H, and H is called a *transitive* inverse subsemigroup of $\mathscr{I}(X)$ if τ_H is the universal relation on its domain $X\tau_H$, i.e. if for all a, b in $X\tau_H$ there exists κ in H for which $a \in \text{dom}(\kappa)$ and $a\kappa = b$.

If $\phi: S \to \mathscr{I}(X)$ is a representation of an inverse semigroup S, we say that ϕ is a *transitive* representation if $S\phi$ is a transitive inverse subsemigroup of $\mathscr{I}(X)$, and that ϕ is an *effective* representation if $S\phi$ is an effective inverse subsemigroup of $\mathscr{I}(X)$. An effective representation "makes use of" all the elements of X.

Let $\{X_i : i \in I\}$ be a family of pairwise disjoint sets and let

$$X = \bigcup\{X_i : i \in I\}.$$

If for each i in I we have a representation $\phi_i: S \to \mathscr{I}(X_i)$ of an inverse semigroup S, then we can define a representation $\phi: S \to \mathscr{I}(X)$ as follows:

$$\text{dom}(s\phi) = \bigcup\{\text{dom}(s\phi_i): i \in I\};$$

If $x \in \text{dom}(s\phi)$ then there is a unique i in I such that $x \in \text{dom}(s\phi_i)$; then

$$x(s\phi) = x(s\phi_i).$$

If, as will increasingly be useful, we regard each $s\phi_i (\in \mathscr{I}(X_i))$ as a subset of $X_i \times X_i$, then we can give the simple definition that

$$s\phi = \bigcup\{s\phi_i : i \in I\}. \tag{8.2}$$

We refer to ϕ as the *sum* of the family $\{\phi_i; i \in I\}$ of representations, and write

$$\phi = \bigoplus_{i \in I} \phi_i.$$

If $I = \{1, 2, \ldots, n\}$ we write

$$\phi = \phi_1 \oplus \phi_2 \oplus \ldots \oplus \phi_n.$$

Because of the definition (8.2) in terms of set-theoretic union, the infinite commutative and associative laws hold for the operation $\oplus$. In particular, if we have two families $\{\phi_i: i \in I\}$ and $\{\psi_i; i \in I\}$ of representations of S, with $\phi_i: S \to \mathscr{I}(X_i)$, $\psi_i: S \to \mathscr{I}(Y_i)$ and with

$$X_i \cap Y_i = \varnothing, \quad (X_i \cup Y_i) \cap (X_j \cup Y_j) = \varnothing$$

for all i, j in I, $i \neq j$, then

$$\bigoplus_{i \in I} (\phi_i \oplus \psi_i) = (\bigoplus_{i \in I} \phi_i) \oplus (\bigoplus_{i \in I} \psi_i) \tag{8.3}$$

If $\phi: S \to \mathscr{I}(X)$ and $\psi: S \to \mathscr{I}(Y)$ are representations, we say that ϕ and ψ are *equivalent* if there exists a bijection $\theta: X \to Y$ with the property that, for each s in S,

$$s\psi = \{(x\theta, x'\theta) \in Y \times Y: (x, x') \in s\phi\}.$$

That is to say,

$$\text{dom}(s\psi) = (\text{dom}(s\phi))\theta,$$

and for all x in dom$(s\phi)$

$$(x(s\phi))\theta = (x\theta)(s\psi).$$

This amounts to saying that the two representations differ "in name only". In practice it is important to be able to replace a representation by an equivalent one when we want to form the sum $\phi_1 \oplus \phi_2$ of representations $\phi_1: S \to \mathscr{I}(X_1)$ and $\phi_2: S \to \mathscr{I}(X_2)$ in the case where X_1 and X_2 are not disjoint. We cannot do this under the rules for addition, but what we can do is to form a representation $\phi_1 \oplus \psi_2$, where $\psi_2: S \to \mathscr{I}(Y_2)$ is a representation equivalent to ϕ_2 and where $Y_2 \cap X_1 = \varnothing$—and for most purposes this is just as good.

The importance of effective and transitive representations lies in the following result:

Theorem 8.4. *An effective representation of an inverse semgiroup S is the sum of a uniquely determined family of effective transitive representations of S.*

Proof. Let $\phi: S \to \mathscr{I}(X)$ be an effective representation of S and let $S\phi = H$. Then τ_H, the transitivity relation of H, is an equivalence relation on X. The

τ_H-classes form a family $\{X_i: i \in I\}$ of pairwise disjoint subsets of X, and

$$\tau_H = \bigcup\{X_i \times X_i: i \in I\}, \quad \bigcup\{X_i: i \in I\} = X.$$

We now define for each i in I the representation $\phi_i: S \to \mathscr{I}(X_i)$ by the rule

$$s\phi_i = s\phi \cap (X_i \times X_i) \qquad (s \in S).$$

That is, we stipulate that the partial mapping $s\phi_i$ is the restriction to $\operatorname{dom}(s\phi) \cap X_i$ of the partial mapping $s\phi$. Since X_i is a transitivity class of $S\phi$ it is then automatic that $\operatorname{ran}(s\phi_i) \subseteq X_i$; in fact

$$\operatorname{ran}(s\phi_i) = \operatorname{ran}(s\phi) \cap X_i.$$

To show that ϕ_i is a representation, observe that for any choice of s and t in S,

$$\begin{aligned}(st)\phi_i &= (st)\phi \cap (X_i \times X_i)\\ &= (s\phi)(t\phi) \cap (X_i \times X_i).\end{aligned}$$

Now $(x, y) \in (s\phi)(t\phi)$ if and only if $(x, z) \in s\phi$ and $(z, y) \in t\phi$ for some z in X. In fact, if x and y are in X_i then it follows (since X_i is a transitivity class of $H = S\phi$) that z also is in X_i. Hence

$$(s\phi)(t\phi) \cap (X_i \times X_i) = [s\phi \cap (X_i \times X_i)][t\phi \cap (X_i \times X_i)]$$

and so we obtain that $(st)\phi_i = (s\phi_i)(t\phi_i)$ as required.

Next, each ϕ_i is transitive and effective, since for each (x, y) in $X_i \times X_i$ there exists (by definition of τ_H) an element $s\phi$ such that $(x, y) \in s\phi$. Thus $(x, y) \in s\phi \cap (X_i \times X_i) = s\phi_i$.

For each s in S,

$$\begin{aligned}&\bigcup\{s\phi_i: i \in I\} = \bigcup\{s\phi \cap (X_i \times X_i): i \in I\}\\ &= (s\phi) \cap \bigcup\{X_i \times X_i: i \in I\}\\ &= s\phi, \quad \text{since} \quad s\phi \subseteq \bigcup\{X_i \times X_i: i \in I\}.\end{aligned}$$

Hence ϕ is the sum of the representations ϕ_i.

Finally, to show that the family $\{\phi_i: i \in I\}$ is unique, suppose that ϕ is the sum of a family $\{\psi_j: j \in J\}$ of effective transitive representations $\psi_j: S \to \mathscr{I}(Y_j)$, where X is the union of the pairwise disjoint sets Y_j. The transitivity classes of the sum ϕ of the ψ_j are the sets Y_j, and for each s in S,

$$(s\phi) \cap (Y_j \times Y_j) = s\psi_j.$$

It thus follows that the sets Y_j are just the sets X_i in some order, and that each ψ_j is equal to the appropriate ϕ_i. ■

The next stage in the investigation is to discover more about effective transitive representations. We begin by describing a particular kind of representation of an inverse semigroup S associated with an inverse subsemigroup H of S that is *closed* (in the sense of the word introduced in Section 2).

The first step in this process is to generalize the group-theoretic notion of "right coset". Let S be an inverse semigroup and let H be a (not necessarily closed) inverse subsemigroup of S. If $s \in S$ then the subset Hs of S may fail to contain s, but it certainly contains s if $ss^{-1} \in H$. We define a *right coset* of H to be a set Hs ($s \in S$) for which $ss^{-1} \in H$. Even if H is closed, the right coset Hs need not be closed. The subset $(Hs)\omega$ ($s \in S, ss^{-1} \in H$) of S will be called a *right ω-coset* of H. Among the right ω-cosets of H is $H\omega$ itself. Let $\mathscr{X}$ be the set of all the right ω-cosets of H.

The following results, strongly reminiscent of group theory, will be useful:

Proposition 8.5. *Let H be an inverse subsemigroup of an inverse semigroup S. If $(Ha)\omega$, $(Hb)\omega$ are right ω-cosets of H, then the following statements are equivalent:*

(i) $(Ha)\omega = (Hb)\omega$; (ii) $ab^{-1} \in H\omega$

(iii) $a \in (Hb)\omega$; (iv) $b \in (Ha)\omega$.

Proof. (i) $\Rightarrow$ (ii). If $(Ha)\omega = (Hb)\omega$, then

$$a = aa^{-1}a \in Ha \subseteq (Ha)\omega = (Hb)\omega$$

and so $a \geqslant hb$ for some h in H. Hence

$$ab^{-1} \geqslant hbb^{-1} \in H$$

and so $ab^{-1} \in H\omega$.

(ii) $\Rightarrow$ (iii). If $ab^{-1} \in H\omega$ then $ab^{-1} \geqslant h$ for some h in H. Hence $a \geqslant ab^{-1}b \geqslant hb$ and so $a \in (Hb)\omega$. Similarly (ii) $\Rightarrow$ (iv).

(iii) $\Rightarrow$ (i). If $a \in (Hb)\omega$ then $a \geqslant hb$ for some h in H. If $s \in (Ha)\omega$ then $s \geqslant ka$ for some k in H and so $s \geqslant khb$. Hence $s \in (Hb)\omega$ and so we have established that $(Ha)\omega \subseteq (Hb)\omega$. To establish the reverse inclusion, notice first that $a \geqslant hb$ implies (by Proposition 2.2 (vi)) that

$$(hb)^{-1}(hb) = (hb)^{-1}a.$$

Hence

$$hb = (hb)(hb)^{-1}(hb) = (hb)(hb)^{-1}a = hbb^{-1}h^{-1}a.$$

If now $s \in (Hb)\omega$, then

$$\begin{aligned} s &\geqslant kb \qquad (k \in H) \\ &\geqslant kh^{-1}hb = kh^{-1}hbb^{-1}h^{-1}a \in Ha \end{aligned}$$

and so $s \in (Ha)\omega$. The proof that (iv) $\Rightarrow$ (i) is similar. ■

Now for each s in S we define an element $s\phi_H$ of $\mathscr{I}(\mathscr{X})$ as follows:

$$\mathrm{dom}(s\phi_H) = \{(Hx)\omega \in \mathscr{X} : (Hxs)\omega \in \mathscr{X}\},$$

$$[(Hx)\omega](s\phi_H) = (Hxs)\omega \qquad ((Hx)\omega \in \mathrm{dom}(s\phi_H)).$$

Alternatively, we can define

$$s\phi_H = \{((Hx)\omega, (Hxs)\omega) : (Hx)\omega, (Hxs)\omega \in \mathscr{X}\}, \tag{8.6}$$

and from this point on it is this more abstract, set-theoretic approach that will be the more useful.

To verify that $s\phi_H$ does indeed belong to $\mathscr{I}(\mathscr{X})$, first notice that it is a well-defined partial mapping of $\mathscr{X}$. If $(Hx)\omega = (Hy)\omega$, with $xx^{-1}, yy^{-1} \in H$, and if $z \in (Hxs)\omega$, then $z \geqslant hxs$ for some h in H. Now $hx \in Hx \subseteq (Hx)\omega = (Hy)\omega$ and so $hx \geqslant h'y$ for some h' in H. Hence $z \geqslant hxs \geqslant h'ys$ and so $z \in (Hys)\omega$. Thus $(Hxs)\omega \subseteq (Hys)\omega$, and the opposite inclusion follows similarly.

Also, $s\phi_H$ is one-one, for if $(Hxs)\omega = (Hys)\omega$, where $(Hx)\omega$, $(Hy)\omega$, $(Hxs)\omega$ and $(Hys)\omega$ are all in $\mathscr{X}$, and if $z \in (Hx)\omega$, then $z \geqslant hx$ for some h in H. Hence

$$z \geqslant hxss^{-1}.$$

Now $hxs \in Hxs \subseteq (Hxs)\omega = (Hys)\omega$ and so $hxs \geqslant h'ys$ for some h' in H. Thus

$$z \geqslant h'yss^{-1} = h'yy^{-1}yss^{-1} = h'yss^{-1}y^{-1}y = h''y,$$

where $h'' \in H$, since $yss^{-1}y^{-1} \in H$ (as a result of the assumption that $(Hys)\omega \in \mathscr{X}$). Thus $z \in (Hy)\omega$ and so we have shown that $(Hx)\omega \subseteq (Hy)\omega$. The opposite inclusion follows in the same way.

We have thus shown that $s\phi_H \in \mathscr{I}(\mathscr{X})$. In fact we can show:

Proposition 8.7. *If H is a closed inverse subsemigroup of an inverse semigroup S, then the mapping $\phi_H : S \to \mathscr{I}(\mathscr{X})$ defined by* (8.6) *is an effective transitive representation of S.*

Proof. To show that ϕ_H is a representation, let $((Hx)\omega, (Hxst)\omega) \in (st)\phi_H$. Then $xx^{-1}, xstt^{-1}s^{-1}x^{-1} \in H$. Since H is closed and since $xss^{-1}x^{-1} \geqslant xstt^{-1}s^{-1}x^{-1} \in H$, we have that

$$((Hx)\omega, (Hxs)\omega) \in s\phi_H, \; ((Hxs)\omega, (Hxst)\omega) \in t\phi_H.$$

Thus $((Hx)\omega, (Hxst)\omega) \in (s\phi_H)(t\phi_H)$, and so we have shown that $(st)\phi_H \subseteq (s\phi_H)(t\phi_H)$. Conversely, if $((Hx)\omega, (Hy)\omega) \in (s\phi_H)(t\phi_H)$ then there exists $(Hz)\omega$ in $\mathscr{X}$ such that

$$((Hx)\omega, (Hz)\omega) \in s\phi_H, \qquad ((Hz)\omega, (Hy)\omega) \in t\phi_H.$$

Hence $(Hz)\omega = (Hxs)\omega$, $(Hy)\omega = (Hxst)\omega$, and so $((Hx)\omega, (Hy)\omega) \in (st)\phi_H$ as required. Thus ϕ_H is a representation.

To show that ϕ_H is effective and transitive, we establish that for any $(Hx)\omega$, $(Hy)\omega$ in $\mathscr{X}$ there exists s in S, namely $s = x^{-1}y$, such that

$$((Hx)\omega, (Hy)\omega) \in s\phi_H.$$

To see that this is so, first note that

$$(xs)(xs)^{-1} = xx^{-1}yy^{-1}xx^{-1} \in H,$$

since xx^{-1} and yy^{-1} are by assumption in H. Thus $(Hxs)\omega \in \mathscr{X}$. Also,

$$(Hxs)\omega = (Hxx^{-1}y)\omega \subseteq (Hy)\omega,$$

since $xx^{-1} \in H$. To show that the reverse inclusion also holds, note that $hy \geqslant hxx^{-1}y$ for all h in H. Hence $Hy \subseteq (Hxx^{-1}y)\omega$ and so (by (2.6) and (2.7))

$$(Hy)\omega \subseteq (Hxx^{-1}y)\omega = (Hxs)\omega. \blacksquare$$

We now show that *every* effective and transitive representation $\psi: S \to \mathscr{I}(X)$ of an inverse semigroup S is equivalent to one of the type ϕ_H.

Proposition 8.8. *Let X be a set and let $\psi: S \to \mathscr{I}(X)$ be an effective and transitive representation of the inverse semigroup S. Let z be an arbitrary fixed element of X and let*

$$H = \{s \in S: (z, z) \in s\psi\}.$$

Then H is a closed inverse subsemigroup of S, and ψ is equivalent to the representation ϕ_H defined by (8.6).

Proof. Certainly H is a subsemigroup, since $(z, z) \in s\psi$ and $(z, z) \in t\psi$ imply that $(z, z) \in (s\psi)(t\psi) = (st)\psi$. Also, since $s^{-1}\psi = (s\psi)^{-1}$ by formula (1.7), we have that

$$s \in H \Rightarrow (z, z) \in s^{-1}\psi \Rightarrow s^{-1} \in H;$$

thus H is an inverse subsemigroup of S. If $k \in H\omega$ then $k \geqslant h$ for some h in H. Since the order relation $\geqslant$ is defined in terms of multiplication we also have that $k\psi \geqslant h\psi$ in $\mathscr{I}(X)$. By Proposition 2.3 we thus have that $k\psi \supseteq h\psi$ (where $k\psi$ and $h\psi$ are now being considered as subsets of $X \times X$). Now $h \in H$ and so, by definition, $(z, z) \in h\psi$. It is now evident that $(z, z) \in k\psi$ and hence that $k \in H$. Thus $H\omega = H$ and so H is closed as required.

To show that ψ is equivalent to ϕ_H, we must begin by defining a bijection θ from X onto the set $\mathscr{X}$ of right ω-cosets of H. If $x \in X$ then since ψ is effective and transitive there exists a_x in S such that $(z, x) \in a_x\psi$. The element a_x of S then necessarily has the property that $a_x a_x^{-1} \in H$, for we have

$$(z, x) \in a_x\psi \quad \text{and} \quad (x, z) \in a_x^{-1}\psi,$$

and so $(z, z) \in (a_x a_x^{-1})\psi$. Thus $(Ha_x)\omega$ is a right ω-coset of H. In fact we may characterize $(Ha_x)\omega$ as follows:

$$(Ha_x)\omega = \{s \in S : (z, x) \in s\psi\}. \tag{8.9}$$

To see this, consider first an element of $(Ha_x)\omega$, i.e. an element s such that $s \geqslant ha_x$ for some h in H. It follows that $s\psi \geqslant (h\psi)(a_x\psi)$, i.e. (by Proposition 2.3) that $s\psi \supseteq (h\psi)(a_x\psi)$, where again we are regarding these elements of $\mathscr{I}(X)$ as subsets of $X \times X$. Now $(z, z) \in h\psi$ and $(z, x) \in a_x\psi$. Hence $(z, x) \in (h\psi)(a_x\psi) \subseteq s\psi$.

Conversely, if $(z, x) \in s\psi$ then it follows from the fact that $(x, z) \in a_x^{-1}\psi$ that $sa_x^{-1} \in H$. Hence $s \in (Ha_x)\omega$ by Proposition 8.5.

The element a_x is not uniquely determined by x. However, if b_x also has the property that $(z, x) \in b_x\psi$, then (8.9) assures us that $b_x \in (Ha_x)\omega$, and hence, by Proposition 8.5, we have

$$(Ha_x)\omega = (Hb_x)\omega.$$

It is, therefore, correct to say that the right ω-coset $(Ha_x)\omega$ is uniquely determined by x.

We now define a mapping $\theta: X \to \mathscr{X}$ by the rule that $x\theta = (Ha_x)\omega$, where a_x is any element of S such that $(z, x) \in a_x\psi$. By the remark in the last paragraph, θ is well-defined. It is, moreover, immediate that θ is one-one, for if $x\theta = x'\theta = (Ha)\omega$ then

$$(z, x) \in a\psi, \qquad (z, x') \in a\psi,$$

from which it immediately follows that $x = x'$, since $a\psi$ is a partial mapping of X. In fact θ is also onto, for if $(Ha)\omega$ is a right ω-coset of H, i.e. if $aa^{-1} \in H$, then

$$(z, z) \in (aa^{-1})\psi = (a\psi)(a^{-1}\psi)$$

and so there exists x in X such that $(z, x) \in a\psi$, $(x, z) \in a^{-1}\psi$. But then $x\theta = (Ha)\omega$.

Suppose finally that x, y in X and s in S are such that $(x, y) \in s\psi$; we must show that $(x\theta, y\theta) \in s\phi_H$. Suppose that $x\theta = (Ha)\omega$, $y\theta = (Hb)\omega$, i.e. that $(z, x) \in a\psi$, $(z, y) \in b\psi$. Then aa^{-1}, $bb^{-1} \in H$ and so $(Ha)\omega$, $(Hb)\omega \in \mathscr{X}$. Also $ass^{-1}a^{-1} \in H$, since $(z, z) \in (ass^{-1}a^{-1})\psi$, and $(z, y) \in (as)\psi$. Hence by (8.9) we have that $(Hb)\omega = (Has)\omega$ and so

$$(x\theta, y\theta) = ((Ha)\omega, (Has)\omega) \in s\phi_H.$$

Conversely, if $(x\theta, y\theta) = ((Ha)\omega, (Has)\omega) \in s\phi_H$, with $(z, x) \in a\psi$ and $(z, y) \in (as)\psi$, then

$$(x, y) \in (a^{-1}as)\psi \subseteq s\psi.$$

Thus $(x, y) \in s\psi$ if and only if $(x\theta, y\theta) \in s\phi_H$ and so the equivalence of ψ and ϕ_H is proved. ■

We summarize the results so far obtained as follows:

Theorem 8.10. *If S is an inverse semigroup then every effective representation of S is uniquely expressible as a sum of effective transitive representations ψ_i $(i \in I)$, each of which is equivalent to ϕ_{H_i} for some closed inverse subsemigroup H_i of S.* ■

We shall establish below (Theorem 8.13) one consequence of this "internalization" of representations that will be useful in Section VII.4. First, if $\psi : S \to \mathscr{I}(X)$ is a representation of an inverse semigroup S and if U is an inverse subsemigroup of S, then the restriction $\psi | U$ of ψ to U is a representation of the inverse semigroup U.

Next, if X is the disjoint union of sets X_i $(i \in I)$ and if for each i in I we have a representation $\psi_i: S \to \mathscr{I}(X_i)$, then

$$\left(\bigoplus_{i \in I} \psi_i\right) \Big| U = \bigoplus_{i \in I} (\psi_i | U); \tag{8.11}$$

for if we write ψ for the sum of the representations ψ_i we have, for each u in U,

$$u(\psi | U) = u\psi = \bigcup\{u\psi_i : i \in I\} \quad \text{(by (8.2))}$$

$$= \bigcup\{u(\psi_i | U) : i \in I\} = u\left[\bigoplus_{i \in I} (\psi_i | U)\right].$$

If a representation $\psi: S \to \mathscr{I}(X)$ is the sum of representations $\psi_1 : S \to \mathscr{I}(Y)$ and $\psi_2 : S \to \mathscr{I}(Z)$, where $Y \cup Z = X$, $Y \cap Z = \varnothing$, we call it *decomposable*. Very often we shall want to write $\psi_1 = \psi_Y$, $\psi_2 = \psi_Z$.

We shall shortly want to consider a situation of the following type. Let X be the disjoint union of Y and Z and let Y and Z be partitioned further as follows:

$$Y = \bigcup\{Y_i : i \in I\}, \quad Z = \bigcup\{Z_i : i \in I\}.$$

Let $\beta_i : S \to \mathscr{I}(Y_i)$ and $\gamma_i : S \to \mathscr{I}(Z_i)$ be representations of the inverse semigroup S (for each i in I). Let $\alpha_i = \beta_i \oplus \gamma_i : S \to \mathscr{I}(Y_i \cup Z_i)$ and let

$$\beta = \bigoplus_{i \in I} \beta_i : S \to \mathscr{I}(Y), \quad \gamma = \bigoplus_{i \in I} \gamma_i : S \to \mathscr{I}(Z).$$

Then by (8.3) we have

$$\beta \oplus \gamma = \bigoplus_{i \in I} \alpha_i : S \to \mathscr{I}(X).$$

Let us call this representation α. Then, using the notation introduced in the last paragraph, we have

$$\left.\begin{aligned} \bigoplus_{i\in I} [(\alpha_i)_{Y_i}] &= \left[\bigoplus_{i\in I} \alpha_i\right]_Y, \\ \bigoplus_{i\in I} [(\alpha_i)_{Z_i}] &= \left[\bigoplus_{i\in I} \alpha_i\right]_Z. \end{aligned}\right\} \tag{8.12}$$

These are a good deal more trivial than they appear: to show the first, simply note that

$$\bigoplus_{i\in I} [(\alpha_i)_{Y_i}] = \bigoplus_{i\in I} \beta_i = \beta = \alpha_Y = \left[\bigoplus_{i\in I} \alpha_i\right]_Y;$$

the second is proved similarly.

It is convenient now to state the result towards which we are aiming:

Theorem 8.13. *Let U be an inverse subsemigroup of an inverse semigroup S, and let $\phi: U \to \mathscr{I}(Y)$ be a representation of U. Then there exists a representation $\psi: S \to \mathscr{I}(Y \cup Z)$ (where $Y \cap Z = \varnothing$) with the property that*

$$\psi | U = (\psi | U)_Y \oplus (\psi | U)_Z$$

and $(\psi | U)_Y = \phi$.

Proof. Suppose for the moment that ϕ is effective. We shall see eventually that this is an unimportant restriction.

We begin by expressing ϕ as a sum of effective transitive representations $\phi_i: U \to \mathscr{I}(Y_i)$, where $Y = \bigcup\{Y_i: i \in I\}$ and the sets Y_i are pairwise disjoint. This can be done by virtue of Theorem 8.4. We shall then show that for each of these transitive representations ϕ_i we can find a representation $\psi_i: S \to \mathscr{I}(Y_i \cup Z_i)$, where $Y_i \cap Z_i = \varnothing$, with the property that

$$(\psi_i | U)_{Y_i} = \phi_i \quad \text{and} \quad \psi_i | U = (\psi_i | U)_{Y_i} \oplus (\psi_i | U)_{Z_i}.$$

We shall show moreover that this can be done in such a way that the sets $Y_i \cup Z_i$ $(i \in I)$ are pairwise disjoint. Having done this we then define Z as $\bigcup\{Z_i: i \in I\}$ and define $\psi: S \to \mathscr{I}(Y \cup Z)$ as the sum of the representations ψ_i $(i \in I)$. Then

$$\psi | U = \left(\bigoplus_{i\in I} \psi_i\right) \Big| U = \bigoplus_{i\in I} (\psi_i | U) \qquad \text{(by (8.11))}$$

$$= \bigoplus_{i\in I} [(\psi_i | U)_{Y_i} \oplus (\psi_i | U)_{Z_i}]$$

$$= \left[\bigoplus_{i\in I} (\psi_i | U)_{Y_i}\right] \oplus \left[\bigoplus_{i\in I} (\psi_i | U)_{Z_i}\right] \qquad \text{(by (8.3))}$$

$$= \left[\bigoplus_{i\in I} (\psi_i | U)\right]_Y \oplus \left[\bigoplus_{i\in I} (\psi_i | U)\right]_Z \qquad \text{(by (8.12))}$$

$$= \left[\left(\bigoplus_{i\in I} \psi_i\right) | U\right]_Y \oplus \left[\left(\bigoplus_{i\in I} \psi_i\right) | U\right]_Z \qquad \text{(by (8.11))}$$

$$= (\psi | U)_Y \oplus (\psi | U)_Z,$$

and

$$(\psi | U)_Y = \bigoplus_{i\in I} (\psi_i | U)_{Y_i} = \bigoplus_{i\in I} \phi_i = \phi.$$

It remains, therefore, to prove that the representation $\psi_i: S \to \mathscr{I}(Y_i \cup Z_i)$ can be found. To avoid unnecessary complications with suffices, we now simplify the notation and consider an effective, transitive representation $\phi: U \to \mathscr{I}(Y)$. From Proposition 8.7 we know that ϕ is equivalent to the representation $\phi_H: U \to \mathscr{I}(\mathscr{Y})$, where, for some arbitrarily chosen z in Y,

$$H = \{u \in U: (z, z) \in u\phi\}.$$

Then H is a closed inverse subsemigroup of U, the set $\mathscr{Y}$ is the set of right ω-cosets of H, and

$$u\phi_H = \{((Ha)\omega, (Hau)\omega): (Ha)\omega, (Hau)\omega \in \mathscr{Y}\}.$$

The inverse subsemigroup H of U is also an inverse subsemigroup of S, but it need not be closed, since it is perfectly possible for us to have $s \geqslant h$ with s in $S \backslash U$ and h in H. (See Exercise 14 below.) For clarity we shall therefore require to make a notational distinction between the closure $A\omega_U$ in U of a subset A of U and its closure $A\omega_S$ in S. Thus $K = H\omega_S$ is a closed inverse subsemigroup of S which will in general contain elements of $S \backslash U$ and so will not coincide with H. Let $\mathscr{W}$ be the set of right ω-cosets of K in S, and consider the effective transitive representation $\phi_K: S \to \mathscr{I}(\mathscr{W})$.

Now define a mapping $\varepsilon: \mathscr{Y} \to \mathscr{W}$ by

$$[(Ha)\omega_U]\varepsilon = (Ka)\omega_S \quad ((Ha)\omega_U \in \mathscr{Y}).$$

Then ε is well-defined, since if $(Ha)\omega_U, (Hb)\omega_U \in \mathscr{Y}$, then

$$\begin{aligned}(Ha)\omega_U = (Hb)\omega_U &\Rightarrow ab^{-1} \in H \quad \text{(by Proposition 8.5)}\\ &\Rightarrow ab^{-1} \in K\\ &\Rightarrow (Ka)\omega_S = (Kb)\omega_S \quad \text{(by Proposition 8.5).}\end{aligned}$$

Moreover, ε does map from $\mathscr{Y}$ into $\mathscr{W}$, since

$$(Ha)\omega_U \in \mathscr{Y} \Rightarrow aa^{-1} \in H \Rightarrow aa^{-1} \in K \Rightarrow (Ka)\omega_S \in \mathscr{W}.$$

Next, ε is one–one, since for any a, b in U for which $aa^{-1}, bb^{-1} \in H$,

$$\begin{aligned}(Ka)\omega_S = (Kb)\omega_S &\Rightarrow ab^{-1} \in K \cap U = H \\ &\Rightarrow (Ha)\omega_U = (Hb)\omega_U.\end{aligned}$$

Let $\mathscr{Z} = \mathscr{W} \backslash \mathscr{Y}\varepsilon$; thus $\mathscr{W}$ is the disjoint union of $\mathscr{Y}\varepsilon$ and $\mathscr{Z}$, where $\mathscr{Y}\varepsilon$ is a set in one-one correspondence with $\mathscr{Y}$. Consider now the restriction $\phi_K | U$. For any u in U,

$$u\phi_K = \{((Ks)\omega_S, (Ksu)\omega_S) \colon ss^{-1} \in K, suu^{-1}s^{-1} \in K\}.$$

Now $s \in U \Rightarrow su \in U$. Conversely, if $su \in U$ then certainly $v = suu^{-1} \in U$ and $vu = su$. Also

$$vv^{-1} = suu^{-1}uu^{-1}s^{-1} = suu^{-1}s^{-1} \in K$$

and

$$vs^{-1} \in K.$$

Hence $(Ks)\omega_S = (Kv)\omega_S$ by Proposition 8.5, and so

$$((Ks)\omega_S, (Ksu)\omega_S) = ((Kv)\omega_S, (Kvu)\omega_S).$$

We conclude that, for each u in U,

$$\begin{aligned}u\phi_K &= \{((Kv)\omega_S, (Kvu)\omega_S) \colon v \in U, vv^{-1} \in H, vuu^{-1}v^{-1} \in H\} \\ &\cup \{((Ks)\omega_S, (Ksu)\omega_S) \colon s \in S\backslash U, ss^{-1} \in K, suu^{-1}s^{-1} \in K\}.\end{aligned}$$

To express the same thing in the language of mappings, we have that the mapping $u\phi_K$ in $\mathscr{I}(\mathscr{W})$ maps elements of $\mathscr{Y}\varepsilon$ to elements of $\mathscr{Y}\varepsilon$ and elements of $\mathscr{Z}$ to elements of $\mathscr{Z}$. That is,

$$\phi_K | U = (\phi_K | U)_{\mathscr{Y}\varepsilon} \oplus (\phi_K | U)_{\mathscr{Z}}.$$

Now the representation $(\phi_K | U)_{\mathscr{Y}\varepsilon} \colon U \to \mathscr{I}(\mathscr{Y}\varepsilon)$ is equivalent to the representation $\phi_H \colon U \to \mathscr{I}(\mathscr{Y})$, since for all u, v, v' in U,

$$\begin{aligned}&([(Hv)\omega_U]\varepsilon, [(Hv')\omega_U]\varepsilon) \in u(\phi_K | U)_{\mathscr{Y}\varepsilon} \\ &\Leftrightarrow ((Kv)\omega_S, (Kv')\omega_S) \in u(\phi_K | U)_{\mathscr{Y}\varepsilon},\end{aligned}$$

$$\text{i.e.} \quad \Leftrightarrow (Kv')\omega_S = (Kvu)\psi_S, vv^{-1} \in K, vuu^{-1}v^{-1} \in K,$$

$$\text{i.e.} \quad \Leftrightarrow v'u^{-1}v^{-1}, vv^{-1}, vuu^{-1}v^{-1} \in K,$$

$$\text{i.e.} \quad \Leftrightarrow v'u^{-1}v^{-1}, vv^{-1}, vuu^{-1}v^{-1} \in H \quad (\text{since } U \cap K = H),$$

$$\text{i.e.} \quad \Leftrightarrow (Hv')\omega_U = (Hvu)\omega_U, vv^{-1} \in H, vuu^{-1}v^{-1} \in H,$$

$$\text{i.e.} \quad \Leftrightarrow ((Hv)\omega_U, (Hv')\omega_U) \in u\phi_H.$$

Since $(\phi_K|U)_{\mathscr{Y}\varepsilon}$ is equivalent to ϕ_H, it is thus equivalent to the original effective transitive representation $\phi: U \to \mathscr{I}(Y)$.

Recall now that the strategy of our proof was to extend each representation in a family $\phi_i: U \to \mathscr{I}(Y_i)$ $(i \in I)$ of effective representations of U to a representation $\psi_i: S \to \mathscr{I}(Y_i \cup Z_i)$ of S in such a way that the sets $Y_i \cup Z_i$ $(i \in I)$ are pairwise disjoint and

$$\psi_i|U = (\psi_i|U)_{Y_i} \oplus (\psi_i|U)_{Z_i}, \quad (\psi_i|U)_{Y_i} = \phi_i. \tag{8.14}$$

Let us now restore the suffix i and observe that what we have actually done is to find a representation $\phi_{K_i}: S \to \mathscr{I}(\mathscr{Y}_i\varepsilon \cup \mathscr{Z}_i)$ such that

$$\phi_{K_i}|U = (\phi_{K_i}|U)_{\mathscr{Y}_i\varepsilon} \oplus (\phi_{K_i}|U)_{\mathscr{Z}},$$

and such that $(\phi_{K_i}|U)_{\mathscr{Y}_i\varepsilon}$ is *equivalent* to ϕ_i (and to ϕ_{H_i}). This is in fact good enough, for we may now for each i in I choose a set Z_i in one–one correspondence with $\mathscr{Z}_i$ and in such a way as to ensure that the sets $Y_i \cup Z_i$ $(i \in I)$ are pairwise disjoint. If we denote by θ the bijection from Y_i onto $\mathscr{Y}_i$ that gives effect to the equivalence between the representations ϕ_i and ϕ_{H_i}, and by χ the bijection from Z_i onto $\mathscr{Z}_i$, we now have a bijection $\lambda: Y_i \cup Z_i \to \mathscr{Y}_i\varepsilon \cup \mathscr{Z}_i$ defined by

$$y\lambda = y\theta\varepsilon \ (y \in Y_i), \qquad z\lambda = z\chi \ (z \in Z_i).$$

The representation $\psi_i: S \to \mathscr{I}(Y_i \cup Z_i)$ defined for each s in S by

$$s\psi_i = \{(w, w') \in (Y_i \cup Z_i)^2 : (w\lambda, w'\lambda) \in \phi_{K_i}\}$$

is then equivalent to ϕ_{K_i} and has precisely the desired properties described in (8.14).

The supposition at the beginning of the proof was that ϕ is effective. If ϕ is not effective then there is a subset Y' of Y with the property that $U\phi \subseteq \mathscr{I}(Y')$ and $\phi: U \to \mathscr{I}(Y')$ is effective. We then carry out our extension routine on this restricted version of ϕ, obtaining a representation $\psi: S \to \mathscr{I}(Y' \cup Z)$, and finally we restore the otiose elements of $Y \backslash Y'$, obtaining a representation $\psi: S \to \mathscr{I}(Y \cup Z)$ with the desired properties. ∎

EXERCISES

1. Show that the $\mathscr{D}$-classes in an inverse semigroup are "square". More precisely, show that there is a bijection from the set of $\mathscr{L}$-classes in a $\mathscr{D}$-class onto the set of $\mathscr{R}$-classes in the same $\mathscr{D}$-class, defined by the rule that L_a maps to R_b if and only if $L_a \cap R_b$ contains an idempotent.

2. In the symmetric inverse semigroup $\mathscr{I}(X)$, show that
(i) $\alpha \mathscr{L} \beta$ if and only if $\mathrm{ran}(\alpha) = \mathrm{ran}(\beta)$;
(ii) $\alpha \mathscr{R} \beta$ if and only if $\mathrm{dom}(\alpha) = \mathrm{dom}(\beta)$;
(iii) $\alpha \mathscr{D} \beta$ if and only if $|\mathrm{dom}(\alpha)| = |\mathrm{dom}(\beta)|$;
(iv) $\mathscr{D} = \mathscr{J}$.

3. If $|X| = n$, show that

$$|\mathscr{I}(X)| = \sum_{r=0}^{n} \binom{n}{r}^2 r!.$$

4. If K is a subset of an inverse semigroup S, show that $(Ks)\omega = ((K\omega)s)\omega$.

5. Let S be an inverse semigroup with semilattice of idempotents E, and let σ be the minimum group congruence on S. Show that the following statements are equivalent:

(i) $(x, y) \in \sigma$; (ii) $(\exists e \in E) xe = ye$;
(iii) $(\exists a \in S) ax = ay$; (iv) $(\exists a \in S) xa = ya$;
(v) $(\exists e, f \in S) ex = fy$; (vi) $(\exists e, f \in S) xe = yf$;
(vii) $xy^{-1} \in E\omega$; (viii) $x^{-1}y \in E\omega$;
(ix) $(Ex)\omega = (Ey)\omega$; (x) $y \in (Ex)\omega$;
(xi) $x \in (Ey)\omega$.

6. (For definitions, see Exercise II.12). Let U be a subsemigroup of an inverse semigroup S.
(i) Show that if U is full and left unitary then U is an inverse subsemigroup of S.
(ii) Show that if U is a left unitary inverse subsemigroup then U is unitary.
(iii) Show that an inverse subsemigroup U is unitary if and only if it is closed.

7. Let S be an inverse semigroup with semilattice of idempotents E. Define an inverse subsemigroup N of S to be *normal* if it is full and closed and if $x^{-1}Nx \subseteq N$ for all x in S.
(i) Show that $E\omega$ is a normal inverse subsemigroup of S.
(ii) Show that an inverse subsemigroup N of S is normal if and only if

$$(Nx)\omega = (xN)\omega$$

for every x in S.
(iii) If N is a normal inverse subsemigroup of S, show that the relation

$$\rho_N = \{(x, y) \in S \times S : xy^{-1} \in N\}$$

is a group congruence on S. Show that the identity of the group S/ρ_N is the ρ_N-class N.
(iv) If τ is a group congruence on S, show that there exists a normal inverse subsemigroup N of S such that $\tau = \rho_N$.

8. Following Munn (1975), we shall call an inverse semigroup *E-unitary* if its semilattice of idempotents is a unitary subsemigroup. Saitô (1965) and McAlister (1973, 1974) have called such inverse semigroups *proper*, and O'Carroll (1974) has

called them *reduced*. Show that the following statements about an inverse semigroup S (with semilattice of idempotents E) are equivalent:

(i) S is E-unitary;
(ii) E is a left unitary subsemigroup of S;
(iii) $E\omega = E$;
(iv) E is a σ-class.

9. (Reilly 1971). Show that an inverse semigroup S is E-unitary if and only if $\sigma \cap \mathscr{R} = 1_S$. [Hint: to show the direct half, observe first that $(x, y) \in \sigma \cap \mathscr{R}$ if and only if $xx^{-1} = yy^{-1}$ and $x^{-1}y \in E$; to show the converse half, notice that $x \in E\omega \Rightarrow x \, \sigma \, xx^{-1}$.]

10. (Part of this has already appeared as Exercise IV.4). The Clifford semigroup $\mathscr{S}(Y; G_\alpha; \phi_{\alpha,\beta})$ is an inverse semigroup with semilattice of idempotents $E = \{e_\alpha : \alpha \in Y\}$ isomorphic to Y.

(i) Show that if $x \in G_\alpha$ and $y \in G_\beta$ then $x \geqslant y$ if and only if $\alpha \geqslant \beta$ and $y = x\phi_{\alpha,\beta}$.
(ii) Show that

$$E\omega = \bigcup \{\ker \phi_{\alpha,\beta} : \alpha, \beta \in Y, \alpha \geqslant \beta\}.$$

(Here we are using 'ker' in the group-theoretic sense: thus $\ker \phi_{\alpha,\beta}$ is the normal subgroup $\{e_\beta\}\phi_{\alpha,\beta}^{-1}$ of G_α.)

(iii) Show that S is E-unitary if and only if the homomorphisms $\phi_{\alpha,\beta}(\alpha, \beta \in Y, \alpha \geqslant \beta)$ are all one–one.

11. If T is an inverse semigroup and θ is a homomorphism from T into the group of units H_1 of T, show that the Bruck-Reilly extension $BR(T, \theta)$ is E-unitary if and only if T is E-unitary and $\ker \theta = \sigma$ (the minimum group congruence on T).

Deduce that a bisimple inverse ω-semigroup $BR(T, \theta)$ is E-unitary if and only if θ is one–one.

12. Let $(E, .)$ be a semilattice and let Γ be a group of (left) automorphisms of E. Define a binary operation on $S = E \times \Gamma$ by

$$(e, \alpha)(f, \beta) = (e(\alpha f), \alpha\beta).$$

(i) Show that S is a semigroup relative to the given operation.
(ii) Show that $(\alpha^{-1}e, \alpha^{-1})$ is an inverse of (e, α) in S, and deduce that S is regular.
(iii) Show that (e, α) is idempotent if and only if $\alpha = 1_E$. Deduce that idempotents commute and hence that S is an inverse semigroup.
(iv) Show that $(e, \alpha) \, \sigma \, (f, \beta)$ (where σ is the minimum group congruence on S) if and only if $e = f$. Deduce that $S/\sigma \simeq \Gamma$.
(v) Show that S is E-unitary.

McAlister (1973, 1974) defines a "P-semigroup" as arising from a generalized version of this construction. He then shows that *every* E-unitary inverse semigroup is isomorphic to some P-semigroup. See also (Munn 1975).

13. Let U be a closed inverse subsemigroup of T, where T is a closed inverse subsemigroup of the inverse semigroup S. Show that U is a closed inverse subsemigroup of S.

14. Let S be the Clifford semigroup consisting of the unions of the two disjoint groups U and V with identities e and f respectively, where $e < f$, and with connecting homomorphism $\phi : V \to U$. If H is a subgroup of U, show that H is a closed inverse subsemigroup of U, but that the closure of H in S is $H \cup H\phi^{-1}$.

15. (Howie 1964). If S is an inverse semigroup and σ is the minimum group congruence on S, show that, for every congruence ξ on S,

$$\xi \vee \sigma = \sigma \circ \xi \circ \sigma.$$

[Hint. The essential point is to prove that the relation $\sigma \circ \xi \circ \sigma$ is transitive. If

$$ea = ep, \quad p \,\xi\, q, \quad \text{and} \quad fq = fb$$

and

$$gb = gr, \quad r \,\xi\, s \quad \text{and} \quad hs = hc,$$

where e, f, g, h are idempotent, show that

$$ia = it, \qquad t \,\xi\, u \qquad \text{and} \quad ju = jc,$$

where $i = efg$, $t = fgp$, $u = fgs$ and $j = hfg$.]

16. (Howie 1964). If H is a subset of a semigroup S, define $H\zeta$, the *centralizer* of H in S, by

$$H\zeta = \{s \in S : (\forall h \in H) \quad hs = sh\}.$$

If S is an inverse semigroup with semilattice of idempotents E and if μ is the maximum idempotent-separating congruence on S, show that

$$\mu = \{(a, b) \in S \times S: \quad (a, b) \in \mathscr{H} \quad \text{and} \quad ab^{-1} \in E\zeta\}.$$

17. Show that an inverse semigroup S is a Clifford semigroup if and only if $S/\mu \simeq E_S$.

18. Let α and β be $\mathscr{H}$-equivalent elements of the symmetric inverse semigroup $\mathscr{I}(X)$. (Thus $\mathrm{dom}(\alpha) = \mathrm{dom}(\beta) = A$ (say), and $\mathrm{ran}(\alpha) = \mathrm{ran}(\beta) = B$.) Show that if 1_C is an idempotent in $\mathscr{I}(X)$ then

$$\alpha^{-1} 1_C \alpha = 1_{(A \cap C)\alpha}, \qquad \beta^{-1} 1_C \beta = 1_{(A \cap C)\beta}.$$

Deduce that $\mathscr{I}(X)$ is fundamental. Deduce further that every inverse semigroup is embeddable in a fundamental inverse semigroup.

19. If X is a set then the set $\mathscr{E} = \mathscr{P}(X)$ of all subsets of X is a semilattice with respect to the operation $\cap$. Show that $T_{\mathscr{E}} \simeq \mathscr{I}(X)$. (Notice that from this it follows that $\mathscr{I}(X)$ is fundamental, as stated in Exercise 18.)

20. Let E be the semilattice $C_\omega \times C_\omega$. Then E may be identified with $\{e_{m,n}: m, n \in N\}$, where

$$e_{m,n}e_{p,q} = e_{\max(m,p),\,\max(n,q)}.$$

(i) Show that the group of automorphisms of E is $\{1_E, \gamma\}$, where

$$e_{m,n}\gamma = e_{n,m} \qquad (m, n \in N).$$

(ii) Show that E is uniform and that if $e = e_{m,n}$ and $f = e_{p,q}$, then $T_{e,f} = \{\alpha, \beta\}$, where

$$e_{r,s}\alpha = e_{r-m+p,\,s-n+q} \qquad (r \geqslant m, s \geqslant n),$$

$$e_{r,s}\beta = e_{s-n+p,\,r-m+q} \qquad (r \geqslant m, s \geqslant n).$$

(iii) Deduce that $\mu \neq \mathscr{H}$ in T_E.

21. (Bruck 1958). Show that any semigroup S can be embedded up to isomorphism in a simple semigroup with identity. [Hint: consider $BR(S^1,\theta)$, where θ maps every element of S^1 to 1.]

Preston (1959) showed how to improve "simple" to "bisimple" in this result. Reilly (1965) showed how to embed an arbitrary inverse semigroup in a bisimple inverse semigroup with identity.

22. By Proposition II.3.1, a $\mathscr{D}$-class cannot contain both regular and irregular elements. Show that this does not apply to $\mathscr{J}$-classes. Specifically, let $T = \{1, x, 0\}$, with $x^2 = 0$, let $\theta: T \to H_1$ be given by $1\theta = x\theta = 0\theta = 1$, and let $S = BR(T, \theta)$.

(i) Show that the $\mathscr{D}$-classes of S are

$$D^1 = \{(m, 1, n): m, n \in N\}, \qquad D^0 = \{(m, 0, n): m, n \in N\},$$

$$D^x = \{(m, x, n): m, n \in N\}.$$

(ii) Show that D^1 and D^0 are regular $\mathscr{D}$-classes, but that D^x is irregular.

(iii) Deduce that the single $\mathscr{J}$-class of the simple semigroup S contains both regular and irregular elements.

23. (Munn 1970c). Let A be the three-element semilattice $\{a, b, z\}$, in which z is the zero element and $ab = z$. Let $E = C_\omega \times A$, with the lexicographic ordering: $(e_m, x) \leqslant (e_n, y)$ if and only if either $m > n$, or $m = n$ and $x \leqslant y$. Show that E (the "lazy tongs" semilattice) is subuniform but not uniform.

24. (Munn 1970c). More generally, if A is an arbitrary semilattice and U is a uniform semilattice, let E be the *ordinal product* $U \circ A$ (in the sense of Birkhoff (1948)); i.e. let E be the set $U \times A$ with the lexicographic order given by the rule that

$(u, a) \leqslant (v, b)$ if and only if either $u < v$, or $u = v$ and $a \leqslant b$. Show that E is subuniform. (It need not be uniform—see the previous exercise.)

25. Let T be the Clifford semigroup consisting of the chain of groups $G_0, \ldots, G_{d-1}$, with identity elements $e_0 > e_1 > \ldots > e_{d-1}$ and with structure homomorphisms $\gamma_i : G_i \to G_{i+1}$ $(i = 0, \ldots, d-2)$. If θ is an endomorphism of T with $S\theta \subseteq G_0$ (as in a Bruck-Reilly extension) and if $\theta | G_{d-1} = \delta : G_{d-1} \to G_0$, show that

$$\theta | G_i = \gamma_i \ldots \gamma_{d-2}\delta \qquad (i = 0, \ldots, d-2).$$

Conversely, if $\delta : G_{d-1} \to G_0$ is a homomorphism, show that $\theta : S \to G_0$ defined by

$$\theta | G_i = \gamma_i \ldots \gamma_{d-2}\delta \qquad (i = 0, \ldots, d-2)$$

$$\theta | G_{d-1} = \delta$$

is an endomorphism of S such that $S\theta \subseteq G_0$. (This enables one to recover Munn's (1968) original version of Theorem 7.8.)

26. Let $S = \langle a \rangle$ be an infinite monogenic semigroup. Show that the relation

$$\rho_n = \{(a^p, a^q) : p \equiv q (\text{mod } n)\}$$

is a group congruence on S. Show conversely that if ρ is a group congruence on S then there exists n $(= \min \{|r - s| : (a^r, a^s) \in \rho \backslash 1_S\})$ such that $\rho = \rho_n$.

Deduce that there is no minimum group congruence on S.

Chapter VI

Orthodox Semigroups

INTRODUCTION

An *orthodox* semigroup is defined as a regular semigroup in which the idempotents form a subsemigroup. The class of orthodox semigroups thus includes both the class of inverse semigroups and the class of bands. Various other special cases have been known for some time: Fantham (1960), Yamada (1963a, b, c; 1965; 1971) and Petrich (1965) have studied the case where the semigroups is also a union of groups, and a recent review article by Clifford (1972b) gives an extremely useful account of the various approaches. Specializing in another direction, Yamada (1967) studied the case where the band of idempotents of the semigroup is *normal* (see Section IV.5). More recently, the structure of orthodox semigroups in general has been clarified (independently) by Yamada (1970) and T. E. Hall (1969, 1970, 1971). Their results have a good deal in common, but since Hall's treatment achieves on the way an interesting generalization of the Munn semigroup T_E encountered in the last chapter, the account given below more closely resembles his than it does Yamada's.

More recently, Hall (1973) has generalized the Munn semigroup still further to the case of a general regular semigroup. This, however, is beyond the scope of the present book. Developments of Hall's ideas are to be found in papers by Grillet (1974a, b, c, d) and Nambooripad (1973; 1975a, b). See also the articles by Clifford (1974, 1975).

For further work on regular semigroups see the review article by Lallement (1972), where brief reference is made in particular to the work of Allen (1971), who approaches the structure problem in a quite different way *via* Rees matrix semigroups.

For a study of the structure of bisimple orthodox semigroups based on the ideas in Reilly (1968) and Reilly and Clifford (1968), see Clifford (1972a).

1. BASIC PROPERTIES OF ORTHODOX SEMIGROUPS

The crucial observation that enables the study of orthodox semigroups to get under way is contained in the following theorem, due to Reilly and Scheiblich (1967):

Theorem 1.1. *If S is a regular semigroup, then the following statements are equivalent:*

(A) *S is orthodox;*

(B) *for any a, b in S, if a' is an inverse of a and b' is an inverse of b, then $b'a'$ is an inverse of ab;*

(C) *if e is idempotent then every inverse of e is idempotent.*

Proof. To show that (A) $\Rightarrow$ (B), suppose that S is orthodox (i.e. that the product of any two idempotents is idempotent) and let $a, b \in S$. Then if a', b' are inverses respectively of a and b, we have that $a'a$ and bb' are idempotent; hence $a'abb'$ and $bb'a'a$ are idempotent by the orthodox property. It now follows that

$$abb'a'ab = aa'abb'a'abb'b = a(a'abb')^2b = aa'abb'b = ab,$$

and that

$$b'a'abb'a' = b'bb'a'abb'a'aa' = b'(bb'a'a)^2a' = b'bb'a'aa' = b'a';$$

that is, $b'a'$ is an inverse of ab.

To show that (B) $\Rightarrow$ (C), let e be idempotent and let x be an inverse of e:

$$xex = x, exe = e.$$

Now xe and ex are both idempotent and so each is an inverse of itself. By property (B) it follows that $(ex)(xe)$ is an inverse of $(xe)(ex)$, i.e. that ex^2e is an inverse of $xe^2x = xex = x$. Hence

$$x = x(ex^2e)x = (xex)(xex) = (xex)^2 = x^2$$

and so x is idempotent.

To show that (C) $\Rightarrow$ (A), let e, f be idempotents, and let x be an inverse of ef. Then

$$(ef)(fxe)(ef) = efxef = ef,$$

and

$$(fxe)(ef)(fxe) = f(xefx)e = fxe,$$

and so ef is an inverse of fxe. But

$$(fxe)^2 = f(xefx)e = fxe.$$

Thus fxe is idempotent and hence ef is idempotent by property (C). ∎

The property (B), which is a generalization of the property

$$(ab)^{-1} = b^{-1}a^{-1}$$

possessed by inverse semigroups, enables us to reproduce in suitably modified form for orthodox semigroups some of the features of the theory of inverse semigroups. The notation $V(a)$ introduced in Section II.4 for the set of inverses of an element a of S is useful here. Using this notation we may rewrite property (B) of an orthodox semigroup as

$$(\forall a, b \in S) \qquad V(b)V(a) \subseteq V(ab). \tag{1.3}$$

A further analogy with inverse semigroups is provided by

Proposition 1.4. *If S is an orthodox semigroup, if e is idempotent and if $a \in S$ then, for every inverse a' of a, the elements $a'ea$ and aea' are both idempotent.*

Proof. To prove that $a'ea$ is idempotent, simply note that

$$(a'ea)^2 = a'eaa'ea = a'eaa'eaa'a = a'(eaa')^2a = a'eaa'a = a'ea;$$

the proof that aea' is idempotent is very similar. ■

Let S be an orthodox semigroup and let us denote the band of idempotents of S by B. The band B is by Theorem IV.3.1 a semilattice Y of rectangular bands E_α ($\alpha \in Y$), where $E_\alpha \cap E_\beta = \varnothing$ if $\alpha \neq \beta$ and where

$$E_\alpha E_\beta \subseteq E_{\alpha\beta} \qquad (\alpha, \beta \in Y). \tag{1.5}$$

If $e \in E_\alpha$ we shall sometimes write $E_\alpha = E(e)$. In fact $E(e)$ is, as we saw in Section IV.3, the $\mathscr{J}^B$-class containing e, but because of the potential confusion between the equivalence $\mathscr{J}^B$ on B and the equivalence $\mathscr{J}^S$ on S it is useful to have the new notation $E(e)$. An alternative, somewhat cumbersome notation would be J_e^B.

Notice that the formula (1.5) "translates" to

$$E(e)E(f) \subseteq E(ef) = E(fe) \qquad (e, f \in B). \tag{1.6}$$

For simplicity we shall denote $\mathscr{J}^B$ by ε. It is the minimum semilattice congruence on B, and the ε-classes are the subsets $E(e)$.

Property (C) of orthodox semigroups assures us that $V(e) \subseteq B$ for every e in B. Now, if $f \in V(e)$ we have

$$e = efe, \qquad f = fef$$

and so certainly $f \in J_e^B = E(e)$. Conversely, if $f \in E(e)$ then f is an inverse of e, since $E(e)$ is a rectangular band and *any* two elements of a rectangular band are mutually inverse. Hence, for every e in B,

$$V(e) = E(e). \tag{1.7}$$

Thus $V(e)$ is determined solely by the nature of the band B.

We record here for future use that for any two elements e, f in a band B,

$$E(efe) = E(fef); \tag{1.8}$$

for the elements efe and fef can be easily seen to be mutually inverse.

If a is a general element of S, we cannot of course expect $V(a)$ to be solely determined by properties of B. The properties of B are, however, highly influential in determining $V(a)$, in the sense that if we know a single inverse a' of a then $V(a)$ is wholly determined by a' and by B. Precisely, we have

Proposition 1.9. *Let S be an orthodox semigroup. If $a \in S$ and $a' \in V(a)$, then*

$$V(a) = E(a'a)a'E(aa').$$

Proof. If $e \in E(a'a)$ and $f \in E(aa')$ then

$$a'aea'a = a'a, \qquad aa'faa' = aa'.$$

Hence

$$a(ea'f)a = aa'a(ea'aa'aa'f)aa'a = a(a'aea'a)a'(aa'faa')a = aa'aa'aa'a = a,$$

and

$$(ea'f)a(ea'f) = (ea'aa'f)aa'aa'a(ea'aa'f)$$
$$= ea'(aa'faa')a(a'aea'a)a'f = ea'aa'aa'aa'f = ea'f.$$

Thus $E(a'a)a'E(aa') \subseteq V(a)$.

Conversely, if $a^* \in V(a)$, then

$$a^* = a^*aa^* = a^*aa'aa^*.$$

Now

$$(a^*a)(a'a)(a^*a) = a^*(aa'a)a^*a = a^*aa^*a = a^*a,$$

and

$$(a'a)(a^*a)(a'a) = a'(aa^*a)a'a = a'aa'a = a'a,$$

and so $a^*a \in E(a'a)$. Similarly $aa^* \in E(aa')$, and so $a^* \in E(a'a)a'E(aa')$ as required. ∎

We can now supplement Theorem 1.1 with a further characterization of orthodox semigroups within the class of regular semigroups:

Theorem 1.10. *A regular semigroup S is orthodox if and only if*

$$(\forall a, b \in S) \qquad [V(a) \cap V(b) \neq \varnothing \Rightarrow V(a) = V(b)].$$

Proof. First, if S is orthodox and if $V(a) \cap V(b) \neq \varnothing$, let $x \in V(a) \cap V(b)$. Then a and b both belong to $V(x)$ and so

$$E(xa) = E(xb), E(ax) = E(bx).$$

(Indeed we even have the stronger statements that

$$R_{xa} = R_{xb}(= R_x), \qquad L_{ax} = L_{bx}(= L_x).)$$

Hence, by Proposition 1.9,

$$V(a) = E(xa)xE(ax) = E(xb)xE(bx) = V(b).$$

Conversely, suppose that S is regular and that the implication given in the statement of the theorem holds in S. Let e, f be idempotents of S and let x be an inverse of ef:

$$efxef = ef, \qquad xefx = x.$$

Then it is easy to verify that fxe and $efxe$ are idempotents and that

$$fxe \in V(fxe) \cap V(efxe).$$

By hypothesis it now follows that $V(fxe) = V(efxe)$. Now again it is straightforward to verify that $ef \in V(fxe)$; hence $ef \in V(efxe)$ and so

$$ef = (ef)(efxe)(ef) = ef(efxef) = (ef)^2.$$

Thus S is orthodox as required. ■

The equivalence relation

$$\gamma = \{(x, y) \in S \times S : V(x) = V(y)\} \tag{1.11}$$

was considered by Yamada (1967) for the case where S is a "generalized inverse" semigroup (i.e. an orthodox semigroup in which the band of idempotents is normal). It was considered in the general orthodox case by T. E. Hall (1969), who obtained the following result:

Theorem 1.12. *If S is an orthodox semigroup then the relation γ defined by* (1.11) *is a congruence on S. It is the smallest inverse semigroup congruence on S.*

Proof. If $(a, b) \in \gamma$ and $c \in S$ then for any x in $V(a)$ $(= V(b))$ and for any c' in $V(c)$ it follows from (1.3) that $xc' \in V(ca) \cap V(cb)$. Hence $V(ca) = V(cb)$ by Theorem 1.10 and so $(ca, cb) \in \gamma$. Similarly $(ac, bc) \in \gamma$, and thus γ is a congruence.

To show that S/γ is an inverse semigroup, notice first that it is certainly regular, since *any* homomorphic image of a regular semigroup is regular. Now, by Lallement's Lemma (II.4.6) any idempotent of S/γ is of the form $e\gamma$, where e is an idempotent of S. For any two idempotents e, f in S, we have

$$\begin{aligned} V(ef) &= E(ef) && \text{(by (1.7))} \\ &= E(fe) && \text{(by (1.6))} \\ &= V(fe) \end{aligned}$$

and so $(ef, fe) \in \gamma$. Thus $(e\gamma)(f\gamma) = (f\gamma)(e\gamma)$ in S/γ, and so S/γ is an inverse semigroup.

Finally, suppose that ρ is a congruence on S such that S/ρ is an inverse semigroup. If $(a, b) \in \gamma$ then $V(a) = V(b)$. Hence for any x in $V(a)$ $(= V(b))$ it is the case that $a\rho$ and $b\rho$ are both inverses in S/ρ of $x\rho$. By uniqueness of inverses in S/ρ we conclude that $a\rho = b\rho$, i.e. that $(a, b) \in \rho$. Thus $\gamma \subseteq \rho$ and so γ is as stated, the *smallest* inverse semigroup congruence on S. ■

It is worth remarking that for each a in the orthodox semigroup S,

$$\begin{aligned} a\gamma &= \{b : V(b) = V(a)\} \\ &= V(a'), \quad \text{where} \quad a' \in V(a) \\ &= V(V(a)) = E(aa')aE(a'a). \end{aligned} \tag{1.13}$$

Two further observations about γ are worth recording at this stage. The first is that

$$\gamma \cap (B \times B) = \varepsilon\,(= \mathscr{J}^B), \tag{1.14}$$

which follows immediately from (1.7). Another way of expressing this fact is S and where $\chi: B/\varepsilon \to S/\gamma$ is a monomorphism defined by

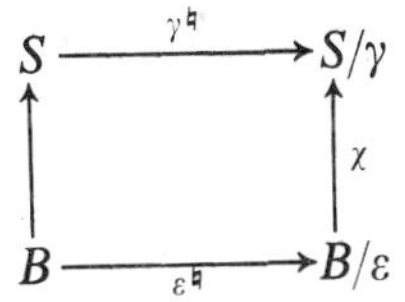

(1.15)

in which the unlabelled arrow denotes the obvious inclusion map from B into S and where $\chi: B/\varepsilon \to S/\gamma$ is a monomorphism defined by

$$(b\varepsilon)\chi = b\gamma.$$

In words, we have that the semilattice of idempotents of the maximum inverse semigroup homomorphic image of an orthodox semigroup S is isomorphic to the maximum semilattice homomorphic image of the band of idempotents of S.

The second observation is that

$$\gamma \cap \mathscr{H} = 1_S. \tag{1.16}$$

To see this, suppose that $(a, b) \in \mathscr{H}$ and that a' is any element of $V(a)$. If we also have $(a, b) \in \gamma$ then $V(a) = V(b)$ and so $a' \in V(b)$. Hence the element a' has two inverses, namely a and b, in the same $\mathscr{H}$-class. It follows by Theorem II.3.5(iii) that $a = b$.

We have seen (Proposition II.4.8) that in a regular semigroup the maximum idempotent-separating congruence μ coincides with $\mathscr{H}^\flat$, the largest congruence contained in $\mathscr{H}$. In the case of inverse semigroups a useful characterization of μ is given by Theorem V.3.2. An analogous characterization in the orthodox case has been given by Meakin (1961) as follows. (See T. E. Hall (1973) for a generalization to regular semigroups.)

Theorem 1.17. *If S is an orthodox semigroup with band of idempotents B and if $\mu = \mathscr{H}^{\flat}$ is the maximum idempotent-separating congruence on S, then $(a, b) \in \mu$ if and only if*

$$(\exists a' \in V(a))(\exists b' \in V(b))(\forall x \in B)[a'xa = b'xb \quad \text{and} \quad axa' = bxb'].$$

Proof. Clearly the given relation—let us temporarily call it ν—is reflexive and symmetric. To show that it is transitive, suppose that $(a, b) \in \nu$ and $(b, c) \in \nu$, so that there exist $a' \in V(a)$, $b', b^* \in V(b)$ and $c^* \in V(c)$ such that, for every x in B,

$$a'xa = b'xb, \qquad b^*xb = c^*xc,$$

$$axa' = bxb', \qquad bxb^* = cxc^*.$$

Now $\bar{a} = b^*ba'bb' \in V(a)$, for if we first observe that

$$bb'a = bb'.a(a'a)a'.a = bb'.b(a'a)b'.a$$

$$= b(a'a)b'.a = a(a'a)a'.a = a,$$

and that

$$a(b^*b)a' = b(b^*b)b' = bb',$$

it follows that

$$a\bar{a}a = a(b^*b)a'.bb'a = bb'a = a,$$

and that

$$\bar{a}a\bar{a} = b^*ba'bb'.a(b^*b)a'.bb' = b^*ba'(bb')^3 = \bar{a}.$$

A similar argument establishes that $\bar{c} = b^*bc^*bb'$ is an inverse of c. Now, for all x in B,

$$\bar{a}xa = b^*b.a'(bb'x)a = b^*b.b'(bb'x)b = b^*bb'xb$$

$$= b^*b.b^*(bb'x)b = b^*b.c^*(bb'x)- = \bar{c}xc,$$

and

$$ax\bar{a} = a(xb^*b)a'.bb' = b(xb^*b)b'.bb' = bxb^*bb'$$

$$= b(xb^*b)b^*.bb' = c(xb^*b)c^*.bb' = cx\bar{c};$$

thus $(a, c) \in \nu$ as required.

If $(a, b) \in \nu$ and $c \in S$, choose c' in $V(c)$. Then, for every x in B,

$$c'(a'xa)c = c'(b'xb)c.$$

Also, since $cxc' \in B$, we have

$$a(cxc')a' = b(cxc')b'$$

for every x in B. Since $c'a' \in V(ac)$ and $c'b' \in V(bc)$ we have thus established that $(ac, bc) \in \nu$. Similarly $(ca, cb) \in \nu$ and so ν is a congruence.

If $(e, f) \in \nu \cap (B \times B)$ then in particular, for some e' in $V(e)$ and f' in $V(f)$,

$$f'ef = e'ee = e'e;$$

hence

$$ef = ee'ef = ef'eff = ef'ef = ee'e = e.$$

Equally,

$$efe' = fff' = ff',$$

and so

$$ef = eff'f = eefe'f = efe'f = ff'f = f.$$

Thus $e = f$ and so ν is idempotent-separating.

Finally, suppose that ρ is an idempotent-separating congruence on S. If $a \,\rho\, b$ then $a \,\mathscr{H}\, b$ and so by Theorem II.4.1 there exist a' in $V(a)$ and b' in $V(b)$ such that $aa' = bb'$ and $a'a = b'b$. Since

$$a' = a'aa' = a'bb' \quad \text{and} \quad a'ab' = b'bb' = b',$$

and since $(a'bb', a'ab') \in \rho$, it follows that $(a', b') \in \rho$. Hence, since ρ is a congruence, we may deduce that, for every x in B,

$$(a'xa, b'xb) \in \rho, \qquad (axa', bxb') \in \rho.$$

But ρ is idempotent-separating, and so

$$a'xa = b'xb, \qquad axa' = bxb'$$

for every x in B. Thus $(a, b) \in \nu$ and so $\rho \subseteq \nu$ as required. ∎

2. THE ANALOGUE OF THE MUNN SEMIGROUP

In the case of an inverse semigroup S with semilattice of idempotents E we were able to find a homomorphism ϕ from S into the symmetric inverse semigroup $\mathscr{I}(E)$ associating the elements a of S with the mappings $\alpha_a : Eaa^{-1} \to Ea^{-1}a$ in $\mathscr{I}(E)$ sending $e (\in Eaa^{-1})$ to $a^{-1}ea$. If we are to find a generalization of this to orthodox semigroups we must begin by finding the appropriate generalization of $\mathscr{I}(E)$.

The Vagner–Preston representation of an inverse semigroup S associates each a in S with a partial one–one mapping $\alpha_a : Saa^{-1} \to Sa^{-1}a$ given by

$$x\alpha_a = xa.$$

The correct generalization of this is not obvious until we remark that, for any x in S,

$$x \in Saa^{-1} \quad \text{if and only if} \quad (xa, x) \in \mathscr{R}.$$

To see that this is so we use the characterization of $\mathscr{R}$ in an inverse semigroup given by Proposition V.1.4, and obtain

$$\begin{aligned}(xa, x) \in \mathscr{R} &\Leftrightarrow xaa^{-1}x^{-1} = xx^{-1}\\ \text{i.e.} \quad &\Leftrightarrow xaa^{-1}x^{-1}x = xx^{-1}x\\ \text{i.e.} \quad &\Leftrightarrow xaa^{-1} = x\\ \text{i.e.} \quad &\Leftrightarrow x \in Saa^{-1}.\end{aligned}$$

This now suggests that for a more general semigroup S we might attempt to find a representation by partial transformations by associating with each a in S an element δ_a of $\mathscr{PT}(S)$ defined by

$$\delta_a = \{(x, y) \in S \times S : y = xa \quad \text{and} \quad (x, y) \in \mathscr{R}\}.$$

Showing that this does give a representation amounts to showing that $\operatorname{dom}(\delta_{ab}) = \operatorname{dom}(\delta_a\delta_b)$, for it is then obvious that $x\delta_{ab} = x\delta_a\delta_b$ for every x in the domain. Now,

$$\begin{aligned}x \in \operatorname{dom}(\delta_a\delta_b) &\Rightarrow (x, xa) \in \mathscr{R} \quad \text{and} \quad (xa, xab) \in \mathscr{R}\\ &\Rightarrow (x, xab) \in \mathscr{R}, \text{ by transitivity}\\ &\Rightarrow x \in \operatorname{dom}(\delta_{ab}).\end{aligned}$$

Conversely, if $x \in \operatorname{dom}(\delta_{ab})$, so that $(x, xab) \in \mathscr{R}$, then there exists u in S^1 such that $x = xabu$ and so certainly $(x, xa) \in \mathscr{R}$. Since $\mathscr{R}$ is an equivalence it now follows that $(xa, xab) \in \mathscr{R}$. Hence $x \in \operatorname{dom}(\delta_a\delta_b)$ as required.

The representation $a \mapsto \delta_a$ is, however, not in general faithful, and it is an important property of the Vagner–Preston representation (though admittedly not of the Munn representation) that it is faithful. In the case of a regular semigroup we can overcome this disadvantage by simultaneously considering the left-right dual of δ_a, i.e. the mapping $a \mapsto \gamma_a$, where $\gamma_a \in \mathscr{PT}(S)$ is defined by

$$\gamma_a = \{(x, y) \in S \times S : y = ax \quad \text{and} \quad (x, y) \in \mathscr{L}\}.$$

This is not strictly a representation, since it has instead the *anti-representation* property that

$$\gamma_a\gamma_b = \gamma_{ba}.$$

For this reason we shall want to think of γ_a as belonging not to the semigroup $\mathscr{PT}(S)$ itself but rather to the dual semigroup $\mathscr{PT}^*(S)$ whose elements are the elements of $\mathscr{PT}(S)$ but with multiplication $*$ defined by

$$\alpha * \beta = \beta\alpha \qquad (\alpha, \beta \in \mathscr{PT}^*(S)).$$

(The juxtaposition on the right denotes the ordinary composition of mappings, i.e. the semigroup operation in $\mathscr{PT}(S)$.) Then $a \mapsto \gamma_a$ is a homomorphism of S into $\mathscr{PT}^*(S)$, since $\gamma_{ab} = \gamma_a * \gamma_b$ for every a, b in S.

We now have the following result, due to Lallement (1967):

Proposition 2.1. *Let S be a regular semigroup. For each a in S define*

$$\gamma_a = \{(x, y) \in S \times S : y = ax \quad \text{and} \quad (x, y) \in \mathscr{L}\},$$

$$\delta_a = \{(x, y) \in S \times S : y = xa \quad \text{and} \quad (x, y) \in \mathscr{R}\}.$$

Then the representation $\alpha: S \to \mathscr{PT}(S) \times \mathscr{PT}^*(S)$ defined by

$$a\alpha = (\delta_a, \gamma_a)$$

is faithful.

Proof. We have already seen that α is a representation. Suppose now that a, b in S are such that $\gamma_a = \gamma_b$, $\delta_a = \delta_b$. If $a' \in V(a)$ then $(a'a, a) \in \gamma_a$. Hence $(a'a, a) \in \gamma_b$ and so

$$a = ba'a.$$

Similarly, $b = ab'b$, where b' is any element of $V(b)$. It follows that $(a, b) \in \mathscr{R}$. By using δ_a we can equally well obtain that $(a, b) \in \mathscr{L}$. Hence there exists u in S such that $b = ua$, and so

$$b = ua = uaa'a = ba'a.$$

But we have already seen that $ba'a = a$, and so we conclude that $a = b$ as required. ■

This, then, is the general background to our approach. In the inverse semigroup case we sacrificed the faithfulness of the Vagner–Preston representation in order to obtain a representation that provided more useful information about the structure of the semigroup. The key lay in the idea of conjugates of idempotents, and this idea, as we have seen, is available in the orthodox case.

If, however, we try in this case to associate with each a in S a mapping $\rho_a: B \to B$ (where B is the band of idempotents of S) by the rule that

$$x\rho_a = a'xa \qquad (x \in B)$$

we immediately face the difficulty that a' is not unique and that a different choice (say a^*) of inverse for a would lead to a potentially different mapping. However, we do have that $a^*xa \mathrel{\mathscr{L}} a'xa$ $(x \in B)$, since

$$\begin{aligned} a^*xaa'xa &= a^*xaa'xaa'a = a^*(xaa')^2a \\ &= a^*xaa'a = a^*xa, \end{aligned}$$

and similarly $a'xaa^*xa = a'xa$. Indeed we can say more: if x and y are $\mathscr{L}$-equivalent elements of B and if $a', a^* \in V(a)$ then $xy = x$, $yx = y$ and so

$$a'xaa^*ya = a'xyaa^*yaa^*a = a'x(yaa^*)^2a$$
$$= a'xyaa^*a = a'xya = a'xa,$$
$$a^*yaa'xa = a^*yxaa'xaa'a = a^*y(xaa')^2a$$
$$= a^*yxaa'a = a^*yxa = a^*ya;$$

thus $a'xa \mathscr{L} a^*ya$. We can therefore define, for each a in S, a mapping $\rho_a: B/\mathscr{L} \to B/\mathscr{L}$ by

$$L_x\rho_a = L_{a'xa} \qquad (x \in B) \tag{2.2}$$

and be sure that the choice of inverse a' on the right-hand side is unimportant. In particular, if $e \in B$, we have

$$L_x\rho_e = L_{exe} \qquad (x \in B). \tag{2.3}$$

Notice that if S is an inverse semigroup (so that B is a semilattice) then $L_x = \{x\}$ and $L_{a'xa} = \{a^{-1}xa\}$, and so formula (2.2) reduces to the standard formula (V.4.1) for inverse semigroups.

By dual arguments we can define $\lambda_a: B/\mathscr{R} \to B/\mathscr{R}$ by

$$R_x\lambda_a = R_{axa'} \qquad (x \in B), \tag{2.4}$$

where a' is an arbitrarily chosen inverse of a. Again, if $e \in B$, we have

$$R_x\lambda_3 = R_{exe} \quad (x \in B). \tag{2.5}$$

From (2.2) and from the fact that $b'a' \in V(ab)$ it is evident that $\rho_{ab} = \rho_a\rho_b$ for all a, b, in S. Dually we obtain that $\lambda_{ab} = \lambda_b\lambda_a$, and so we have

Proposition 2.6. *Let S be an orthodox semigroup with band of idempotents B. The mapping* $\xi: S \to \mathscr{T}(B/\mathscr{L}) \times \mathscr{T}^*(B/\mathscr{R})$ *defined by*

$$a\xi = (\rho_a, \lambda_a) \qquad (a \in S)$$

is a homomorphism. ■

If the analogy with inverse semigroups is to be continued then ker ξ ought to be μ, the maximum idempotent-separating congruence on S. To see that this is indeed the case, suppose first that $(e, f) \in \ker \xi \cap (B \times B)$. Then $\lambda_e = \lambda_f$ and so in particular $R_e\lambda_e = R_e\lambda_f$. Applying formula (2.5), we obtain that $R_e = R_{fef}$, i.e. that $e \mathscr{R} fef$. Similarly, by considering ρ_e and ρ_f, we may show that $e \mathscr{L} fef$. Since $\mathscr{H}$ is trivial on any band we deduce that $e = fef$. By symmetry we may also show that $f = efe$. Hence

$$e = fef = f.fef = fe = efe.e = efe = f.$$

Thus ker ξ is an idempotent-separating congruence and so ker $\xi \subseteq \mu$.

Conversely, if $(a, b) \in \mu$ then for suitably chosen inverses a' and b' we have by Theorem 1.17 that

$$R_x\lambda_a = R_{axa'} = R_{bxb'} = R_x\lambda_b,$$
$$L_x\rho_a = L_{a'xa} = L_{b'xb} = L_x\rho_b$$

for all x in B. Hence $\lambda_a = \lambda_b$ and $\rho_a = \rho_b$, giving $(a, b) \in \ker \xi$ as required.

For convenience we restate Proposition 2.6 in its new improved form:

Theorem 2.7. *Let S be an orthodox semigroup with band of idempotents B, and let ξ be the mapping from S into $\mathscr{T}(B/\mathscr{L}) \times \mathscr{T}^*(B/\mathscr{R})$ defined by*

$$a\xi = (\rho_a, \lambda_a),$$

where ρ_a, λ_a are given by formulae (2.2) and (2.4). Then ξ is a homomorphism whose kernel is the maximum idempotent-separating congruence μ on S. ■

The result we have achieved is not quite as exact an analogue of Theorem V.4.9 as we might wish, because the semigroup $\mathscr{T}(B/\mathscr{L}) \times \mathscr{T}^*(B/\mathscr{R})$ used for the representation corresponds to $\mathscr{I}(E)$ rather than to T_E. To obtain a true analogue we must identify an orthodox subsemigroup of $\mathscr{T}(B/\mathscr{L}) \times \mathscr{T}^*(B/\mathscr{R})$ containing the image of S under ξ and having band of idempotents isomorphic to B.

If we start with a band B on its own, the obvious analogue of the ideal $Ee = \{x \in E : x \leqslant e\}$ in the semilattice E is the sub-band $eBe = \{x \in E : x \leqslant e\}$ of B. It is convenient to denote eBe by $\langle e \rangle$. Notice that for all x, y in B,

$$\langle x \rangle = \langle y \rangle \Rightarrow x = y. \tag{2.8}$$

We define

$$\mathscr{U} = \{(e, f) \in B \times B : \langle e \rangle \simeq \langle f \rangle\}$$

and write $W_{e,f}$ for the set of all isomorphisms from $\langle e \rangle$ onto $\langle f \rangle$. Notice that

$$e\alpha = f \tag{2.9}$$

for all α in W_{ef}. Also,

Proposition 2.10. *If $g \in \langle e \rangle$ and $\alpha \in W_{e,f}$, then*

$$\langle g \rangle\alpha = \langle g\alpha \rangle.$$

Proof. If $x \in \langle g \rangle\alpha$ then $x = (gbg)\alpha$ for some b in B. Hence

$$x = (gbg)\alpha = (gebeg)\alpha \qquad (\text{since } g \leqslant e)$$
$$= (g\alpha)[(ebe)\alpha](g\alpha) \in \langle g\alpha \rangle.$$

Conversely, if $y \in \langle g\alpha \rangle$ then $y = (g\alpha)b(g\alpha)$ for some b in B. Hence

$$
\begin{aligned}
y &= (g\alpha)b(g\alpha) = (g\alpha)fbf(g\alpha) && (\text{since } g\alpha \leqslant f)\\
&= (g\alpha)[(ece)\alpha](g\alpha) \text{ for some } c \text{ in } B && (\text{since } \alpha \text{ is onto})\\
&= (geceg)\alpha \in \langle g \rangle\alpha. \blacksquare
\end{aligned}
$$

Next, if we have an orthodox semigroup S with band of idempotents B we may in an entirely routine manner obtain a generalization of the result for inverse semigroups described at the beginning of Section V.4:

Proposition 2.11. Let S be an orthodox semigroup with band of idempotents B. If $a \in S$ and $a' \in V(a)$ then the maps $x \mapsto a'xa$, $y \mapsto aya'$ are mutually inverse isomorphisms from $\langle aa' \rangle$ onto $\langle a'a \rangle$ and from $\langle a'a \rangle$ onto $\langle aa' \rangle$ respectively. ■

Recalling now from Proposition II.4.9 that two idempotents e, f in a regular semigroup are $\mathscr{D}$-equivalent if and only if there exists a in S and a' in $V(a)$ such that $aa' = e$ and $a'a = f$, we deduce that in any orthodox semigroup S with band of idempotents B,

$$\mathscr{D}^S \cap (B \times B) \subseteq \mathscr{U}. \tag{2.12}$$

It is again a routine matter to verify the following results:

Proposition 2.13. *Let e, f be the elements of a band B. If $E(e) = E(f)$ then the maps*

$$\theta_f : x \mapsto fxf, \qquad \theta_e : y \mapsto eye$$

are mutually inverse isomorphisms from $\langle e \rangle$ onto $\langle f \rangle$ and from $\langle f \rangle$ onto $\langle e \rangle$ respectively. ■

As a consequence,

$$\varepsilon \subseteq \mathscr{U}. \tag{2.14}$$

Notice that this is different from the semilattice situation, where $\mathscr{U}$ may (in the anti-uniform case) be the equality relation.

If B is a band, if $(e, f) \in \mathscr{U}$ and if $\alpha \in W_{e,f}$ we may define $\alpha_l \in \mathscr{I}(B/\mathscr{L})$ and $\alpha_r \in \mathscr{I}(B/\mathscr{R})$ by the formulae

$$L_x\alpha_l = L_{x\alpha}, \qquad R_x\alpha_r = R_{x\alpha} \qquad (x \in \langle e \rangle). \tag{2.15}$$

These formulae do in fact define one–one mappings, since (e.g.)

$$L_x = L_y \quad \Leftrightarrow \quad xy = x, yx = y$$
$$\text{i.e.} \quad \Leftrightarrow \quad (x\alpha)(y\alpha) = x\alpha, (y\alpha)(x\alpha) = y\alpha$$
$$\text{i.e.} \quad \Leftrightarrow \quad L_{x\alpha} = L_{y\alpha}.$$

Moreover, it is easy to verify that $(\alpha_l)^{-1} = (\alpha^{-1})_l$ and $(\alpha_r)^{-1} = (\alpha^{-1})_r$; thus we may use the notations $\alpha_l^{-1}, \alpha_r^{-1}$ without ambiguity.

Now let S be an orthodox semigroup with band of idempotents B, let $a \in S$ and let $a' \in V(a)$. Denoting aa' by e and $a'a$ by f, we observe that the mapping $\rho_a \in \mathscr{T}(B/\mathscr{L})$ defined by (2.2) may be expressed as $\rho_e\theta_l$, where θ is the mapping in $W_{e,f}$ given by

$$x\theta = a'xa \qquad (x \in \langle e \rangle).$$

(Here the multiplication of ρ_e in $\mathscr{T}(B/\mathscr{L})$ and $\theta_l \in \mathscr{I}(B/\mathscr{L})$ takes place in the common "supersemigroup" $\mathscr{PT}(B/\mathscr{L})$ and results in an element of $\mathscr{T}(B/\mathscr{L})$.) Equally, the mapping $\lambda_a \in \mathscr{T}(B/\mathscr{R})$ defined by (2.4) may be expressed as $\lambda_f\theta_r^{-1}$. The crucial point to notice here is that the underlying mapping θ is the same in each case. The range of the mapping $\xi: a \mapsto (\lambda_a, \rho_a)$ is thus contained in the subset

$$W_B = \{(\rho_e\alpha_l, \lambda_f\alpha_r^{-1}): \alpha \in W_{e,f}, (e,f) \in \mathscr{U}\}, \tag{2.16}$$

of $\mathscr{T}(B/\mathscr{L}) \times \mathscr{T}^*(B/\mathscr{R})$.

In fact W_B is precisely the analogue of T_E we have been seeking. We shall call it the *Hall semigroup* of the band B. Its most important properties are listed in the next theorem:

Theorem 2.17. *Let B be a band, and let W_B be defined by formula* (2.16). *Then*

(i) *W_B is a subsemigroup of $\mathscr{T}(B/\mathscr{L}) \times \mathscr{T}^*(B/\mathscr{R})$;*

(ii) *W_B is orthodox, with band of idempotents $B^* = \{(\rho_e, \lambda_e): e \in B\}$ isomorphic to B;*

(iii) *if B^* is identified with B, then, in W_B,*

$$\mathscr{D} \cap (B \times B) = \mathscr{U}.$$

Proof. To prove (i), we consider the product in the semigroup $\mathscr{T}(B/\mathscr{L}) \times \mathscr{T}^*(B/\mathscr{R})$ of two elements $(\rho_e\alpha_l, \lambda_f\alpha_r^{-1})$ and $(\rho_g\beta_l, \lambda_h\beta_r^{-1})$ in W_B. Thus (e,f), $(g,h) \in \mathscr{U}$, and $\alpha \in W_{e,f}$, $\beta \in W_{g,h}$. We begin by examining the product $\rho_e\alpha_l\rho_g\beta_l$.

Notice first that the range of $\rho_e\alpha_l$ is $\{L_{fxf}: x \in B\}$, and hence that the range of $\rho_e\alpha_l\rho_g\beta_l$ is $\{L_{(gfxfg)\beta}: x \in B\}$. Now

$$gfBfg = gfgfBfgfg = gfg(fBf)gfg \subseteq gfgBgfg = gf(gBg)fg \subseteq gfBg,$$

and so, in our previous notation,

$$gfBfg = \langle gfg \rangle.$$

Hence, by Proposition 2.10, the range of $\rho_e\alpha_l\rho_g\beta_l$ is the subset $\{L_{jxj} : x \in B\}$ of $B/\mathscr{L}$, where $j = (gfg)\beta$. Notice that $\beta|\langle gfg \rangle$ is an isomorphism from $\langle gfg \rangle$ onto $\langle j \rangle$, with inverse $\beta^{-1}|\langle j \rangle$).

Now observe that since $E(fgf) = E(gfg)$ (by (1.8)) we have by Proposition 2.13 that

$$\theta_{gfg} : \langle fgf \rangle \to \langle gfg \rangle, \qquad \theta_{fgf} : \langle gfg \rangle \to \langle fgf \rangle$$

are mutually inverse isomorphisms. If we write $(fgf)\alpha^{-1}$ as i, we have that $\alpha|\langle i \rangle$ is an isomorphism from $\langle i \rangle$ onto $\langle fgf \rangle$, with inverse $\alpha^{-1}|\langle fgf \rangle$. Let

$$\gamma = (\alpha|\langle i \rangle)(\theta_{gfg}|\langle fgf \rangle)(\beta|\langle gfg \rangle) \tag{2.18}$$

be the isomorphism from $\langle i \rangle$ onto $\langle j \rangle$ formed by taking the composition of the three isomorphisms described. Then, as we shall show,

$$\rho_e\alpha_l\rho_g\beta_l = \rho_i\gamma_l.$$

If $L_x \in B/\mathscr{L}$, then $L_x\rho_e\alpha_l\rho_g\beta_l = L_{(g[(exe)\alpha]g)\beta}$. On the other hand,

$$L_x\rho_i\gamma_l = L_{(ixi)\gamma},$$

and

$$\begin{aligned}(ixi)\gamma &= (iexei)\gamma \qquad (\text{since } i \leqslant e)\\ &= (iexei)\alpha(\theta_{gfg}|\langle fgf \rangle)(\beta|\langle gfg \rangle)\\ &= fgf[(exe)\alpha]fgf(\theta_{gfg}|\langle fgf \rangle)(\beta|\langle gfg \rangle)\\ &= (gfgfgf[(exe)\alpha]fgfgfg)\beta\\ &= (gf[(exe)\alpha]fg)\beta\\ &= (g[(exe)\alpha]g)\beta \qquad (\text{since } (exe)\alpha \in fBf).\end{aligned}$$

Thus $\rho_e\alpha_l\rho_g\beta_l = \rho_i\gamma_l$ as required. A similar argument shows that

$$(\lambda_f\alpha_r^{-1}) * (\lambda_h\beta_r^{-1}) = \lambda_h\beta_r^{-1}\lambda_f\alpha_r^{-1} = \lambda_j\gamma_r^{-1}.$$

Thus the product of the elements $(\rho_e\alpha_l, \lambda_f\alpha_r^{-1})$, $(\rho_g\beta_l, \lambda_h\beta_r^{-1})$ of W_B is the element $(\rho_i\gamma_l, \lambda_j\gamma_r^{-1})$, which also lies in W_B.

To prove (ii) we first establish that every element of W_B has an inverse in W_B. If $a = (\rho_e\alpha_l, \lambda_f\alpha_r^{-1}) \in W_B$, consider the element $a' = (\rho_f\alpha_l^{-1}, \lambda_e\alpha_r)$ (also in W_B). Then

$$aa'a = (\rho_e\alpha_l\rho_f\alpha_l^{-1}\rho_e\alpha_l, \qquad \lambda_f\alpha_r^{-1}\lambda_e\alpha_r\lambda_f\alpha_r^{-1}).$$

Since ρ_f acts identically on the range of $\rho_e\alpha_i$ and λ_e acts identically on the range of $\lambda_f\alpha_r^{-1}$, it easily follows that $aa'a = a$. Similarly $a'aa = a'$.

To show that W_B is orthodox, let us begin by considering the conditions under which the element $a = (\rho_e\alpha_l, \lambda_f\alpha_r^{-1})$ of W_B is idempotent. For this to be the case we require in the first instance that $\rho_e\alpha_l\rho_e\alpha_l = \rho_e\alpha_l$, which implies that $\rho_e\alpha_l$ acts identically on any element L_{fxf} $(x \in B)$ lying in its range. In particular,

$$L_f = L_f\rho_e\alpha_l = L_{efe}\alpha_l.$$

Since it is also the case that

$$L_e\alpha_l = L_{e\alpha} = L_f,$$

we conclude from the one–one property of α_l that $L_e = L_{efe}$. Hence efe is a right identity in L_e, and so

$$e = e(efe = efe)$$

Similarly, from the fact that

$$\lambda_f\alpha_r^{-1}\lambda_f\alpha_r^{-1} = \lambda_f\alpha_r^{-1}$$

we may deduce that $fef = f$. Thus $E(e) = E(f)$.

Now, from the fact that $\rho_e\alpha_l$ acts identically on its range we may further deduce that

$$\begin{aligned} L_x\rho_e\alpha_l &= L_{exe}\alpha_l = L_{(efe)x(efe)}\alpha_l \\ &= L_{fexef}\rho_e\alpha_l = L_{fexef}. \end{aligned}$$

It is convenient to present the next phase of the argument as a lemma:

Lemma 2.19. *If* x, e, f *are elements of a band, then* $L_{fexef} = L_{efxef}$.

Proof. The crucial observation is that if $a. b$ are elements of a band B then $ab \mathscr{L} bab$. One half of this is trivial, and the other half follows from the observation that $ab = a(bab)$. Taking $a = x$ and $b = ef$ we obtain that $xef \mathscr{L} efexf$. Again, taking $a = x$ and $b = e$ we obtain that $xe \mathscr{L} exe$. Hence $xef \mathscr{L} exef$, since $\mathscr{L}$ is a right congruence. Finally, taking $a = exe$ and $b = f$ we obtain that $exef \mathscr{L} fexef$. It now follows by the symmetry and transitivity of $\mathscr{L}$ that $fexef \mathscr{L} efxef$ as required. ■

Returning now to the main proof, we make use of the lemma to deduce that

$$L_x\rho_e\alpha_l = L_{efxef} = L_x\rho_{ef} \qquad (x \in B).$$

A similar argument shows that

$$R_x\lambda_f\alpha_r^{-1} = R_{efxfe} = R_{efxef} = R_x\lambda_{ef} \qquad (x \in B).$$

It follows that $a = (\rho_{ef}, \lambda_{ef})$.

Conversely, it is easy to verify that any element (ρ_e, λ_e) $(e \in B)$ is an idempotent element of W_B. Thus the set of idempotents of W_B is

$$B^* = \{(\rho_e, \lambda_e): e \in B\}.$$

It is now clear that W_B is orthodox, for if $e, f \in B$ then

$$L_x\rho_e\rho_f = L_{fexef} = L_{efxef} = L_x\rho_{ef},$$

while

$$R_x\lambda_f\lambda_e = R_{efxfe} = R_{efxef} = R_x\lambda_{ef};$$

thus, in W_B,

$$(\rho_e, \lambda_e)(\rho_f, \lambda_f) = (\rho_{ef}, \lambda_{ef}).$$

This shows, moreover that the mapping $\psi: B \to B^*$ defined by

$$e\psi = (\rho_e, \lambda_e) \qquad (e \in B)$$

is a homomorphism. It clearly maps onto B^*. In fact,

Lemma 2.20. *The mapping ψ is an isomorphism from B onto B^*.*

Proof. It remains to show that ψ is one–one. If $e\psi = f\psi$ $(e, f \in B)$, then $\lambda_e = \lambda_f$ and $\rho_e = \rho_f$. Hence $R_e\lambda_e = R_e\lambda_f$ and so $e \mathscr{R} fef$. Equally $L_e\rho_e = L_e\rho_f$ and so $e \mathscr{L} fef$. Hence $e = fef$. By symmetry we may also show that $f = efe$. Hence

$$e = fef = f.fef = fe = efe.e = efe = f. \blacksquare$$

It remains to prove the final part (iii) of Theorem 2.17. By (2.12) it is always the case that $\mathscr{D} \cap (B \times B) \subseteq \mathscr{U}$. Conversely, if $(e, f) \in \mathscr{U}$, then to each α in $W_{e,f}$ there correspond elements $a = (\rho_e\alpha_l, \lambda_f\alpha_r^{-1})$ and $a' = (\rho_f\alpha_l^{-1}, \lambda_e\alpha_r)$ of W_B. We saw earlier that a and a' are mutually inverse. Also, recalling that

$$\rho_e\alpha_l\rho_f = \rho_e\alpha_l, \qquad \lambda_e\alpha_r\lambda_f = \lambda_e\alpha_r,$$

we see that

$$aa' = (\rho_e\alpha_l\rho_f\alpha_l^{-1}, \lambda_e\alpha_r\lambda_f\alpha_r^{-1}) = (\rho_e, \lambda_e).$$

A similar argument establishes that $a'a = (\rho_f, \lambda_f)$ and so, by Proposition II.4.9, the idempotents (ρ_e, λ_e) and (ρ_f, λ_f) are $\mathscr{D}$-equivalent in W_B. If we agree to identify B^* with B (in accordance with the isomorphism ψ), this amounts to saying that $(e,f) \in \mathscr{D} \cap (B \times B)$. Hence in W_B we have $\mathscr{D} \cap (B \times B) = \mathscr{U}$ as required. $\blacksquare$

The Hall semigroup W_B is, as we have seen, closely analogous to the Munn semigroup T_E of a semilattice. It effectively reduces to T_E in the case where the band B is a semilattice:

Proposition 2.21. *If E is a semilattice, then $W_E \simeq T_E$.*

Proof. Since $L_x = \{x\} = R_x$ for every x in E, and since $W_{e,f}$ coincides with $T_{e,f}$, we may identify W_E with

$$\{(\theta_e\alpha, \theta_f\alpha^{-1}) : \alpha \in T_{e,f}, (e,f) \in \mathscr{U}\},$$

where as before θ_g $(g \in E)$ is defined as the mapping $x \mapsto gxg$ of E into itself. The product in W_E of two elements $(\theta_e\alpha, \theta_f\alpha^{-1})$ and $(\theta_g\beta, \theta_h\beta^{-1})$ is of the form $(\theta_i\gamma, \theta_j\gamma^{-1})$, where $i = (fgf)\alpha^{-1}$ and γ is given in accordance with formula (2.18). But in the semilattice E we have $fgf = gfg = fg$; hence $i = (fg)\alpha^{-1}$, $j = (fg)\beta$ and formula (2.18) reduces to

$$\gamma = (\alpha|\langle i\rangle)(\beta|\langle fg\rangle) \qquad (\in T_{i,j}).$$

Observe now that γ is none other than the product in T_E of α and β. It is then a routine matter to verify that the mapping $\Psi: T_E \to W_E$ defined by

$$\alpha\Psi = (\theta_e\alpha, \theta_f\alpha^{-1}) \qquad (\alpha \in T_{e,f})$$

is an isomorphism.■

3. UNIFORM AND ANTI-UNIFORM BANDS

By analogy with the case of semilattices, we call a band B *uniform* if the equivalence

$$\mathscr{U} = \{(e,f) \in B \times B : \langle e\rangle \simeq \langle f\rangle\}$$

is the universal relation on B, i.e. if $eBe \simeq fBf$ for all e, f in B. We have seen (2.12) that in any orthodox semigroup having B as band of idempotents

$$\mathscr{D}^S \cap (B \times B) \subseteq \mathscr{U},$$

and so we immediately obtain

Proposition 3.1. *The band of idempotents of a bisimple orthodox semigroup is uniform.* ■

In fact, since we have now produced, for a given band B, an orthodox semigroup W_B for which

$$\mathscr{D} \cap (B \times B) = \mathscr{U},$$

and since any regular semigroup in which all the idempotents are $\mathscr{D}$-equivalent is easily seen to be bisimple, we can assert

Theorem 3.2. *A band B is uniform if and only if there exists a bisimple orthodox semigroup S whose band of idempotents is isomorphic to B.*■

In the case of a semilattice E we used the term 'anti-uniform' to describe the situation where $\mathscr{U} = 1_E$. In the present situation such a definition will not do, since we have observed (2.14) that $\varepsilon \subseteq \mathscr{U}$ in any band B, where ε is the minimum semilattice congruence. It seems reasonable, therefore, to describe a band B as *anti-uniform* if $\mathscr{U} = \varepsilon$.

We can now achieve an analogue for orthodox semigroups of Theorem V.5.2. First we establish

Proposition 3.3. *An orthodox semigroup S with band of idempotents G is a union of groups if and only if $\mathscr{D}^S \cap (B \times B) = \varepsilon$, where ε is the minimum semilattice congruence on the band B.*

Proof. If S is an orthodox semigroup in which

$$\mathscr{D}^S \cap (B \times B) = \varepsilon,$$

consider an arbitrary element a of S and an arbitrary inverse a' of a. Since $(aa'\ a'a) \in \mathscr{D}^S$ we deduce that $(aa', a'a) \in \varepsilon$. Thus

$$e = aa' \in E(a'a), \qquad f = a'a \in E(aa').$$

Hence, by Proposition 1.9,

$$a^* = ea'f\ (=a(a')^3(a) \in E(a'a)a'E(aa') = V(a).$$

Now

$$\begin{aligned} aa^* &= aea'f = a(a'aea'a)a'f = a(fef)a'f \\ &= afa'f \quad (\text{since } (e, f) \in \varepsilon) \\ &= aa'aa'f = ef, \end{aligned}$$

and

$$\begin{aligned} a^*a &= ea'fa = ea'(aa'faa')a = ea'(efe)a \\ &= ea'ea \quad (\text{since } (e, f) \in \varepsilon) \\ &= ea'aa'a = ef. \end{aligned}$$

Hence $ef \in H_a$. Since ef is idempotent it follows that H_a is a group. Thus every $\mathscr{H}$-class of S is a group and so S is a union of groups.

Conversely, if S is orthodox and is a union of groups, consider two $\mathscr{D}^S$-equivalent idempotents e, f in S. By Proposition II.3.6 there exists an element a in S such that $aa' = e, a'a = f$. But H_a is a group with identity g (say) and so the element a has an inverse a^* such that $aa^* = a^*a = g$. But aa' and aa^* are both $\mathscr{R}^S$-equivalent to a. Since $\mathscr{R}^S \cap (B \times B) = \mathscr{R}^B$ by Proposition II.4.5, we may deduce that

$$(aa', aa^*) \in \mathscr{R}^B \subseteq \mathscr{J}^B = \varepsilon.$$

Similarly

$$(a'a, a^*a) \in \varepsilon.$$

That is, we have $(e, g) \in \varepsilon$, $(f, g) \in \varepsilon$, from which it immediately follows that $(e, f) \in \varepsilon$. We have shown that

$$\mathscr{D}^S \cap (B \times B) \subseteq \varepsilon.$$

Since $\varepsilon = \mathscr{D}^B \subseteq \mathscr{D}^S \cap (B \times B)$ we therefore have the required result.∎

We now easily deduce

Theorem 3.4. *A band B has the property that every orthodox semigroup band of idempotents whose isomorphic to B is a union of groups if and only if it is anti-uniform.*

Proof. It is evident from (2.12) that if B is anti-uniform then $\mathscr{D} \cap (B \times B) \subseteq \mathscr{U} = \varepsilon$ in any orthodox semigroup S having B as band of idempotents. Hence, by the last proposition, S is a union of groups.

Conversely, if B is not anti-uniform then $\varepsilon \subset \mathscr{U}$ (properly). Since in W_B we have shown that $\mathscr{D} \cap (B \times B) = \mathscr{U}$, it follows that $\mathscr{D} \cap (B \times B) \neq \varepsilon$ in this semigroup and so W_B is not a union of groups.∎

4. THE STRUCTURE OF ORTHODOX SEMIGROUPS

The Hall semigroup W_B also plays an important part in the structure theory to be described in this section. Many of the ideas involved are in Yamada's (1970) paper, but the treatment we shall give is drawn from the rather more complete theory in Hall (1971).

Let S be an orthodox semigroup with band of idempotents B. The mapping $\xi: a \mapsto (\rho_a, \lambda_a)$ of S into W_B is not in general one–one. Indeed we have seen that its kernel is μ (Theorem 2.7). However, since by (1.16) we have

$$\gamma \cap \mu \subseteq \gamma \cap \mathscr{H} = 1_S,$$

we do have that the homomorphism η of S into $W_B \times S/\gamma$ defined by

$$a\eta = ((\rho_a, \lambda_a), a\gamma) \qquad (a \in S) \tag{4.1}$$

is one–one.

The monomorphism η is not onto, but if we can identify its range we may have a useful pointer towards a structure theorem for orthodox semigroups in terms of bands and inverse semigroups. If γ_1 is the minimum inverse semigroup congruence on W_B *then* $\xi\gamma_1^{\natural}$ is a homomorphism from S into the inverse semigroup W_B/γ_1 which must factor through S/γ in accordance with the commutative diagram

$$\begin{array}{ccc} S & \xrightarrow{\ \xi\ } & W_B \\ \downarrow{\scriptstyle \gamma^\natural} & & \downarrow{\scriptstyle \gamma_1^\natural} \\ S/\gamma & \xrightarrow[\ \theta\ ]{} & W_B/\gamma_1 \end{array} \tag{4.2}$$

The homomorphism θ is uniquely defined by the diagram, and we have

Lemma 4.3. *The homomorphism θ is idempotent-separating. Its range contains all the idempotents of W_B/γ_1.*

Proof. Let $e\gamma, f\gamma$ by idempotents in S/γ (where $e, f \in B$) and suppose that $(e\gamma)\theta = (f\gamma)\theta$. Then $e\xi\gamma_1 = f\xi\gamma_1$; i.e. $(e\xi, f\xi) \in \gamma_1 \cap (B^* \times B^*)$. By (1.14) it follows that the idempotents $e\xi, f\xi$ in W_B are $\mathscr{J}$-equivalent in B^*. Since $\xi|B$ is an isomorphism of B onto the band B^* of idempotents of W_B, it follows that e and f are $\mathscr{J}$-equivalent in B. Hence, again by (1.14) it follows that $e\gamma = f\gamma$.

Any idempotent in W_B/γ_1 is by Lallement's Lemma (II.4.6) expressible as $(\rho_e, \lambda_e)\gamma_1$, where (ρ_e, λ_e) is an idempotent in W_B. That is to say, it is expressible as $e\xi\gamma_1$ for some idempotent e in S. The commutativity of the diagram (4.2) then enables us to express our idempotent as $(e\gamma)\theta$; hence every idempotent in W_B/γ_1 lies in the range of θ.■

Now, any element $((\rho_a, \lambda^a), a\gamma)$ in the range of η has the property that

$$(\rho_a, \lambda_a)\gamma_1 = a\xi\gamma_1 = a\gamma\theta = (a\gamma)\theta.$$

Conversely, we shall show that if $(x, a\gamma) \in W_B \times S/\gamma$ is such that

$$x\gamma_1 = (a\gamma)\theta,$$

then $(x, a\gamma) = b\eta$ for some b in S. In other words, we shall establish

Proposition 4.4. *Let S be an orthodox semigroup with band of idempotents B. The mapping η defined by* (4.1) *is an isomorphism of S onto*

$$\{(x, a\gamma) \in W_B \times S/\gamma : x\gamma_1^\natural = (a\gamma)\theta\},$$

the spined product of W_B and S/γ with respect to $W_B/\gamma_1, \gamma_1^\natural$ and θ.

Proof. It remains to show that η is onto. If $(x, a\gamma) \in W_B \times S/\gamma$ is such that $x\gamma_1 = (a\gamma)\theta$, notice first that

$$x\gamma_1 = (a\gamma)\theta = (\rho_\alpha, \lambda_a)\gamma_1$$

so that $V(x) = V(\rho_a, \lambda_a)$ in W_B (by 1.11)). Now for any inverse c of a in S it is easy to verify that $(\rho_c, \lambda_c) \in V(\rho_a, \lambda_a)$ in W_B. Hence $(\rho_c, \lambda_c) \in V(x)$ and so both $(\rho_c, \lambda_c)x$ and $x(\rho_c, \lambda_c)$ are idempotents in W_B. That is, there exist e, f in B such that

$$(\rho_c, \lambda_c)x = (\rho_e, \lambda_e), \qquad x(\rho_c, \lambda_c) = (\rho_f, \lambda_f)$$

As a consequence we have that

$$(\rho_e, \lambda_e)\,\mathscr{R}\,(\rho_c, \lambda_c), \qquad (\rho_c, \lambda_c)\,\mathscr{L}\,(\rho_f, \lambda_f)$$

in W_B. That is, in W_B,

$$e\xi\,\mathscr{R}\,e\xi, \qquad c\xi\,\mathscr{L}c\,f\xi.$$

Examining the first of these, we first use Proposition II.4.5 to deduce that $e\xi$ and $c\xi$ are $\mathscr{R}$-equivalent in $S\xi$. Thus there exist u, v in S such that

$$e\xi = (c\xi)(u\xi), \qquad c\xi = (e\xi)(v\xi),$$

i.e. such that $(e, cu) \in \xi \circ \xi^{-1}$, $(c, ev) \in \xi \circ \xi^{-1}$. Now $\xi \circ \xi^{-1} = \mu \subseteq \mathscr{H} \subseteq \mathscr{R}$ and so in S there exist elements x and y such that

$$e = cux, \qquad c = evy.$$

We conclude that $e\,\mathscr{R}\,c$ in S. Similarly, $c\,\mathscr{L}\,f$ in S. (We could have shortened this part of the argument by quoting Exercise II.6).

The familiar eggbox argument based on Theorem II.3.5 now assures us that the $\mathscr{H}$-class $L_e \cap R_f$ contains an inverse b of c. It follows that in W_B the element(ρ_b, λ_b) is an inverse of (ρ_c, λ_c) and that it is $\mathscr{L}$-equivalent to (ρ_e, λ_e) and $\mathscr{R}$-equivalent to (ρ_f, λ_f). Since x also has these properties and since by Theorem II.3.5 (iii) no $\mathscr{H}$-class contains more than one inverse of a given element, we conclude that $x = (\rho_b, \lambda_b)$. Notice also that $b\gamma$ and $a\gamma$ are both inverses of $c\gamma$ in the inverse semigroup S/γ. Hence $b\gamma = a\gamma$ and so

$$(x, a\gamma) = ((\rho_b, \lambda_b), b\gamma) = b\eta. \blacksquare$$

It now seems natural to make the following construction. Let B be a band and let T be an inverse semigroup whose semilattice of idempotents is isomorphic to B/ε. Let γ_1 be the minimum inverse semigroup congruence on the Hall semigroup W_B of B. Then W_B/γ_1 is an inverse semigroup whose semilattice of idempotents is by (1.14) isomorphic to B/ε. Let $\psi\colon T \to W_B/\gamma_1$ be an idempotent-separating homomorphism whose range contains all the idempotents of W_B/γ_1. Then we shall denote the spined product

$$S = \{(x, t) \in W_B \times T : x\gamma_1^\natural = t\psi\} \tag{4.5}$$

of W_B and T with respect to W_B/γ_1, $\gamma_1^\natural$ and ψ by $\mathscr{H}(B, T, \psi$ and call it the *Hall–Yamada semigroup determined by the band B, the inverse semigroup T and the idempotent-separating homomorphism* ψ.

Theorem 4.6. *Let B be a band and let T be an inverse semigroup whose semilattice of idempotents is isomorphic to B/ε. Let γ_1 be the minimum inverse semigroup congruence on the Hall semigroup W_B and let ψ be an idempotent-separating homorphism from T into W_B/γ_1 whose range contains all the idempotents of W_B/γ_1. Then the Hall–Yamada semigroup $S = \mathscr{H}(B, T, \psi)$ is an orthodox semigroup whose band of idempotents is isomorphic to B. If γ is the minimum inverse semigroup congruence on S, then $S/\gamma \simeq T$.*

Conversely, if S is an orthodox semigroup whose band of idempotents is B then there exists an idempotent-separating homomorphism $\theta: S/\gamma \rightarrow W_B/\gamma_1$ whose range contains all the idempotents of W_B/γ_1 and such that S is isomorphic to $\mathscr{H}(B, S/\gamma, \theta)$.

Proof. Notice that the second (converse) half of this theorem has already been proved, being merely a restatement of Lemma 4.3 and Proposition 4.4.

To prove the direct half we being by showing that $S = \mathscr{H}(B, T, \psi)$ is regular. It is of course obvious that $W_B \times T$ is regular: indeed we can say that the set of inverses of an element (x, t) in $W_B \times T$ is $V(x) \times \{t^{-1}\}$. If the element (x, t) is in S, i.e. if $t\psi = t\gamma_1^\natural$, then for every x' in $V(x)$ the elements $x'\gamma_1^\natural$ and $t^{-1}\psi$ are both inverses of the single element $x\gamma_1^\natural = t\psi$ of the inverse semigroup W_B/γ_1; hence $x'\gamma_1^\natural = t^{-1}\psi$ and so $(x', t^{-1}) \in S$. Thus S is a regular subsemigroup of $W_B \times T$, and we have shown moreover that the set of inverses of an element (x, t) of S is $V(x) \times \{t^{-1}\}$. That S is orthodox follows immediately from the fact that W_B and T are orthodox and from the fact that an element (x, t) of $W_B \times T$ is idempotent if and only if x is an idempotent of W_B and t is an idempotent of T.

Let us denote the band of idempotents of S by $\bar{B}$. We know that the idempotents of W_B form a band B^* isomorphic to B; indeed we know that the mapping $\xi | B: e \mapsto (\rho_e, \lambda_e)$ is an isomorphism from B onto B^*. Denoting the inverse of $\xi | B$ by $\kappa: B^* \rightarrow B$, let us now define a mapping $\zeta: \bar{B} \rightarrow B$ by

$$(x, t)\zeta = x\kappa \qquad ((x, t) \in \bar{B}).$$

It is obvious that ζ is a homomorphism. To see that it maps *onto* B, notice that for any e in B the element $(\rho_e, \lambda_e)\gamma_1^\natural$ is an idempotent in W_B/γ_1 and so there is a unique idempotent g in T such that $g\psi = (\rho_e, \lambda_e)\gamma_1^\natural$. Then $((\rho_e, \lambda_e), g) \in \bar{B}$ and has image e under ζ.

To show that ζ is one–one, let us suppose that the elements (x, t), (y, u) in $\bar{B}$ are such that $(x, t)\zeta = (y, u)\zeta$. Then $x\kappa = y\kappa$ and so (since κ is an isomorphism) $x = y$. Hence certainly $x\gamma_1^\natural = y\gamma_1^\natural$, and so $t\psi = u\psi$ by the definition (4.5) of S. But t, u are idempotents of T and so, since ψ is idempotent-separating, $t = u$. Thus $(x, t) = (y, u)$, and we conclude that ζ is an isomorphism from the band $\bar{B}$ of idempotents of S onto B.

It is easy to see that the mapping $\pi : (x, t) \mapsto t$ is a homomorphism from S into the inverse semigroup T. In fact π maps *onto* T, since $\gamma_1^\natural$ maps W_B onto W_B/γ_1 and so for every t in T there is an element x in W_B such that $x\gamma_1^\natural = t\psi$, i.e. such that $(x, t) \in S$. If γ is the minimum inverse semigroup congruence on S, it follows that $\gamma \subseteq \pi \circ \pi^{-1}$ and that there is a homomorphism α from S/γ onto T such that the diagram

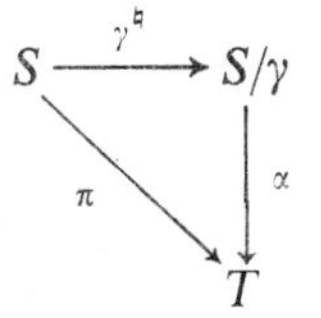

$$
\begin{array}{ccc}
S & \xrightarrow{\;\gamma^\natural\;} & S/\gamma \\
 & \searrow{\scriptstyle \pi} & \downarrow{\scriptstyle \alpha} \\
 & & T
\end{array}
\qquad (4.7)
$$

commutes.

Now, we saw earlier that the set of inverses of the element (x, t) of S is $V(x) \times \{t^{-1}\}$, where $V(x)$ is the set of inverses of the element x in W_B. Hence, using the characterization (1.11) of γ, we have that

$$
\gamma = \{((x, t), (y, y)) \in S \times S : V(x) \times \{t^{-1}\} = V(y) \times \{u^{-1}\}\}
$$
$$
= \{((x, t), (y, u)) \in S \times S : t = u \text{ and } V(x) = V(y)\}.
$$

But $t = u$ implies $t\psi = u\psi$, which in turn implies $x\gamma_1 = y\gamma_1$ (since (x, t), $(y, u) \in S$). Thus, using the characterization (1.11) of γ_1, we have that if $t = u$ then it *necessarily follows* that $V(x) = V(y)$. Thus

$$
\gamma = \{((x, t), (y, u)) \in S \times S : t = u\} = \pi \circ \pi^{-1}
$$

and so the mapping $\alpha : S/\gamma \to T$ in the diagram (4.7) is an isomorphism.∎

EXERCISES

1. Show by an example that the inclusion (1.3) for an orthodox semigroup can be strict. [Hint: consider the left normal band $\{a, b, c, d\}$ with Cayley table

	a	b	c	d
a	a	a	a	a
b	a	b	b	a
c	a	c	c	a
d	d	d	d	d

and compute $V(a)$, $V(b)$, $V(ab)$, $V(b) \, V(a)$.]

2. (T. E. Hall 1969). Show that a semigroup S is orthodox if and only if each of its principal factors is orthodox, i.e. (in the finite case) if and only if each of its principal factors is an orthodox completely 0-simple semigroup.

(Hall (1969) goes on to describe necessary and sufficient conditions for a Rees matrix semigroup $\mathcal{M}^\circ[G; I, \Lambda; P]$ to be orthodox.)

3. (T. E. Hall 1970). If a is an element of an orthodox semigroup S with band of idempotents B and if $a' \in V(a)$, show that

$$aV(a) = aa'E(aa') = R^B_{aa'}, \qquad V(a)a = E(a'a)a'a = L^B_{a'a}.$$

4. (T. E. Hall 1970; see also Yamada 1967, 1970). Let S be an orthodox semigroup with band of idempotents B, and let a, b be elements of S. Using the characterization

$$a\gamma = E(aa')aE(a'a)$$

(formula (1.13)) for $a\gamma$, show that the equality of the triples

$$(aV(a), a\gamma, V(a)a), \qquad (bV(b), b\gamma, V(b)b)$$

in $B/\mathcal{R} \times S/\gamma \times B/\mathcal{L}$ implies the equality of a and b.

5. Let $B = Z^1$, where $Z = \{b, c\}$ is the two element left zero semigroup. Show that B is anti-uniform. Compute W_B and show that $W_B \neq B^*$. (Specifically, show that if α denotes the element

$$\begin{pmatrix} 1 & b & c \\ 1 & c & b \end{pmatrix}$$

of $W_{1,1}$, then the element $(\rho_1\alpha_1, \lambda_1\alpha_r^{-1})$ does not belong to B^*.) Verify that W_B is a union of groups.

6. Let $B = \mathcal{S}(Y, E_\alpha, \phi_{\alpha,\beta})$ be a normal band. (See Theorem IV.5.14) Show that if $e_\alpha \in E_\alpha$ then

$$\langle e_\alpha \rangle = \{e_\alpha \phi_{\alpha,\beta} : \beta \leqslant \alpha\},$$

and deduce that $\langle e_\alpha \rangle$ is isomorphic to the ideal $Y\alpha$ of the semilattice Y. Deduce that B is a uniform [anti-uniform] band if and only if Y is a uniform [anti-uniform] semilattice.

7. (T. E. Hall 1971). By way of contrast with the last example, construct a band B in the following way. For $n = 0, 1, 2, \ldots$ let E_n be a finite rectangular band, and suppose that

$$i \neq j \Rightarrow |E_i \neq |E_j|, \qquad E_i \cap E_j = \varnothing.$$

Let $B = \bigcup_{n=0}^{\infty} E_n$. If $x \in E_m \subset B$ and $y \in E_n \subset B$, define

n = 0

$$xy = \begin{cases} x \text{ if } m > n \\ y \text{ if } m < n \\ xy \text{ (in } E_n) \text{ if } m = n. \end{cases}$$

Show that B is an anti-uniform band but that $B/\varepsilon \simeq C_\omega$ which is a uniform semilattice.

8. Show that a band is both uniform and anti-uniform if and only if it is rectangular.

9. Given a band B and an inverse semigroup T whose semilattice of idempotents is isomorphic to B/ε, it is reasonable (in view of Theorem 4.6) to ask whether there will always exist an orthodox semigroup S whose band of idempotents is B and for which $S/\gamma \simeq T$. Using the band B described in Exercise 7 and taking T as the Munn semigroup of B/ε, show that the question has a negative answer.

10. The following examples, due to Hall (1967), show that $\mathscr{H}(B, T, \psi)$ depends not only on B and T, but also on ψ.

Let $S_1 = \{1, b, c, x\}$ and $S_2 = \{1, b, c, y\}$ be semigroups with Cayley tables

S_1	1	b	c	x
1	1	b	c	x
b	b	b	b	b
c	c	c	c	c
x	x	c	b	1

S_2	1	b	c	y
1	1	b	c	y
b	b	b	b	b
c	c	c	c	c
y	y	b	c	1

(i) Verify that S_1 and S_2 are semigroups. The easiest way of doing this is to check that S_1 is represented by matrices

$$1 = \begin{bmatrix} 1 & 0 \\ 0 & 1 \end{bmatrix}, b = \begin{bmatrix} 1 & 0 \\ 0 & 0 \end{bmatrix}, c = \begin{bmatrix} 1 & 0 \\ 1 & 0 \end{bmatrix}, x = \begin{bmatrix} 1 & 0 \\ 1 & -1 \end{bmatrix},$$

while S_2 is represented by matrices

$$1 = \begin{bmatrix} 1 & 0 & 0 \\ 0 & 1 & 0 \\ 0 & 0 & 1 \end{bmatrix}, b = \begin{bmatrix} 1 & 0 & 0 \\ 0 & 0 & 0 \\ 0 & 0 & 0 \end{bmatrix}, c = \begin{bmatrix} 1 & 0 & 0 \\ 1 & 0 & 0 \\ 0 & 0 & 0 \end{bmatrix}, y = \begin{bmatrix} 1 & 0 & 0 \\ 0 & 1 & 0 \\ 0 & 0 & -1 \end{bmatrix}.$$

(ii) Verify that S_1 and S_2 are orthodox, with band of idempotents $B = \{1, \mathrm{b}, \mathrm{c}\}$.

(iii) Show that if γ_i denotes the minimum inverse semigroup congruence on S_i ($i = 1, 2$) then γ_1 has classes $\{1\}, \{x\}, \{b, c\}$, while γ_2 has classes $\{1\}, \{y\}, \{b, c\}$. Show that $S_1/\gamma_1 \simeq S/\gamma_2 \simeq T$, where $T = \{1, 0, a\}$ is the inverse semigroup with Cayley table

	1	0	a
1	1	0	a
0	0	0	0
a	a	0	1

(iv) Show that the $\mathscr{H}$-classes in S_1 are $\{1, x\}, \{b\}, \{c\}$, that the $\mathscr{H}$-classes in S_2 are $\{1, y\}, \{b\}, \{c\}$, that $\mathscr{H}$ is a congruence on S_2 but that $\mathscr{H}$ is *not* a congruence on S_1. Deduce that S_1 and S_2 are not isomorphic.

11. Let S be an orthodox semigroup with a rectangular band B of idempotents. Use Theorem 4.6 to deduce that $S \simeq B \times G$, where G is a group. (This can also be deduced by more elementary means: see Exercise III.12.)

Chapter VII

Semigroup Amalgams

INTRODUCTION

A (*semigroup*) *amalgam* will be defined more carefully below, but may conveniently be thought of as an indexed family $\{S_i : i \in I\}$ of semigroups intersecting in a common subsemigroup U; thus $S_i \cap S_j = U$ if $i \neq j$. Effectively $\bigcup\{S_i : i \in I\}$ is a partial semigroup; that is to say, if x and y are members of the set, then the product xy may be defined; and if $x, y, z \in \bigcup\{S_i : i \in I\}$ then $(xy)z = x(yz)$ whenever both sides are meaningful. The central question concerning a semigroup amalgam is whether or not the partial semigroup $\bigcup\{S_i : i \in I\}$ is embeddable in a semigroup, i.e. whether or not there exists a semigroup T containing $\bigcup\{S_i : i \in I\}$ in which the product of two elements is *always* defined and in which previously defined multiplications within $\bigcup\{S_i : i \in I\}$ take place as before. It was shown by Schreier (1927) that a *group* amalgam is always embeddable in a group. This simple positive answer will not suffice for semigroups, and the purpose of this chapter is to determine various sufficient conditions under which semigroup amalgams are embeddable. The chapter ends with an account of a very recent result of T. E. Hall (1975) to the effect that Schreier's theorem extends to the case of inverse semigroups.

1. FREE PRODUCTS

Given an indexed family $\{S_i : i \in I\}$ of disjoint semigroups, we show how to form a semigroup $F = \Pi^* \{S_i : i \in I\}$, the *free product* of the family $\{S_i : i \in I\}$. First, let us introduce a useful notational device: if $a \in \bigcup\{S_i : i \in I\}$ then there is a unique k in I such that $a \in S_k$; we shall refer to k as the *index* of a and write $k = \sigma(a)$.

Now let F consist of all finite "strings"

$$(a_1, a_2, \ldots, a_m),$$

where $m\,(\geqslant 1)$ is an integer, where $a_r \in \bigcup\{S_i : i \in I\}$ for $r = 1, \ldots, m$ and where $\sigma(a_r) \neq \sigma(a_{r+1})$ for $r = 1, \ldots, m-1$. A binary operation is defined on F by the rule that

$$\left.\begin{aligned}(a_1, a_2, \ldots, a_m)(b_1, b_2, \ldots, b_n) &= (a_1, \ldots, a_m, b_1, \ldots, b_n) \\ &\quad \text{if } \sigma(a_m) \neq \sigma(b_1); \\ (a_1, a_2, \ldots, a_m)(b_1, b_2, \ldots, b_n) &= (a_1, \ldots, a_{m-1}, a_m b_1, b_2, \ldots, b_n) \\ &\quad \text{if } \sigma(a_m) = \sigma(b_1).\end{aligned}\right\} \quad (1.1)$$

If $\mathbf{a} = (a_1, \ldots, a_m)$, $\mathbf{b} = (b_1, \ldots, b_n)$, $\mathbf{c} = (c_1, \ldots, c_p)$ are three members of F, it is a routine matter to check that $(\mathbf{ab})\mathbf{c} = \mathbf{a}(\mathbf{bc})$. The four cases

(i) $\sigma(a_m) \neq \sigma(b_1)$, $\quad \sigma(b_n) \neq \sigma(c_1)$,

(ii) $\sigma(a_m) = \sigma(b_1)$, $\quad \sigma(b_n) \neq \sigma(c_1)$,

(iii) $\sigma(a_m) \neq \sigma(b_1)$, $\quad \sigma(b_n) = \sigma(c_1)$,

(iv) $\sigma(a_m) = \sigma(b_1)$, $\quad \sigma(b_n) = \sigma(c_1)$

must be distinguished, but no difficulties are encountered. Indeed it is worth remarking that the verification of associativity here is considerably easier than in the corresponding group-theoretic construction—see (Kurosh 1956). It must be emphasized that if the semigroup S_i are all groups then the semigroup $F = \Pi^*\{S_i : i \in I\}$ we are here describing is *not* the free product of the groups S_i as normally understood in group theory. Indeed, it is not even a group. This is because in the *group* free product all the identity elements of the individual groups are identified, while in the *semigroup* free product they remain distinct.

We have established that F is a semigroup. Among the elements of F are strings of length one, such as (s_i), where $s_i \in S_i$. In fact F is generated by these strings of length one, since

$$(a_1, a_2, \ldots, a_m) = (a_1)(a_2)\ldots(a_m)$$

for every $(a_1, a_2, \ldots, a_m)$ in F. It is customary to leave out the brackets in writing down a string of length one, and hence to regard F as consisting of finite non-empty *words* $a_1 a_2 \ldots a_m$ in the *alphabet* $\bigcup\{S_i : i \in I\}$. Multiplication of words $a_1 a_2 \ldots a_m$ and $b_1 b_2 \ldots b_n$ in F is then defined simply by juxtaposition if $\sigma(a_m) \neq \sigma(b_1)$:

$$(a_1 a_2 \ldots a_m)(b_1 b_2 \ldots b_n) = a_1 a_2 \ldots a_m b_1 b_2 \ldots b_n;$$

while if $\sigma(a_m) = \sigma(b_1) = i$ (say), then

$$(a_1 a_2 \ldots a_m)(b_1 b_2 \ldots b_n) = a_1 a_2 \ldots a_{m-1}\, c b_2 \ldots b_n,$$

where c is the product in S_i of a_m and b_1.

If the context ensures that no confusion will arise, we abbreviate $\Pi^*\{S_i: i \in I\}$ to $\Pi^* S_i$. Also, if I is finite—say $I = \{1, 2, \ldots, n\}$—we often write $S_1 * S_2 * \ldots * S_n$ for the free product $\Pi^*\{S_i: i \in I\}$.

The crucial property of free products is contained in the following result:

Proposition 1.2. *Let $F = \Pi^*\{S_i: i \in I\}$ be the free product of a family $\{S_i: i \in I\}$ of disjoint semigroups. Then for each i in I there is a monomorphism $\theta_i: S_i \to F$. If T is a semigroup for which a homomorphism $\psi_i: S_i \to T$ exists for each i in I then there is a unique homomorphism $\gamma: F \to T$ with the property that the diagram*

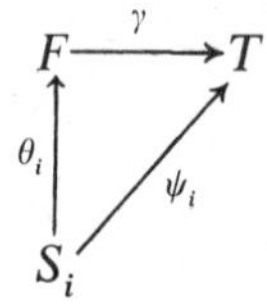

(1.3)

commutes for every i in I.

Proof. The monomorphism $\theta_i: S_i \to F$ is given by

$$s_i\theta_i = (s_i) \qquad (s_i \in S_i);$$

that is, it associates the element s_i with the one-letter word (s_i) (usually written simply as s_i) in F. It is evident that θ_i is a monomorphism. We shall very often want to identify $s_i\theta_i$ with s_i.

If T and $\{\psi_i: i \in I\}$ are given, we define $\gamma: F \to T$ by

$$(a_1 a_2 \ldots a_m)\gamma = (a_1\psi_{\sigma(a_1)})(a_2\psi_{\sigma(a_2)}) \ldots (a_m\psi_{\sigma(a_m)}).$$

The expression on the right is a product of elements of T and so is an element of T; thus γ maps F into T. To show that γ is a homomorphism, let

$$a = a_1 a_2 \ldots a_m, \qquad b = b_1 b_2 \ldots b_n$$

be elements of F. Then, if $\sigma(a_m) \neq \sigma(b_1)$,

$$\begin{aligned} (ab)\gamma &= (a_1 a_2) \ldots a_m b_1 b_2 \ldots b_n)\gamma \\ &= (a_1\psi_{\sigma(a_1)}) \ldots (a_m\psi_{\sigma(a_m)})\,(b_1\psi_{\sigma(b_1)}) \ldots (b_n\psi_{\sigma(b_n)}) \\ &= [(a_1\psi_{\sigma(a_1)}) \ldots (a_m\psi_{\sigma(a_m)})]\,[(b_1\psi_{\sigma(b_1)}) \ldots (b_n\psi_{\sigma(b_n)})] \\ &= (a\gamma)(b\gamma). \end{aligned}$$

If $\sigma(a_m) = \sigma(b_1) = i$ (say) and $a_m b_1 = c$ in S_i,

$$\begin{aligned} (ab)\gamma &= (a_1 \ldots a_{m-1}\, c\, b_2 \ldots b_n)\gamma \\ &= (a_1\psi_{\sigma(a_1)}) \ldots (a_{m-1}\psi_{\sigma(a_{m-1})})\,(c\psi_i)\,(b_2\psi_{\sigma(b_2)}) \ldots (b_n\psi_{\sigma(b_n)}) \end{aligned}$$

$$= (a_1\psi_{\sigma(a_1)})\dots(a_{m-1}\psi_{\sigma(a_{m-1})})(a_m\psi_i)(b_1\psi_i)(b_2\psi_{\sigma(b_2)})\dots(b_n\psi_{\sigma(b_n)})$$

(since ψ_i is a homomorphism)

$$= [(a_1\psi_{\sigma(a_1)})\dots(a_m\psi_{\sigma(a_m)})][(b_1\psi_{\sigma(b_1)})\dots(b_n\psi_{\sigma(b_n)})]$$

$$= (a\gamma)(b\gamma).$$

To see that the diagram (1.3) commutes for every i, observe that the definitions of θ_i and γ imply that

$$s_i(\theta_i\gamma) = (s_i\theta_i)\gamma = (s_i)\gamma = s_i\psi_i$$

for every s_i in S_i. Hence $\theta_i\gamma = \psi_i$ as required.

The uniqueness of γ follows from the fact that F is generated by words of length one. If γ is to make the diagram (1.3) commute then we *must* have $(s_i)\gamma = s_i\psi_i$ for every s_i in S_i and for every i in I. Then, if γ is to be a homomorphism we *must* have

$$(a_1a_2\dots a_m)\gamma = (a_1)\gamma\,.\,(a_2)\gamma\dots(a_m)\gamma$$
$$= (a_1\psi_{\sigma(a_1)})(a_2\psi_{\sigma(a_2)})\dots(a_m\psi_{\sigma(a_m)})$$

for every $a_1a_2\dots a_m$ in F; that is, γ must be exactly as we defined it. ■

The property described in Proposition 1.2 does in fact *characterize* the free product. More precisely:

Proposition 1.4. *Let* $\{S_i : i \in I\}$ *be a family of semigroups and let* H *be a semigroup such that:*

(i) *there exists a monomorphism* $\alpha_i: S_i \to H$ *for each* i *in* I;

(ii) *if* T *is a semigroup for which a homomorphism* $\beta_i: S_i \to T$ *exists for each* i *in* I *then there is a unique homomorphism* $\delta: H \to T$ *such that*

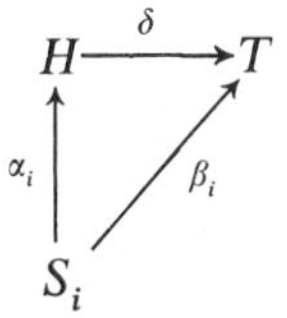

commutes for every i *in* I.

Then H *is isomorphic to* $F = \Pi^*\{S_i : i \in I\}$.

Proof. The property of F established in Proposition 1.2 requires, when applied to the case when $T = F$ and $\psi_i = \theta_i$ $(i \in I)$, the existence of a unique homomorphism $F \to F$ making the diagram

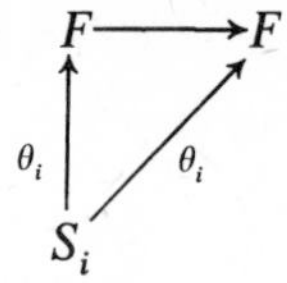

commute for every i in I. It is evident that the identity mapping $1_F : F \to F$ has this property; hence by uniqueness it is the *only* homomorphism from F into F having the property. Equally, by the assumed property of H, the identity mapping 1_H is the only homomorphism from H into H with the property that the diagram

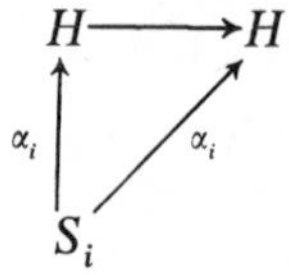

commutes for every i in I. If now we apply Proposition 1.2 with $T = H$ and $\psi_i = \alpha_i$ $(i \in I)$ we obtain a homomorphism $\gamma: F \to H$ such that the diagram

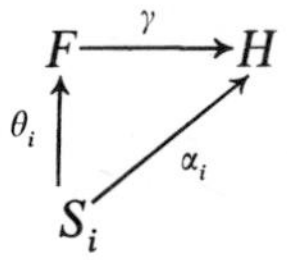

(1.5)

commutes for every i in I. Again, by the assumed property of H with $T = F$ and $\beta_i = \theta_i$ $(i \in I)$ we obtain a homomorphism $\delta: H \to F$ such that the diagram

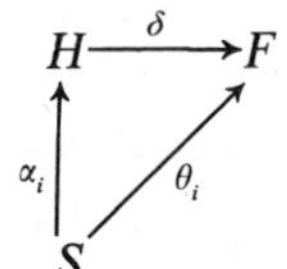

(1.6)

commutes for every i in I. It follows that

$$\theta_i\gamma\delta = \alpha_i\delta = \theta_i, \qquad \alpha_i\delta\gamma = \theta_i\gamma = \alpha_i \qquad (i \in I);$$

that is, if we "tack together" the diagram (1.5) and (1.6) in both of the possible ways we obtain commutative diagrams

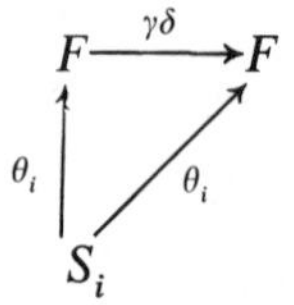

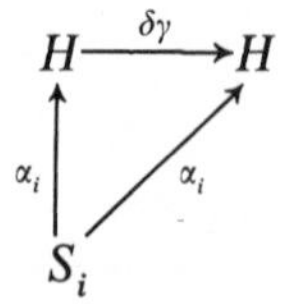

for every i in I. From our earlier remarks we now immediately deduce that $\gamma\delta = 1_F$, $\delta\gamma = 1_H$. Thus γ and δ are mutually inverse isomorphisms and so $H \simeq F$ as required.■

The conclusions of Proposition 1.2 and 1.4 can be summarized by saying that $F = \Pi^* S_i$ is the unique *coproduct*, in the sense of category theory, of the objects S_i. (See, for example, Mitchell 1965.)

We turn now to the more complex situation that faces us when we consider a semigroup amalgam. The intuitive definition in the introduction is not now precise enough, and so we begin by giving a more careful definition as follows. A (*semigroup*) *amalgam* $\mathfrak{A} = [\{S_i : i \in I\}; U; \{\phi_i : i \in I\}]$ consists of a semigroup U (called the *core* of the amalgam), a family $\{S_i : i \in I\}$ of semigroups disjoint from each other and from U, and a family of monomorphisms $\phi_i : U \to S_i$ $(i \in I)$. We shall simplify the notation to $\mathfrak{A} = [S_i; U \cdot \phi_i]$ or to $\mathfrak{A} = [S_i; U]$ when the context allows.

We shall say that the amalgam $\mathfrak{A}$ is *embedded* in the semigroup T if there exist a monomorphism $\lambda : U \to T$ and for each i in I a monomorphism $\lambda_i : S_i \to T$ with the properties:

(a) $\phi_i \lambda_i = \lambda$ for each i in I;
(b) $S_i \lambda_i \cap S_j \lambda_j = U\lambda$ for all i, j in I such that $i \neq j$.

The earlier intuitive definition differs from this only in that all the monomorphisms are regarded as inclusion mappings.

The *free product* $\Pi^*_U S_i$ of the amalgam $\mathfrak{A}$ is a quotient semigroup of the ordinary free product $\Pi^* S_i$ in which for each i and j in I the image $u\phi_i$ of an element u of U in S_i is identified with its image $u\phi_j$ in S_j. More precisely, if as before we denote by θ_i the natural monomorphism from S_i into $\Pi^* S_i$, we define $P = \Pi^*_U S_i$ to be $(\Pi^* S_i)/\rho$, where ρ is the congruence on $\Pi^* S_i$ generated by the subset

$$\mathbf{R} = \{(u\phi_i\theta_i,\ u\phi_j\theta_j) : u \in U,\ i, j \in I\} \tag{1.7}$$

of $(\Pi^* S_i) \times (\Pi^* S_i)$.

It is clear that for each i in I there is a homomorphism $\mu_i = \theta_i \rho^\natural$ from S_i into P. It is clear, moreover, from the definition of ρ that for every u in U and for all i, j in I,

$$u\phi_i\mu_i = u\phi_i\theta_i\rho = u\phi_j\theta_j\rho = u\phi_j\mu_j;$$

thus there exists a homomorphism $\mu : U \to P$ such that $\phi_i\mu_i = \mu$ for every i in I. The amalgam $\mathfrak{A}$ is thus embedded in P if

(1.8) each μ_i is one-one;

(1.9) $S_i\mu_i \cap S_j\mu_j \subseteq U\mu$ for all i, j in I such that $i \neq j$.

Notice that for all i, j in I,

$$S_i\mu_i \supseteq U\phi_i\mu_i = U\mu, \quad S_j\mu_j \supseteq U\phi_j\mu_j = U\mu;$$

thus $S_i\mu_i \cap S_j\mu_j \supseteq U\mu$ for every amalgam $\mathfrak{A}$, and so the condition (1.9) is equivalent to the apparently stronger condition

$$S_i\mu_i \cap S_j\mu_j = U\mu.$$

If the conditions (1.8) and (1.9) hold, we shall say that the amalgam $\mathfrak{A} = [S_i; U; \phi_i]$ is *naturally embedded in its free product* $\Pi_U^* S_i$. We shall see shortly that if an amalgam is embeddable in a semigroup at all then it is naturally embedded in its free product. First, however, we have

Proposition 1.10. *If* $\mathfrak{A} = [\{S_i : i \in I\}; U; \{\phi_i : i \in I\}]$ *is a semigroup amalgam, then the free product* $P = \Pi_U^* S_i$ *of* $\mathfrak{A}$ *is the "pushout" of the diagram* $\{U \to S_i\}_{i \in I}$. *That is:*

(*a*) *there exists for each i in I a homomorphism* $\mu_i : S_i \to P$ *such that the diagram* $\{U \to S_i \to P\}_{i \in I}$ *commutes, i.e. such that* $\phi_i\mu_i = \phi_j\mu_j$ *for all i, j in I;*

(*b*) *if Q is a semigroup for which homomorphisms* ν_i: $S_i \to Q$ $(i \in I)$ *exist such that* $\phi_i\nu_i = \phi_j\nu_j$ *for all i, j in I, then there exists a unique homomorphism* $\delta: P \to Q$ *such that the diagram*

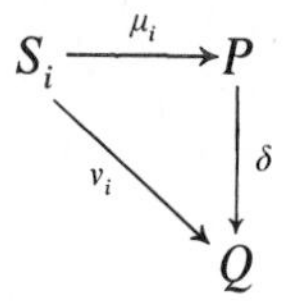

commutes for each i in I.

Proof. Property (a) has already been observed. To see that (b) holds, we first obtain by Proposition 1.2 a unique homomorphism $\gamma : F \to Q$ such that

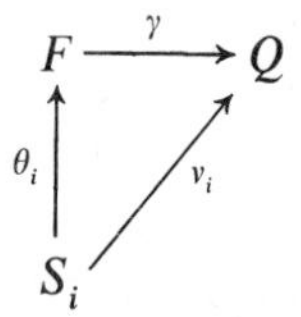

is a commutative diagram for each i in I. If $i, j \in I$, then, for all u in U,

$$u\phi_i\theta_i\gamma = u\phi_i\nu_i = u\phi_j\nu_j = u\phi_j\theta_j\gamma;$$

hence $(u\phi_i\theta_i, u\phi_j\theta_j) \in \gamma \circ \gamma^{-1}$. Recalling formula (1.7), we deduce that $\mathbf{R} \subseteq \gamma \circ \gamma^{-1}$. Hence, since $\gamma \circ \gamma^{-1}$ is a congruence,

$$\rho = \mathbf{R}^\# \subseteq \gamma \circ \gamma^{-1}.$$

By Theorem I.5.4 it follows that the homomorphism $\gamma: F \to Q$ factors through $P = F/\rho$; i.e. there is a unique homomorphism $\delta: P \to Q$ such that the diagram

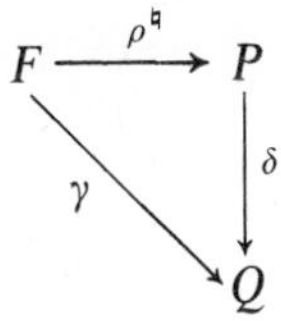

commutes. It now follows from the definition of μ_i and from the commutativity of these two diagrams that

$$\mu_i \delta = \theta_i \rho^\natural \delta = \theta_i \gamma = \nu_i$$

for each i in I, i.e. that the diagram

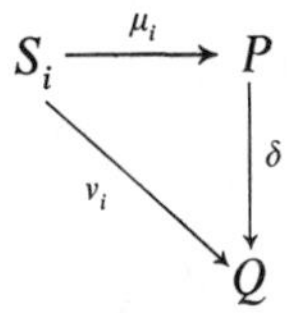

commutes.■

From this we can readily deduce

Theorem 1.11. *The semigroup amalgam*

$$\mathfrak{A} = [\{S_i : i \in I\}; U; \{\phi_i : i \in I\}]$$

is embeddable in a semigroup if and only if it is naturally embedded in its free product.

Proof. One way round this is obvious. Suppose therefore that $\mathfrak{A}$ is embedded in a semigroup T, i.e. that there exist monomorphisms $\lambda: U \to T$, $\lambda_i: S_i \to T\ (i \in I)$ such that $\phi_i \lambda_i = \lambda$ for every i in I and such that $S_i \lambda_i \cap S_j \lambda_j = U\lambda\ (i \neq j)$. By Proposition 1.10(b) there exists a unique homomorphism $\delta: P \to T$ such that

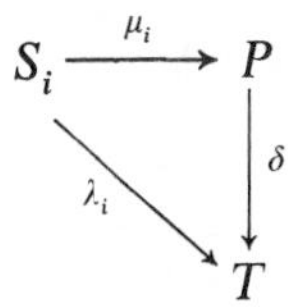

is a commutative diagram for each i in I. It follows that $\mu_i: S_i \to P$ is a monomorphism for every i in I, since if $x, y \in S_i$ then

$$x\mu_i = y\mu_i \Rightarrow x\mu_i\delta = y\mu_i\delta \Rightarrow x\lambda_i = y\lambda_i \Rightarrow x = y.$$

Also, if $i \neq j$ and if $x = s_i\mu_i = s_j\mu_j \in S_i\mu_i \cap S_j\mu_j$, then $x\delta = s_i\mu_i\delta = s_i\lambda_i \in S_i\lambda_i$ and similarly $x\delta \in S_j\lambda_j$. Then $x\delta \in S_i\lambda_i \cap S_j\lambda_j = U\lambda$. That is, there exists u in U such that $x\delta = u\lambda$, i.e. such that $s_i\lambda_i = u\phi_i\lambda_i$. Since λ_i is a monomorphism, it follows that $s_i = u\phi_i$. Hence $x = s_i\mu_i = u\phi_i\mu_i \in U\mu$. We have shown that both (1.8) and (1.9) hold; hence $\mathfrak{A}$ is naturally embedded in P as required.■

2. DOMINIONS AND ZIGZAGS

We turn now to an idea which appears at first sight to have no connection at all with amalgams, but which turns out on closer examination to be intimately connected with them. First, if U is a subsemigroup of a semigroup S, then, following Isbell (1965), we say that U *dominates* an element d of S if for all semigroups T and for all homomorphisms $\beta, \gamma: S \to T$,

$$[(\forall u \in U)\ u\beta = u\gamma] \Rightarrow d\beta = d\gamma.$$

More informally, U dominates d if any two homomorphisms of S that coincide on elements of U coincide also on d. The set of elements dominated by U is called the *dominion* of U in S and is written $\mathrm{Dom}_S(U)$. It is evident that $U \subseteq \mathrm{Dom}_S(U)$. Also, $\mathrm{Dom}_S(U)$ is a subsemigroup of S, since if $d, d' \in \mathrm{Dom}_S(U)$ and $\beta, \gamma: S \to T$ are such that $u\beta = u\gamma$ for every u in U, then $d\beta = d\gamma$ and $d'\beta = d'\gamma$, from which it follows that

$$(dd')\beta = (d\beta)(d'\beta) = (d\gamma)(d'\gamma) = (dd')\gamma;$$

thus $dd' \in \mathrm{Dom}_S(U)$.

The function $U \to \mathrm{Dom}_S(U)$ is a *closure* operation in the sense that is usual in algebra—see Exercise 1. If $\mathrm{Dom}_S(U) = U$ we say that U is a *closed* subsemigroup of S, while if $\mathrm{Dom}_S(U) = S$ we say that U is a *dense* subsemigroup of S. Both these types of semigroup can arise, and it can also happen that

$$U \subset \mathrm{Dom}_S(U) \subset S.$$

(See Exercise 2.)

It is of course potentially very hard to discover whether or not an element belongs to the dominion of a subsemigroup, since the definition of dominion involves phrases such as "for all semigroups" and "for all homomorphisms". For this reason it is important to characterize the notion of dominion in some more accessible way. With this end in view, consider a semigroup S and a subsemigroup U of S, and let S' be an isomorphic copy of S, disjoint from S. We denote the isomorphism by $\alpha: S \to S'$ and write $s\alpha$ as s'. Let $U' = \{u' : u \in U\}$. We now examine the amalgam $\mathfrak{A} = [\{S, S'\}; U; \{\iota, \alpha | U\}]$, where ι is the inclusion mapping of U into S, and show that it is what is sometimes called *weakly embeddable*. With reference to the free product of the amalgam, this means that property (1.8) is satisfied but not necessarily property (1.9). Stating the result formally, we have

Proposition 2.1. *Let S be a semigroup and let U be a subsemigroup of S. Let S' be a semigroup disjoint from S and let $\alpha: S \to S'$ be an isomorphism. Let $P = S *_U S'$ be the free product of the amalgam*

$$\mathfrak{A} = [\{S, S'\}; U; \{\iota, \alpha | U\}],$$

where ι is the inclusion mapping of U into S. Then the natural mappings $\mu: S \to P$, $\mu': S' \to P$ defined by

$$s\mu = (s\theta)\rho \quad (s \in S), \qquad s'\mu' = (s'\theta')\rho \quad (s' \in S')$$

are both monomorphisms.

Proof. In the statement of the proposition, θ and θ' are the standard monomorphisms from S and S' respectively into the free product $S * S'$, and

$$\rho = \{(u\theta, u'\theta') : u \in U\}.$$

Certainly we have a commutative diagram

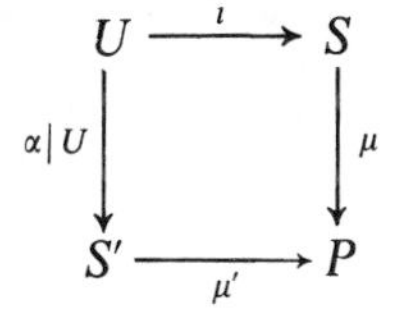

We also have a commutative diagram

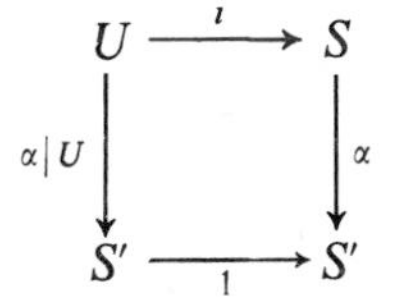

and so, by the pushout property of the first diagram (more precisely, by part (b) of Proposition 1.10) there is a unique homomorphism $\delta: P \to S'$ such that

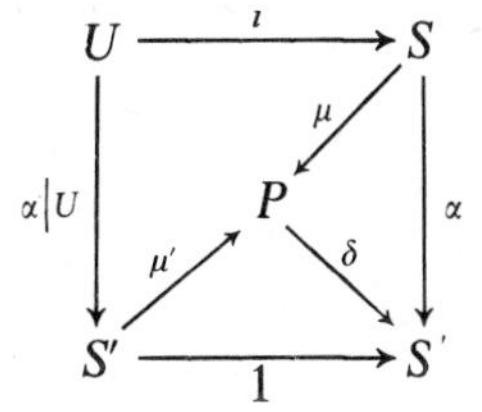

is a commutative diagram. From this it readily follows that μ and μ' are monomorphisms; for if $s_1, s_2 \in S$ then

$$s_1\mu = s_2\mu \Rightarrow s_1\mu\delta = s_2\mu\delta \Rightarrow s_1\alpha = s_2\alpha \Rightarrow s_1 = s_2;$$

and if $s'_1, s'_2 \in S'$, then

$$s'_1\mu' = s'_2\mu' \Rightarrow s'_1\mu'\delta = s'_2\mu'\delta \Rightarrow s'_1 = s'_2. \blacksquare$$

It is natural now to ask whether the amalgam $\mathfrak{A} = [\{S, S'\}; U; \{\iota, \alpha|U\}]$ has the second property (1.9) required for embeddability. In general it does not. In fact Isbell (1965) proved the following result:

Theorem 2.3. *Let U be a subsemigroup of a semigroup S. Let S' be a semigroup disjoint from S and let $\alpha: S \to S'$ be an isomorphism. Let $P = S *_U S'$, the free product of the amalgam*

$$\mathfrak{A} = [\{S, S'\}; U; \{\iota, \alpha|U\}],$$

where ι is the inclusion mapping of U into S, and let μ, μ' be the natural monomorphisms from S, S', respectively, into P. Then

$$(S\mu \cap S'\mu')\mu^{-1} = \mathrm{Dom}_S(U).$$

Proof. The commutativity of the diagram

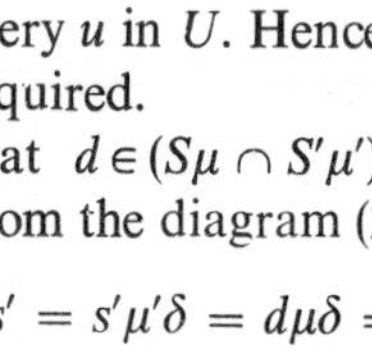

ensures that the monomorphisms $\mu: S \to P$ and $\alpha\mu': S \to P$ have the property that $u\mu = u\alpha\mu'$ for every u in U. Hence if $d \in \mathrm{Dom}_S(U)$ then $d\mu = d\alpha\mu'$ and so $d\mu \in S\mu \cap S'\mu'$ as required.

Conversely, suppose that $d \in (S\mu \cap S'\mu')\mu^{-1}$. Then there exists $s' \in S'$ such that $s'\mu' = d\mu$. Now from the diagram (2.2) it follows that

$$s' = s'\mu'\delta = d\mu\delta = d\alpha.$$

Now let $\beta: S \to T$ and $\gamma: S \to T$ be homomorphisms such that $u\beta = u\gamma$ for every u in U. Then the diagram

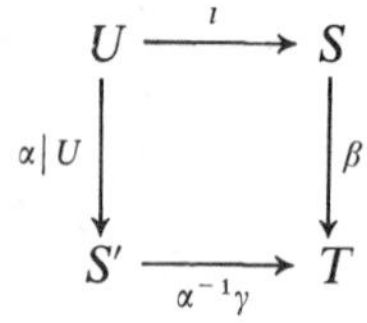

commutes and so, by Proposition 1.10(b), there exists a unique $\xi: P \to T$ such that the diagram

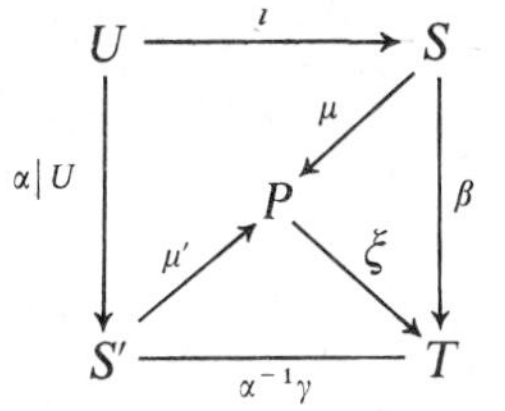

commutes. Hence

$$d\beta = d\mu\,\xi = s'\mu'\xi = d\alpha\mu'\xi = d\alpha\alpha^{-1}\gamma = d\gamma$$

and so $d \in \mathrm{Dom}_S(U)$.■

While Theorem 2.3 provides a very interesting characterization of the dominion, it may still be very hard to determine whether or not an element d in S belongs to $(S\mu \cap S'\mu')\mu^{-1}$. A more usable characterization of the dominion has been provided by Isbell (1965) in terms of what he calls *zigzags*. Isbell's proof involves a degree of 'handwaving' that is (to use his own word) objectionable. A successful amplification of his approach has been provided by Philip (1974), but the quite different approach of Stenström (1971) has many advantages and it is his version in essence that we shall now consider.

If M is a set and S is a semigroup with identity 1, we shall say that M is a *right S-system* if there is a mapping $(x, s) \mapsto xs$ from $M \times S$ into M with the properties that

$$(xs)t = x(st) \qquad (x \in M, s, t \in S)$$

and $x1 = x$ $(x \in M)$. Equally, M is a *left S-system* if there is a mapping $(s, x) \mapsto sx$ from $S \times M$ into M such that

$$s(tx) = (st)x \qquad (x \in M, s, t \in S)$$

and $1x = x$ $(x \in M)$. If S and T are semigroups with identity we say that M is an *(S, T)-bisystem* if it is a left S-system, a right T-system and if, for all $s \in S$, $t \in T$ and $x \in M$,

$$(sx)t = s(xt).$$

These definitions are of course closely modelled on the definitions of right modules, left modules and bimodules. Clifford and Preston (1967, Chapter 11) use the term "operand" rather than "system".

If M is a right S-system and N is a left S-system, let τ be the equivalence relation on $M \times N$ generated by the subset

$$\{((xs, y), (x, sy)) : x \in M, s \in S, y \in N\}$$

of $(M \times N)^2$. The set $(M \times N)/\tau$ will be denoted by $M \otimes_S N$ and will be called the *tensor product over S* of the two S-systems. The equivalence class $(x, y)\tau$ determined by an element (x, y) of $M \times N$ will be written $x \otimes y$. From the definition of τ it is immediate that

$$xs \otimes y = x \otimes sy$$

for all $x \in M$, $s \in S$ and $y \in N$.

With M and N as described, the tensor product $M \otimes_S N$ is merely a set, but if M and N have extra structure the tensor product inherits some of this. Precisely, if M is a (T, S)-bisystem and N is an (S, U)-bisystem, then $M \otimes_S N$ becomes a (T, U)-bisystem if we define

$$t(x \otimes y) = tx \otimes y, \qquad (x \otimes y)u = x \otimes yu$$

for $t \in T$, $u \in U$, $x \otimes y \in M \otimes_S N$. The verification that this is so is entirely routine and is omitted.

If P and Q are right S-systems, we say that a mapping $\alpha : P \to Q$ is a *morphism* of *right S-systems* if, for every x in P and every s in S,

$$(xs)\alpha = (x\alpha)s.$$

Similar definitions apply to left S-systems and (S, T)-bisystems.

Let us now suppose that S is a semigroup not necessarily having an identity, and that U is a subsemigroup of S. Let S^* be the semigroup obtained from S by adjoining an extra identity element 1 *whether or not it already has one*, and let $U^* = U \cup \{1\}$; then U^* is a subsemigroup of S^*. We may in an obvious way regard S^* as either a left- or a right U^*-system and so may form the tensor product $A = S^* \otimes_{U^*} S^*$.

Theorem 2.5. (Stenström 1971). *If U is a subsemigroup of a semigroup S and if $d \in S$, then $d \in \mathrm{Dom}_S(U)$ if and only if $d \otimes 1 = 1 \otimes d$ in the tensor product $A = S^* \otimes_{U^*} S^*$.*

Proof. Suppose first that $d \in S$ and that $d \otimes 1 = 1 \otimes d$ in A. The tensor product A is $(S^* \times S^*)/\tau$, where τ is the equivalence relation on $S^* \times S^*$ generated by

$$\mathbf{T} = \{((xu, y), (x, uy)) : x, y \in S^*, u \in U^*\}.$$

Let T be a semigroup and let $\beta : S \to T$, $\gamma : S \to T$ be homomorphisms coinciding on U. If we define $1\beta = 1\gamma = 1$ we may regard β and γ as homomorphisms of S^* into T^* coinciding on U^*. Define $\psi : S^* \times S^* \to T^*$ by

$$(x, y)\psi = (x\beta)(x\gamma) \qquad ((x, y) \in S^* \times S^*).$$

Then $\mathbf{T} \subseteq \psi \circ \psi^{-1}$, since for all x, y in S^* and for all u in U^*

$$(xu, y)\psi = [(xu)\beta](y\gamma) = (x\beta)(u\beta)(y\gamma) = (x\beta)(u\gamma)(y\gamma)$$
$$= (x\beta)[(uy)\gamma] = (x, uy)\psi.$$

Hence $\tau = \mathbf{T}^e \subseteq \psi \circ \psi^{-1}$, since $\psi \circ \psi^{-1}$ is an equivalence relation. (See Section I.4.) Hence there is a unique well-defined mapping $\chi : (S^* \times S^*)/\tau \to T^*$ such that $\tau^\natural \chi = \psi$; i.e. there is a mapping $\chi : A \to T^*$ given by

$$(x \otimes y)\chi = (x\beta)(y\gamma) \qquad (x \otimes y \in A).$$

We are assuming that $d \otimes 1 = 1 \otimes d$ in A. Hence $(d \otimes 1)\chi = (1 \otimes d)\chi$; i.e. $d\beta = d\gamma$. Thus $d \in \mathrm{Dom}_S U$.

To prove the converse we begin by making a construction due to Silver (1967) and adapted by Stenström (1971) to the case of semigroups. Since we may clearly regard S^* as an (S^*, U^*)-bisystem and as a (U^*, S^*)-bisystem, we may regard the tensor product A as having an (S^*, S^*)-bisystem structure defined by

$$s(x \otimes y) = sx \otimes y, \qquad (x \otimes y)s = x \otimes ys$$

$(s, x, y \in S^*)$. Let $(\mathbf{Z}(A), +)$ be the free abelian group on the set A, i.e. the set of all finite linear combinations $\sum z_i a_i$ of elements with integral coefficients z_i, with the obvious addition. [More formally, we could regard $\mathbf{Z}(A)$ as the set of all mappings $\zeta : A \to \mathbf{Z}$ such that $a\zeta = 0$ for all but finitely many a in A, and define $\zeta + \tau$, $-\zeta$ by

$$a(\zeta + \tau) = a\zeta + a\tau, \qquad a(-\zeta) = -a\zeta \qquad (a \in A).]$$

The abelian group $\mathbf{Z}(A)$ inherits a natural (S^*, S^*)-bisystem structure from A if we define

$$s(\sum z_i a_i) = \sum z_i(sa_i), \qquad (\sum z_i a_i)s = \sum z_i(a_i s)$$

for all $s \in S^*$ and $\sum z_i a_i \in \mathbf{Z}(A)$. Notice also that if $x, y \in \mathbf{Z}(A)$ and $s \in S^*$, then

$$s(x + y) = sx + sy, \qquad (x + y)s = xs + ys. \tag{2.6}$$

Next, we define a binary operation on $S^* \times \mathbf{Z}(A)$ by

$$(p, x)(q, y) = (pq, py + xq). \tag{2.7}$$

The operation is associative, since for every (p, x), (q, y) and (r, z) in $S^* \times \mathbf{Z}(A)$.

$$[(p, x)(q, y)](r, z) = (pq, py + xq)(r, z)$$
$$= ((pq)r, (pq)z + (py + xq)r)$$
$= (p(qr), p(qz) + p(yr) + x(qr))$, by associativity in S^*, by (2.6) and by the bisystem laws for $\mathbf{Z}(A)$,

$= (p(qr), p(qz + yr) + x(qr))$, by (2.6),

$$= (p, x)(qr, qz + yr)$$

$$= (p,x)[(q, y)(r, z)].$$

Thus $S^* \times \mathbf{Z}(A)$ is a semigroup, and it is easy to verify that it has identity element $(1, 0)$.

We now consider two homomorphisms β and γ from S^* into $S^* \times \mathbf{Z}(A)$, and show that they coincide on elements of U. First, define β by

$$s\beta = (s, 0) \qquad (s \in S^*);$$

it is obvious that β is a semigroup homomorphism. Less obviously, if $\gamma : S^* \to S^* \times \mathbf{Z}(A)$ is defined by

$$s\gamma = (s, s(1 \otimes 1) - (1 \otimes 1)s) \qquad (s \in S^*)$$

then γ too is a semigroup homomorphism; for if we denote $1 \otimes 1$ by a for brevity, we easily verify from (2.7), (2.6) and the bisystem laws for $\mathbf{Z}(A)$ that

$$(s, sa - as)(t, ta - at) = (st, s(ta - at) + (sa - as)t) = (st, (st)a - a(st)).$$

If $u \in U^*$, then

$$u(1 \otimes 1) = u \otimes 1 = 1u \otimes 1 = 1 \otimes u1 = (1 \otimes 1)u.$$

and so $u\beta = u\gamma$. Removing identities, we thus have two homomorphisms β and γ from S into $S^* \times \mathbf{Z}(A)$ such that $u\beta = u\gamma$ for all u in U. If $d \in \mathrm{Dom}_S(U)$ we must therefore have that $d\beta = d\gamma$, i.e. that $d \otimes 1 = 1 \otimes d$. ■

When we come to investigate what this result amounts to in more elementary terms, we observe that $d \otimes 1 = 1 \otimes d$ only if the elements $(d, 1)$ and $(1, d)$ of $S^* \times S^*$ are equivalent under the relation τ generated by the subset

$$\mathbf{T} = \{((xu, y), (x, uy)); x, y \in S^*, u \in U^*\},$$

i.e. if and only if $(1, d)$ is connected to $(d, 1)$ by a finite sequence of steps either of the form

$$(xu, y) \to (x, uy) \tag{2.8}$$

or of the form

$$(x, uy) \to (xu, y). \tag{2.9}$$

If we have two successive steps

$$(xu, y) \to (x, uy) = (zv, uy) \to (z, vuy)$$

of the former type we can achieve the same effect with a single step of this type:

$$(xu, y) = (zvu, y) \to (z, vuy).$$

A similar remark applies to consecutive steps of the latter type, and so we may assume that steps of the two types occur alternately in the sequence connecting $(1, d)$ to $(d, 1)$.

Since 1 is an "extra" identity having no factors in S, the first step must be of the form (2.9):

$$(1, d) = (1, uy) \rightarrow (u, y).$$

The last step must also be of this type:

$$(x, u) \rightarrow (xu, 1) = (d, 1).$$

Hence the statement $d \otimes 1 = 1 \otimes d$ is equivalent to the statement that $(1, d)$ is connected to $(d, 1)$ by a sequence of steps as follows:

$$\begin{aligned}(1, d) &= (1, u_0 y_1) \rightarrow (u_0, y_1)\\ &= (x_1 u_1, y_1) \rightarrow (x_1, u_1 y_1)\\ &= (x_1, u_2 y_2) \rightarrow (x_1 u_2, y_2)\\ &= \ldots.\\ &= (x_i u_{2i-1}, y_i) \rightarrow (x_i, u_{2i-1} y_i)\\ &= (x_i, u_{2i} y_{i+1}) \rightarrow (x_i u_{2i}, y_{i+1})\\ &= \ldots.\\ &= (x_m, u_{2m}) \rightarrow (x_m u_{2m}, 1) = (d, 1),\end{aligned} \tag{2.10}$$

where $u_0, \ldots, u_{2m} \in U^*$, $x_1, \ldots, x_m$, $y_1, \ldots, y_m \in S^*$, and where

$$d = u_0 y_1, \qquad u_0 = x_1 u_1$$

$$u_{2i-1} y_i = u_{2i} y_{i+1}, \qquad x_i u_{2i} = x_{i+1} u_{2i+1} \qquad (i = 1, \ldots, m-1)$$

$$u_{2m-1} y_m = u_{2m}, \qquad x_m u_{2m} = d.$$

Now, we may as well assume that $u_0, \ldots, u_{2m} \in U$, since a step of type (2.8) or (2.9) in which $u = 1$ is of no account. If any x_i is 1, let x_k be the *last* x_i that is equal to 1; thus we have a subsequence of (2.10) as follows:

$$(1, d) \rightarrow \ldots \rightarrow (1, u_{2k} y_{k+1}) \tag{2.11}$$

(but ending $(1, u_{2m})$ if $k = m$). Now it is clear that if (p, q) and (r, s) are connected by a sequence of steps of the forms (2.8) and (2.9) then $pq = rs$. In the present instance this gives $d = u_{2k} y_{k+1}$ (or $d = u_{2m}$); hence the subsequence (2.11) merely connects $(1, d)$ to $(1, d)$ and so may be left out. What remains is a sequence of the form (2.10) in which no x_i is 1.

A similar argument shows that if y_k is the *first* y_i that is equal to 1 then the subsequence

$$(x_k u_{2k-1}, 1) \rightarrow \ldots \rightarrow (d, 1)$$

can be left out. We may therefore assume that in the sequence (2.10) no x_i is 1 and no y_i is 1.

We can now obtain Isbell's Zigzag Theorem in its original form. First, if U is a subsemigroup of a semigroup S, a system of equalities

$$d = u_0 y_1, \qquad u_0 = x_1 u_1$$

$$u_{2i-1} y_i = u_{2i} y_{i+1}, \qquad x_i u_{2i} = x_{i+1} u_{2i+1} \qquad (i = 1, 2, \ldots, m-1)$$

$$u_{2m-1} y_m = u_{2m}, \qquad x_m u_{2m} = d \tag{2.12}$$

is called a *zigzag of length m in S over U with value d*. By the *spine* of the zigzag we mean the ordered $(2m+1)$-tuple $(u_0, u_1, \ldots, u_{2m})$.

Theorem 2.13. (Isbell's Zigzag Theorem). If U is a *semigroup S and if* $d \in S$, *then* $d \in \mathrm{Dom}_S(U)$ *if and only if either* $d \in U$ *or there exists a zigzag in S over U with value d.*■

We conclude this section with one application of the Zigzag Theorem. This and other applications are to be found in (Howie and Isbell 1967). A semigroup S is called *absolutely closed* if $\mathrm{Dom}_W(S) = S$ for *every* semigroup W containing S.

Theorem 2.14. *Inverse semigroups are absolutely closed.*

Proof. Let U be an inverse semigroup and let S be a semigroup containing U as a subsemigroup. If $d \in \mathrm{Dom}_S(U)$ then there exists a zigzag (2.12) in S over U with value d. The method of proof is to show that the zigzag (2.12) can be replaced by a zigzag

$$d = u_0 y_1, \qquad u_0 = z_1 u_1,$$

$$u_{2i-1} y_i = u_{2i} y_{i+1}, \qquad z_i u_{2i} = z_{i+1} u_{2i+1} \qquad (i = 1, \ldots, m-1)$$

$$u_{2m-1} y_m = u_{2m}, \qquad z_m u_{2m} = d \tag{1.15}$$

which is *left-inner*, i.e. for which $z_1, \ldots, z_m$ all belong to U. From this it immediately follows that $d = z_m u_{2m} \in U$.

In defining the elements $z_1, \ldots, z_m$, it is convenient to begin by defining elements $v_1, \ldots, v_m$ in U as follows:

$$v_1 = u_1^{-1}, \qquad v_{i+1} = v_i u_{2i} u_{2i+1}^{-1} \qquad (i = 1, \ldots, m-1).$$

Now define $z_i = u_0 v_i$ $(i = 1, \ldots, m)$. The elements z_i evidently all belong to U. Also,

$$u_0 = x_1 u_1 = x_1 u_1 u_1^{-1} u_1 = u_0 u_1^{-1} u_1 = z_1 u_1,$$

and so the top right-hand equality of (2.15) is satisfied. To show that the other equalities hold, it is helpful first to prove by induction that

$$z_i = x_i v_i^{-1} v_i \qquad (i = 1, \ldots, m). \tag{2.16}$$

This is obvious for $i = 1$. If we assume it for $i = r$, then

$$\begin{aligned} z_{r+1} &= u_0 v_{r+1} = u_0 v_r u_{2r} u_{2r+1}^{-1} \\ &= z_r u_{2r} u_{2r+1}^{-1} = x_r v_r^{-1} v_r u_{2r} u_{2r+1}^{-1} \quad \text{(by hypothesis)} \\ &= x_r (v_r^{-1} v_r)(u_{2r} u_{2r}^{-1}) u_{2r} u_{2r+1}^{-1} \\ &= x_r u_{2r} u_{2r}^{-1} v_r^{-1} v_r u_{2r} u_{2r+1}^{-1} \quad \text{(by commuting idempotents)} \\ &= x_{r+1} u_{2r+1} u_{2r}^{-1} v_r^{-1} v_r u_{2r} u_{2r+1}^{-1} \quad \text{(by 2.12)} \\ &= x_{r+1} v_{r+1}^{-1} v_{r+1}, \end{aligned}$$

exactly as required. It now follows that for $i = 1, \ldots, m - 1$,

$$\begin{aligned} z_i u_{2i} &= x_i v_i^{-1} v_i u_{2i} = x_i (v_i^{-1} v_i)(u_{2i} u_{2i}^{-1}) u_{2i} \\ &= x_i u_{2i} u_{2i}^{-1} v_i^{-1} v_i u_{2i} = x_{i+1} u_{2i+1} u_{2i}^{-1} v_i^{-1} v_i u_{2i} \\ &= x_{i+1} u_{2i+1} (u_{2i+1}^{-1} u_{2i+1})(u_{2i}^{-1} v_i^{-1} v_i u_{2i}) \\ &= x_{i+1} u_{2i+1} u_{2i}^{-1} v_i^{-1} v_i u_{2i} u_{2i+1}^{-1} u_{2i+1} \\ &= x_{i+1} v_{i+1}^{-1} v_{i+1} u_{2i+1} = z_{i+1} u_{2i+1}. \end{aligned}$$

It remains to verify the final equality in (2.15). This is now immediate, since

$$d = u_0 y_1 = z_1 u_1 y_1 = z_2 u_3 y_1 = z_2 u_4 y_2 = \ldots = z_m u_{2m-1} y_m = z_m u_{2m}. \blacksquare$$

Other types of absolutely closed semigroups are correctly identified in (Howie and Isbell 1967), but it may be remarked that their proof that full transformation semigroups are absolutely closed is incorrect. A correct proof has been given by Scheiblich and Moore (1973).

3. THE EMBEDDING OF AMALGAMS

In Section 1 we showed that an amalgam is embeddable if and only if it is naturally embedded (in the sense of (1.8) and (1.9)) in its free product. We shall now use this approach to derive some sufficient conditions for embeddability. The results are due to Howie (1962), but the combinatorial devices we shall use here are different from and hopefully more efficient than those in the original paper.

In writing the elements of a free product as words $a_1a_2 \ldots a_m$ rather than as strings $(a_1, a_2, \ldots, a_m)$ we introduce a mild ambiguity, since unless we have clearly specified the semigroups to which $a_1, \ldots, a_m$ belong we cannot be certain whether (say) $a_1a_2a_3a_4$ stands for the string (a_1, a_2, a_3, a_4) or (say) for the string $(a_1a_2a_3, a_4)$. If we know that $\sigma(a_1) \neq \sigma(a_2)$, $\sigma(a_2) \neq \sigma(a_3)$, $\sigma(a_3) \neq \sigma(a_4)$, then $a_1a_2a_3a_4$ stands unambiguously for the string (a_1, a_2, a_3, a_4); if $\sigma(a_1) = \sigma(a_2) = \sigma(a_3) \neq \sigma(a_4)$ then it stands unambiguously for $(a_1a_2a_3, a_4)$. It will be convenient to continue to write words $a_1a_2 \ldots a_m$ in this ambiguous way. By a *subword* of $w = a_1a_2 \ldots a_n$ we shall mean a word $a_r \ldots a_s$, where $1 \leqslant r \leqslant s \leqslant n$. A maximal subword of constant index will be called a *syllable* of w. Thus, for example, if the word $w = a_1a_2a_3a_4a_5$ in $S_1 * S_2$ stands for the string (a_1a_2, a_3, a_4a_5), with $\sigma(a_1) = \sigma(a_2) = \sigma(a_4) = \sigma(a_5) = 1$ and $\sigma(a_3) = 2$, then the syllables of w are a_1a_2, a_3, a_4a_5 and are elements of S_1, S_2 and S_1 respectively.

For simplicity we shall consider an amalgam $[S, T; U]$ of two semigroups. We shall assume that U is a subsemigroup of S and that ϕ is an isomorphism from U onto a subsemigroup V of T. We shall systematically identify S and T with their natural images (see Proposition 1.2) in the free product $S * T$. Elements of U will be denoted by symbols such as u, u', u_i etc., and we shall adopt the very convenient notational convention whereby v, v', v_i etc. will stand respectively for $u\phi, u'\phi, u_i\phi$ etc.

In the new notation, the free product $S *_U T$ of the amalgam $[S, T; U]$ is the quotient of $F = S * T$ by ρ, where ρ is the congruence on F generated by

$$\mathbf{R} = \{(u, u\phi) : u \in U\}.$$

We shall examine the congruence ρ by considering elementary $\mathbf{R}$-transitions, as defined in Section I.5. These are of three types, designated respectively *increasing, decreasing* and *stable*. An *increasing* step increases the number of syllables in the word, and is of one of the following three types ($w_1, w_2 \in F^1$):

$$w_1sw_2 \to w_1s_1vs_2w_2, \tag{3.1}$$

where $s \in S$ is a syllable and $s = s_1us_2$;

$$sw_1 \to vs_1w_1, \tag{3.2}$$

where $s \in S$ is a syllable and $s = us_1$;

$$w_1s \to w_1s_1v, \tag{3.3}$$

where $s \in S$ is a syllable and $s = s_1u$. Analogous types of course exist with S and T interchanged.

A *stable* step is one that leaves the number of syllables in the word unchanged. Stable steps affecting words of more than one syllable are essentially all of the following type ($w_1, w_3 \in F^1$):

$$w_1 s t w_2 \dashrightarrow w_1 s_1 t_1 w_2, \tag{3.4}$$

where $s \in S$ and $t \in T$ are syllables and where, for some u in U,

$$s = s_1 u, \qquad t_1 = vt.$$

The roles of S and T can be interchanged, and the "direction of movement" of u can be reversed, but the differences are not of great importance. A stable step affecting a word of one syllable is necessarily of one or other of the simple types

$$u \rightarrow v \quad \text{or} \quad v \rightarrow u. \tag{3.5}$$

A *decreasing* step is one that decreases the number of syllables in the word. Decreasing steps are of three types, the reverses respectively of the three types (3.1), (3.2) and (3.3) of increasing step.

Our objective is to prove that under suitable circumstances the amalgam $[S, T; U]$ is naturally embedded in its free product $P = S *_U T = (S * T)/\rho$. That is to say, writing (1.8) and (1.9) in our simplified notation, we want to show that the mappings $\mu : S \rightarrow P$ and $\nu : T \rightarrow P$ defined by

$$s\mu = s\rho \ (s \in S), \qquad t\nu = t\rho \ (t \in T)$$

are both one–one, and that $S\mu \cap T\nu \subseteq U\mu$. In combinatorial terms, what we must show is:

(3.6) if, for s, s' in S,

$$s \rightarrow \ldots \rightarrow s'$$

by a sequence of elementary **R**-transitions in F, then $s = s'$;

(3.7) if, for t, t' in T,

$$t \rightarrow \ldots \rightarrow t'$$

by a sequence of elementary **R**-transitions in F, then $t = t'$;

(3.8) if, for s in S and t in T,

$$s \rightarrow \ldots \rightarrow t$$

by a sequence of elementary **R**-transitions in F, then $s \in U$ and $t = u\phi$.

We cannot of course hope to prove these three assertions in general, but in view of the embeddability of group amalgams it seemed to Graham Higman in 1960 reasonable to conjecture that a semigroup amalgam $[S, T; U]$ might be embeddable if its core U was a group. This turned out to be the case. The method of proof, moreover, suggested a more general conjecture. To explain this we require a definition that is new as far as the text is concerned, though we have encountered it previously in the exercises. (See Exercises II.12

and V.6–V.12.) If U is a subsemigroup of a semigroup S we shall say that U is *unitary* if S if

$$(\forall u \in U)(\forall s \in S) \quad us \in U \Rightarrow s \in U$$

and

$$(\forall u \in U)(\forall s \in S) \quad su \in U \Rightarrow s \in U.$$

It is easy to see that a subgroup of a group has the unitary property. (See also Exercise 7.) It is even the case that a subgroup U of a semigroup S is a unitary subsemigroup *provided the identity of the subgroup is an identity for the whole semigroup.* To see this, notice that if $u \in U$ and $us = u' = U$, then

$$s = 1s = u^{-1}us = u^{-1}u' \in U,$$

while if $su = u'' \in U$ then

$$s = s1 = suu^{-1} = u''u^{-1} \in U.$$

The italicized proviso cannot in general be removed—see Exercise 8. If U, with identity e, is a subgroup of a semigroup S, then for every u in U and s in S it is easy to show that

$$u(ese) \in U \Rightarrow ese \in U$$

and that

$$(ese)u \in U \Rightarrow ese \in U.$$

Thus U is unitary in eSe.

The search for a class of subsemigroups that includes both unitary subsemigroups and those having identity e that are unitary in eSe leads us to a definition in terms of left and right translations of the semigroup. As in Section IV.3 we shall adopt the notational convention of writing left translations on the left and right translations on the right. We say that a subsemigroup U of a semigroup S is *almost unitary* if there exist mappings $\lambda : S \to S$, $\rho : S \to S$ such that

(AU1) $\lambda^2 = \lambda$, $\rho^2 = \rho$;
(AU2) $\lambda(st) = (\lambda s)t$, $(st)\rho = s(t\rho)$ for all s, t in S—i.e., λ is a *left translation* and ρ is a *right translation*;
(AU3) $\lambda(s\rho) = (\lambda s)\rho$ for every s in S—i.e., λ and ρ *commute*; thus the notation $\lambda s\rho$ is unambiguous;
(AU4) $s(\lambda t) = (s\rho)t$ for all s, t in S—i.e., λ and ρ are *linked*;
(AU5) $\lambda|U = \rho|U = 1_U$;
(AU6) U is unitary in $\lambda S\rho$.

We call λ and ρ the *associated mappings* of U. The case where U is unitary arises when $\lambda = \rho = 1_S$, while the case where U has identity e and is unitary in eSe arises when λ is taken as $s \mapsto es$ and ρ as $s \mapsto se$.

At first sight the definition "almost unitary" looks substantially more general than "unitary in eSe", but this is not really the case.

Proposition 3.9. *Let U be an almost unitary subsemigroup of a semigroup S, with associated mappings λ and ρ. If U has an identity element e, then U is unitary in eSe.*

Proof. Certainly $U \subseteq eSe$. Also, since $e \in U$ we have $\lambda e = e$, $e\rho = e$ by property (AU5). Hence, for all s in S,

$$ese = (\lambda e)s(e\rho) = \lambda(ese)\rho \in \lambda S\rho.$$

Thus $eSe \subseteq \lambda S\rho$ and so U, being unitary in $\lambda S\rho$, is *a fortiori* unitary in eSe. ■

We remark that this does not mean that λ and ρ necessarily coincide with the mappings $s \mapsto es$, $s \mapsto se$. The subsemigroup $\lambda S\rho$ may well be strictly larger than eSe. (See Exercise 9).

A further remark showing the connection between "almost unitary" and "unitary in eSe" is as follows:

Proposition 3.10. *A subsemigroup U of a semigroup S is almost unitary in S if and only if it is possible to adjoin an idempotent e to S with the properties:*

(i) $eS \subseteq S$, $Se \subseteq S$;

(ii) *e is an identity element for $U \cup \{e\}$;*

(iii) *$U \cup \{e\}$ is unitary in $e(S \cup \{e\})e$.*

Proof. If an idempotent e can be adjoined with the listed properties we define $\lambda : S \to S$ and $\rho : S \to S$ by the rules

$$\lambda s = es, \qquad s\rho = se \quad (s \in S)$$

respectively. Property (i) ensures that λs and $s\rho$ belong to S and properties (AU1) to (AU6) follow easily. Conversely, if U is almost unitary in S we consider an element $e \notin S$ and extend the multiplication on S to $S \cup \{e\}$ by the rules:

$$es = \lambda s, \qquad se = s\rho \quad (s \in S), \qquad e^2 = e.$$

The given properties (AU1) to (AU4) now ensure that the multiplication on $S \cup \{e\}$ is associative:

$$e(es) = e^2 s = es, \qquad (se)e = se^2 = se,$$

$$e(st) = (es)t, \qquad (st)e = s(te), \qquad e(se) = (es)e,$$

$$s(et) = (se)t.$$

Property (AU5) ensures that e is an identity element for $U \cup \{e\}$, while (AU6) gives that $U \cup \{e\}$ is unitary in $e(S \cup \{e\})e$.■

The purpose of the elaborate definition of an almost unitary subsemigroup was to allow both for the most general sort of subgroup and also for unitary subsemigroups. The justification of the definition lies in the following theorem:

Theorem 3.11. (Howie 1962). *The amalgam* $[S, T; U]$ *is embeddable if* U *is almost unitary in* S *and* $V\ (=U\phi)$ *is almost unitary in* T.

Before proving this, we remark that there is no great difficulty in extending the result to a more general amalgam $[\{S_i : i \in I\}; U; \{\phi_i : i \in I\}]$ in which $U\phi_i$ is almost unitary in S_i for each i in I.

The method of proof adopted here differs considerably both from the method in (Howie 1962) and the neater method in (Clifford and Preston 1967). It owes a good deal to the 'proper words' idea in the Clifford and Preston proof, but the remainder of the argument is much shorter.

If we use the letters λ and ρ for the mappings associated with the almost unitary subsemigroup U of S, then strictly we ought to use different letters for the mappings associated with the almost unitary subsemigroup V of T. To do so is inconvenient, however, and since both pairs of mappings operate identically on the core of the amalgam no real confusion arises from the use of a single pair (λ, ρ) of letters.

We begin by defining a word $w = a_1 a_2 \ldots a_n$ in $S * T$ with $n(\geqslant 2)$ syllables $a_1, \ldots, a_n$ as *internal* if

$$a_1 = a_1\rho, \qquad a_n = \lambda a_n \quad \text{and} \quad a_i = \lambda a_i \rho \qquad (2 \leqslant i \leqslant n-1).$$

(Clifford and Preston (1967) use the term *proper*.) We say that w is *left-internal* if it is internal and if in addition $a_1 = \lambda a_1$, that w is *right-internal* if it is internal and if in addition $a_n = a_n\rho$, and that w is *bi-internal* if it is both left internal and right internal. *All* words of one syllable are defined as internal; a one-syllable word a is defined to be *left-internal, right-internal* or *bi-internal* according as $\lambda a = a$, $a\rho = a$ or $\lambda a\rho = a$.

Because λ and ρ are idempotent, it is clear that for any $w = a_1 a_2 \ldots a_n$ in $S * T$ the word

$$w^* = (a_1\rho)(\lambda a_2\rho)\ldots(\lambda a_{n-1}\rho)(\lambda a_n) \tag{3.12}$$

is internal. Moreover, the words

$$(\lambda a_1\rho)(\lambda a_2\rho)\ldots(\lambda a_{n-1}\rho)(\lambda a_n), \qquad (a_1\rho)(\lambda a_2\rho)\ldots(\lambda a_{n-1}\rho)(\lambda a_n\rho)$$

and

$$(\lambda a_1\rho)(\lambda a_2\rho)\ldots(\lambda a_{n-1}\rho)(\lambda a_n\rho)$$

are respectively left-internal, right-internal and bi-internal.

It is useful at this stage to record some elementary consequences of the definition "almost unitary". If U is an almost unitary subsemigroup of a semigroup S, with associated mappings λ and ρ, then for all u in U and all x, y in S,

$$\left.\begin{aligned} \lambda(xu) &= \lambda(xu)\rho = (\lambda x\rho)u, \qquad (ux)\rho = \lambda(ux)\rho = u(\lambda x\rho), \\ ux &= \lambda(ux) = u(\lambda x), \qquad xu = (xu)\rho = (x\rho)u, \\ \lambda(xuy) &= (\lambda x\rho)u(\lambda y), \qquad (xuy)\rho = (x\rho)u(\lambda y\rho), \\ xuy &= (x\rho)u(\lambda y), \qquad \lambda(xuy)\rho = (\lambda x\rho)u(\lambda y\rho). \end{aligned}\right\} \tag{3.13}$$

The proofs are routine: for example,

$$\begin{aligned} \lambda(xu) &= (\lambda x)u && \text{by (AU2)}, \\ &= (\lambda x)(u\rho) && \text{by (AU5)} \\ &= ((\lambda x)u)\rho && \text{by (AU2)}, \\ &= \lambda(xu)\rho && \text{by (AU2) and (AU3)}, \end{aligned}$$

$$\begin{aligned} \text{while } \lambda(xu) = (\lambda x)u &= (\lambda x)(\lambda u) \text{ by (AU5), while} \\ &= (\lambda x\rho)u \quad \text{by (AU4).} \end{aligned}$$

If

$$w_0 \to w_1 \to \ldots \to w_n \tag{3.14}$$

is a sequence of elementary **R**-transitions beginning on an internal word w_0 (e.g. beginning on a monosyllabic word), then we shall show that it is possible to *internalize* the sequence to a sequence

$$w_0 \to w_1^* \to \ldots w_n^*$$

in which each w_i^* is the internal word obtained from w_i in accordance with formula (3.12). We have $w_0 = w_0^*$, since w_0 is already internal.

To prove that this *internalization* is possible, let $w_i = a_1a_2 \ldots a_n$ (where each a_j is a syllable) and consider the various essentially different ways in which the step $w_i \to w_{i+1}$ can take place:

(Inc 1) $w_i = a_1 \ldots a_k \ldots a_n = a_1 \ldots a_k'ua_k'' \ldots a_n$

$\to a_1 \ldots a_k'va_k'' \ldots a_n = w_{i+1}$,

(Inc 2) $w_i = a_1a_2 \ldots a_n = ua_1'a_2 \ldots a_n \to va_1'a_2 \ldots a_n = w_{i+1}$,

(Inc 3) $w_i = a_1 \ldots a_{n-1}a_n = a_1 \ldots a_{n-1}a_n'u \to a_1 \ldots a_{n-1}a_n'v = w_{i+1}$

(and three similar possibilities involving a change from V to U);

$$\text{(Stab 1) } w_i = a_1 \dots a_k a_{k+1} \dots a_n = a_1 \dots a'_k u \,.\, a_{k+1} \dots a_n$$
$$\to a_1 \dots a'_k \,.\, v a_{k+1} \dots a_n = a_1 \dots a'_k a'_{k+1} \dots a_n = w_{i+1}$$

(and similar possibilities involving movement from right to left and/or change from V to U);

$$\text{(Stab 2) } n = 1;\ w_i = a_1 = u \to v = w_{i+1}$$
$$\text{or } w_i = a_1 = v \to u = w_{i+1};$$

$$\text{(Dec 1) } w_i = a_1 \dots a_{k-1} a_k a_{k+1} \dots a_n = a_1 \dots a_{k-1} u a_{k+1} \dots a_n$$
$$\to a_1 \dots a_{k-1} v a_{k+1} \dots a_n = a_1 \dots a_{k-2} a'_k a_{k+2} \dots a_n = w_{i+1},$$
$$\text{where } a'_k = a_{k-1} v a_{k+1},$$

$$\text{(Dec 2) } w_i = a_1 a_2 \dots a_n = u a_2 \dots a_n$$
$$\to v a_2 \dots a_n = a'_1 a_3 \dots a_n = w_{i+1},$$
$$\text{where } a'_1 = v a_2,$$

$$\text{(Dec 3) } w_i = a_1 \dots a_{n-1} a_n = a_1 \dots a_{n-1} u \to a_1 \dots a_{n-1} v$$
$$= a_1 \dots a_{n-2} a'_n = w_{i+1}, \text{ where } a'_n = a_{n-1} v$$

(and three similar possibilities involving change from V to U).

We now show that in all cases there is an elementary **R**-transition of the same type transforming w_i^* to w_{i+1}^*.

In case (Inc 1),

$$w_i^* = (a_1\rho)\dots(\lambda a_k\rho)\dots(\lambda a_n) = (a_1\rho)\dots(\lambda(a'_k u a''_k)\rho)\dots(\lambda a_n)$$
$$= (a_1\rho)\dots(\lambda a'_k\rho)u(\lambda a''_k\rho)\dots(\lambda a_n) \quad \text{by (3.13)},$$
$$\to (a_1\rho)\dots(\lambda a'_k\rho)v(\lambda a''_k\rho)\dots(\lambda a_n) = w_{i+1}^*,$$

since $\lambda v\rho = v$. Small modifications are required if $k = 1$ or $k = n$.

In case (Inc 2),

$$w_i^* = (a_1\rho)(\lambda a_2\rho)\dots(\lambda a_n) = ((ua'_1)\rho)(\lambda a_2\rho)\dots(\lambda a_n)$$
$$= u(\lambda a'_1\rho)(\lambda a_2\rho)\dots(\lambda a_n) \quad \text{by (3.13)},$$
$$\to v(\lambda a'_1\rho)(\lambda a_2\rho)\dots(\lambda a_n) = w_{i+1}^*.$$

A small modification is required if $n = 1$.

Case (Inc 3) is similar.

In case (Stab 1)

$$\begin{aligned} w_i^* &= (a_1\rho)\dots(\lambda a_k\rho)(\lambda a_{k+1}\rho)\dots(\lambda a_n) \\ &= (a_1\rho)\dots(\lambda(a_k'u)\rho)(\lambda a_{k+1}\rho)\dots(\lambda a_n) \\ &= (a_1\rho)\dots((\lambda a_k'\rho)u)(\lambda a_{k+1}\rho)\dots(\lambda a_n) \quad \text{(by (3.13))} \\ &\to (a_1\rho)\dots(\lambda a_k'\rho)(v(\lambda a_{k+1}\rho))\dots(\lambda a_n) \\ &= (a_1\rho)\dots(\lambda a_k'\rho)(\lambda(va_{k+1})\rho)\dots(\lambda a_n) \quad \text{(by 3.13))} \\ &= (a_1\rho)\dots(\lambda a_k'\rho)(\lambda a_{k+1}'\rho)\dots(\lambda a_n) = w_{i+1}^*. \end{aligned}$$

Small modifications are required if $k = 1$ and/or $k + 1 = n$.

Case (Stab 2) is trivial, since $w_i = w_i^*$ and $w_{i+1} = w_{i+1}^*$ in this case.

In case (Dec 1),

$$\begin{aligned} w_i^* &= (a_1\rho)\dots(\lambda a_{k-1}\rho)(\lambda a_k\rho)(\lambda a_{k+1}\rho)\dots(\lambda a_n) \\ &= (a_1\rho)\dots(\lambda a_{k-1}\rho)u(\lambda a_{k+1}\rho)\dots(\lambda a_n) \\ &\to (a_1\rho)\dots[(\lambda a_{k-1}\rho)v(\lambda a_{k+1}\rho)]\dots(\lambda a_n) \\ &= (a_1\rho)\dots[\lambda(a_{k-1}va_{k+1})\rho]\dots(\lambda a_n) \quad \text{(by (3.13))} \\ &= (a_1\rho)\dots(\lambda a_k'\rho)\dots(\lambda a_n) = w_{i+1}^*. \end{aligned}$$

Small modifications are required if $k - 1 = 1$ and/or $k + 1 = n$.

In case (Dec 2),

$$\begin{aligned} w_i^* &= (a_1\rho)(\lambda a_2\rho)\dots(\lambda a_n) = u(\lambda a_2\rho)\dots(\lambda a_n) \\ &\to [v(\lambda a_2\rho)]\dots(\lambda a_n) = (va_2)\rho\dots(\lambda a_n) \qquad \text{(by (3.13))} \\ &= (a_1'\rho)\dots(\lambda a_n) = w_{i+1}^*. \end{aligned}$$

Small modifications are required if $n = 2$.

Case (Dec 3) is similar.

We have now shown that an internalized sequence

$$w_0 \to w_i^* \to \dots \to w_n^* \tag{3.15}$$

can be obtained from the original sequence (3.14), where w_i^* is obtained from w_i by means of formula (3.12). The interesting case is the one in which w_n (as well as w_0) is internal, so that $w_n^* = w_n$. The internalized sequence then connects w_0 and w_n, but has the advantage over the original sequence (3.14) that all the words appearing in the sequence are internal.

We remark for future use that if in such a case we internalize the reverse

$$w_n \to w_{n-1} \to \dots \to w_0$$

of the sequence (3.14), the resulting sequence is the reverse of the internalized version

$$w_0 \to w_1^* \to \ldots \to w_{n-1}^* \to w_n$$

of (3.14).

Returning now to the original sequence (3.14), we next observe that if w_0 is left-internal then an argument that is virtually a repetition of the previous internalization argument shows that we can produce a *left-internalized* sequence

$$w_0 \to w_1^\# \to \ldots \to w_n^\#,$$

in which each $w_i^\#$ is the left-internal word obtained from $w_i = a_1 a_2 \ldots a_n$ by the rule

$$w_i^\# = (\lambda a_1 \rho) \ldots (\lambda a_{n-1} \rho)(a_n \rho).$$

In this statement we can replace "left-internal" by "right-internal" throughout if we use

$$w_i^\# = (a_1 \rho)(\lambda a_2 \rho) \ldots (\lambda a_n \rho),$$

and by 'bi-internal' if we use

$$w_i^\# = (\lambda a_1 \rho)(\lambda a_2 \rho) \ldots (\lambda a_n \rho).$$

The effect of this is that if in the internalized sequence (3.15) there is a w_i^* that is left-internal, then we can further internalize the sequence, modifying all words subsequent to w_i^* in such a way as to make them left-internal. Similar remarks apply with "left-internal" replaced by "right-internal" or "bi-internal". We can therefore treat the original sequence (3.14) as follows: first internalize to obtain (3.15); then find the first word w_i^* (if any) in (3.15) that is left-internal, and left-internalize the sequence from that point on; then find the first word (if any) in the new sequence that is right-internal, and right-internalize from that point on. We obtain a sequence

$$w_0 \to \overline{w}_1 \to \ldots \to \overline{w}_n$$

which we describe as the *fully internalized* sequence obtained from (3.14).

We can now state the crucial combinatorial lemma that leads us to the desired theorem:

Lemma 3.16. *If in a fully internalized sequence*

$$w_0 \to \overline{w}_1 \to \ldots \to \overline{w}_n$$

a stable or decreasing step follows an increasing step, then either the two steps can be carried out in the opposite order or there exists a shorter fully internalized sequence connecting w_0 *to* $\overline{w}_n$.

Proof. The increasing step may be taken as

$$\text{(A) } wsz = ws_1u_1s_2z \to ws_1v_1s_2z \qquad (w, z \in F^1),$$

or as

$$\text{(B) } sz = u_1s_2z \to v_1s_2z \qquad (z \in F^1).$$

These are in essence the only cases, but of course the change may be from V to U and the "edge" case (B) may affect the right-hand rather than the left-hand edge of the word. We are assuming that the s appearing in the initial word is a syllable.

If the next step in the sequence affects either w or z or the left-hand side of s_1 or the right-hand side of s_2, then clearly the order of the two steps can be interchanged with no change either in the ultimate effect of the sequence or in the fully internalized property. The remaining cases can be listed as follows:

$$\text{(AS1) } ws_1v_1s_2z = ws_3u_2v_1s_2z \to ws_3v_2v_1s_2z;$$

$$\text{(AS2) } ws_1v_1s_2z = ws_1v_1u_2s_3z \to ws_1v_1v_2s_3z;$$

$$\text{(AS3) } ws_1v_1s_2z = ws_1v_2ts_2z \to ws_1u_2ts_2z;$$

$$\text{(AS4) } ws_1v_1s_2z = ws_1tv_2s_2z \to ws_1tu_2s_2z;$$

$$\text{(AD1) } ws_1v_1s_2z \to ws_1u_1s_2z;$$

$$\text{(AD2) } ws_1v_1s_2z = w'ts_1v_1s_2z \qquad (w' \in F^1, t \in T)$$

$$= w'tu_2v_1s_2z \to w' \,.\, tv_2v_1 \,.\, s_2z;$$

$$\text{(AD3) } s_1v_1s_2z = u_2v_1s_2z \to v_2v_1 \,.\, s_2z \qquad (w = 1);$$

$$\text{(AD4) } ws_1v_1s_2z = ws_1v_1s_2tz' \qquad (z' \in F^1, t \in T)$$

$$= ws_1v_1u_2tz' \to ws_1 \,.\, v_1v_2t \,.\, z';$$

$$\text{(AD5) } ws_1v_1s_2 = ws_1v_1u_2 \to ws_1 \,.\, v_1v_2 \qquad (z = 1);$$

$$\text{(BS1) } v_1s_2z = v_1u_2s_3z \to v_1v_2s_3z;$$

$$\text{(BS2) } v_1s_2z = tv_2s_2z \to tu_2s_2z;$$

$$\text{(BD1) } v_1s_2z \to u_1s_2z;$$

$$\text{(BD2) } v_1s_2z = v_1s_2tz' \qquad (z' \in F^1, t \in T)$$

$$= v_1u_2tz' \to v_1v_2t \,.\, z';$$

$$\text{(BD3) } v_1s_2 = v_1u_2 \to v_1v_2 \qquad (z = 1).$$

In each case we shall see that the pair of steps can be replaced either by a single step or by no step at all. The fully internalized property is preserved in all cases.

In the case (A, AS1) we replace the two steps by the single step

$$wsz = ws_1u_1s_2z = ws_3(u_2u_1)s_2z \to ws_3v_2v_1s_2z.$$

The case (A, AS2) is similar.

In the case (A, AS3) we have $t = \lambda t\rho$ since the sequence is internalized. Since $v_2t = v_1 \in V$, it follows by property (AU6) of almost unitary semigroups that $t \in V$—say $t = v_3$. The replacement of the two given steps is then as follows:

$$wsz = ws_1u_1s_2z = ws_1u_2u_3s_2z \to ws_1u_2v_3s_2z = ws_1u_2tz,$$

for the equation $v_1 = v_2v_3$ in V implies the corresponding equation $u_1 = u_2u_3$ in the isomorphic semigroup U. The case (A, AS4) is similar.

The case (A, AD1) is trivial, since the steps cancel each other.

In the case (A, AD2) the replacement is

$$wsz = w'ts_1u_1s_2z = w'tu_2u_1s_2z \to w'tv_2v_1s_2z.$$

In the case (A, AD3), the replacement is

$$sz = s_1u_1s_2z = u_2u_1s_2z \to v_2v_1s_2a.$$

The cases (A, AD4) and (A, AD5) are similar.

In the case (B, BS1) we replace the given steps by the single step

$$sz = u_1s_2z = u_1u_2s_3z \to v_1v_2s_3z.$$

The case (B, BS2) is the crucial one. The word v_1sz is certainly internal, and is in fact left-internal, since $\lambda v_1 = v_1$; hence, since the sequence is assumed to be fully internalized, $\lambda t\rho = t$. Since $tv_2 = v_1 \in V$ it follows by the almost unitary property that $t \in V$. Writing $t = v_3$, we can now replace the given steps by the single step

$$sz = u_1s_2z = u_3u_2s_2z \to v_3u_2s_2z = tu_2s_2z.$$

The case (B, BD1) is trivial, since the steps cancel each other.

In the case (B, BD2) the required replacement is

$$sz = u_1s_2tz' = u_1u_2tz' \to v_1v_2tz'.$$

In the case (B, BD3) we replace by

$$s = u_1s_2 = u_1u_2 \to v_1v_2.$$

All the essentially different cases have now been covered, and so Lemma 3.16 is proved.■

As a consequence of Lemma 3.16 we have

Lemma 3.17. *Let w and z be words of one syllable in $S * T$ (i.e. elements of $S \cup T$), and let*

$$w \to w_1 \to \ldots \to w_{n-1} \to z \tag{3.18}$$

be a sequence of elementary **R**-*transitions in $S * T$. Then the corresponding fully internalized sequence*

$$w \to \bar{w}_1 \to \ldots \to \bar{w}_{n-1} \to \bar{z} \tag{3.19}$$

can be replaced by a shorter fully internalized sequence connecting w and $\bar{z}$ unless either:

(a) $n = 0$; *or*

(b) $n = 1, \quad w = u \in U, \quad \bar{z} = z = u\phi$; or

(c) $n = 1, \quad w = v \in V, \quad \bar{z} = z = v\phi^{-1}$.

Proof. If $n = 1$ then the sequence must be of one or other of the two forms (b) and (c). If $n > 1$ and if the sequence (3.18) has no increasing steps, it must be of one or other of the forms

$$u \to v \to u \to v \to \ldots, \qquad v \to u \to v \to u \to \ldots,$$

and so can certainly be shortened. In these cases the sequence (3.18) and the fully internalized sequence (3.19) are identical.

Let us assume now that $n > 1$ and that the sequence (3.18)—and hence also the sequence (3.19)—has at least one increasing step. The last such step cannot be the last step in the sequence, since $\bar{z}$ is monosyllabic. If we consider the last increasing step (in (3.19)) and the step immediately following it, then Lemma 3.16 allows us to reverse the order of the steps or to reduce the length of the sequence. If it is the former alternative that occurs, we apply Lemma 3.16 again to the last increasing step of the modified sequence. Indeed we apply Lemma 3.16 repeatedly until we obtain a reduction in the length of the sequence connecting w to $\bar{z}$. This must happen eventually, since if the former (order-reversal) alternative were to persist indefinitely we would eventually have moved the last increasing step to the end of the sequence—and we have seen that this is impossible.■

The next lemma is now obvious:

Lemma 3.20. *Let w and z be words of one syllable in $S * T$ and let*

$$w \to w_1 \to \ldots \to w_{n-1} \to z$$

be a sequence of elementary **R**-*transitions in* $S * T$. *If*

$$w \to \bar{w}_1 \to \dots \bar{w}_{n-1} \to \bar{z}$$

is the corresponding fully internalized sequence, then either:

(a) $w = \bar{z}$; *or*

(b) $w = u \in U$ *and* $z = \bar{z} = u\phi$; or

(c) $w = v \in V$ *and* $z = \bar{z} = v\phi^{-1}$. ■

We can now quickly complete the proof of Theorem 3.11. In accordance with the plan of action outlined in (3.6), (3.7) and (3.8), we first consider a sequence

$$x \to w_1 \to \dots \to w_{n-1} \to y, \tag{3.21}$$

where $x, y \in S$. Of the three possibilities allowed by Lemma 3.20 only the first can arise here. Hence $x = \bar{y}$. We now distinguish four cases:

(i) if the internalized sequence

$$x \to w_1^* \to \dots \to w_{n-1}^* \to y \tag{3.22}$$

contains no right- or left-internal w_i^*, then $\bar{y} = y$;

(ii) if the internalized sequence (3.22) contains a left-internal w_i^* but no right-internal w_j^*, then $\bar{y} = \lambda y$;

(iii) if the internalized sequence (3.22) contains a right-internal w_i^* but no left-internal w_j^*, then $\bar{y} = y\rho$;

(iv) if the internalized sequence (3.22) contains a left-internal w_i^* and a right-internal w_j^*, then $\bar{y} = \lambda y\rho$.

We can now apply exactly the same analysis to the reverse sequence

$$y \to w_{n-1} \to \dots \to x$$

and conclude that $y = \bar{x}$, where $\bar{x}$ is one of x, λx, $x\rho$ and $\lambda x\rho$. Moreover, whichever one of the four eventualities (i), (ii), (iii), (iv) applies to the sequence (3.22) applies also to the reversed internalized sequence

$$y \to w_{n-1}^* \to \dots \to w_1^* \to x,$$

and so we have either:

(i) $x = y$ and $y = x$; or
(ii) $x = \lambda y$ and $y = \lambda x$; or
(iii) $x = y\rho$ and $y = x\rho$; or
(iv) $x = \lambda y\rho$ and $y = \lambda x\rho$.

In all cases it follows that $x = y$. For example, in case (iv),

$$x = \lambda y\rho = \lambda(\lambda x\rho)\rho = \lambda x\rho \quad \text{(by (AU1))}$$
$$= y.$$

Thus (3.6) is proved. A similar argument establishes (3.7).

To prove (3.8), suppose now that in the sequence (3.21) we have $x \in S$ and $y \in T$. Of the three possibilities allowed by Lemma 3.20 only the second can arise here. Hence $x \in U$ and $y = \bar{y} = x\phi$, exactly as required by (3.8).

This completes the proof of Theorem 3.11. For further results on amalgams and on free products of amalgams see Howie (1962; 1963a, b, c; 1964a, b; 1968; 1975), Grillet and Petrich (1970), B. H. Neumann (1970), and Ljapin (1969, 1970, 1974).

4. INVERSE SEMIGROUP AMALGAMS

In this final section we establish the following theorem, due to T. E. Hall (1975), concerning an *inverse semigroup amalgam*, that is to say, a semigroup amalgam

$$\mathfrak{A} = [\{S_i : i \in I\}; U; \{\phi_i : i \in I\}] \tag{4.1}$$

in which S_i ($i \in I$) and U are all inverse semigroups.

Theorem 4.2. *An inverse semigroup amalgam is embeddable in an inverse semigroup.*

The proof we shall give is different in plan from that used by Hall, although Hall does indicate that the approach of the proof below is an alternative to his. Representation theory is used at a crucial stage in the proof, and here the basic idea is Hall's, although it appears below in a modified form that makes reference to the results of Section V.8.

Following Hall, we shall establish the theorem only for an amalgam $[S, T; U]$ of two semigroups. In the case where I is finite it is a routine matter to extend the result to any amalgam of type (4.1), and even if I is infinite the result can be extended using Zorn's Lemma. Details of the approach can be found in (Howie 1968).

The first hint that inverse semigroup amalgams might be more amenable than general semigroup amalgams comes from Theorem 2.14. The fact that inverse semigroups are absolutely closed implies in particular by Theorem 2.3 that if U is an inverse subsemigroup of an inverse semigroup S and α: $S \to S'$ is an isomorphism, then the amalgam $[\{S, S'\}; U; \{\iota, \alpha|U\}]$ is embeddable in a semigroup. If we can show that it is embeddable in an inverse semigroup then we shall have shown that in certain special circumstances (where the two semigroups S and S' are isomorphic) an inverse semigroup amalgam is embeddable in an inverse semigroup. And this will in fact be an important step on the way to the more general result.

We begin by describing the construction for inverse semigroups that is analogous to the free product of semigroups discussed in Section 1. The description follows that of Preston (1973). If $\{S_i: i \in I\}$ is a family of disjoint inverse semigroups, then for each word $w = a_1 a_2 \dots a_n$ in the free product $F = \Pi^* S_i$ we define the *formal inverse* w' of w by

$$w' = a_n^{-1} a_{n-1}^{-1} \dots a_1^{-1}.$$

It is clear that for all w, z in F,

$$w''(=(w')') = w, \qquad (wz)' = z'w'. \tag{4.3}$$

Let τ be the congruence on F generated by the subset

$$\mathbf{T} = \{(ww'w, w): w \in F\} \cup \{(ww'zz', zz'ww'): w, z \in F\} \tag{4.4}$$

of $F \times F$. Define $\mathfrak{I}^* S_i$, the *free inverse product* of $\{S_i: i \in I\}$, by

$$\mathfrak{I}^* S_i = (\Pi^* S_i)/\tau.$$

We shall show that the name "free inverse product" is reasonable by showing that $\mathfrak{I}^* S_i$ is the *coproduct* (see (Mitchell 1965)) of $\{S_i: i \in I\}$ in the category of inverse semigroups. First we must show that $\mathfrak{I}^* S_i$ is an inverse semigroup. Certainly it is regular, since the very definition of τ implies that $w'\tau$ is an inverse of $w\tau$ for any w in F. We now show that idempotents commute in F/τ. Let $e\tau$, $f\tau$ be idempotents in F/τ. If e' is the formal inverse of e, then

$$\begin{aligned} e'\tau &= (e'ee')\tau = (e'e\,.\,ee')\tau \\ &= (ee'\,.\,e'e)\tau, \end{aligned}$$

since $e'' = e$ and since, by definition,

$$(ee'\,.\,e'e'', e'e''\,.\,ee') \in \mathbf{T} \subseteq \tau.$$

Hence

$$\begin{aligned} e'\tau &= (e^2e'\,.\,e'e^2)\tau = (e\tau)[(ee'\,.\,e'e)\tau](e\tau) \\ &= (e\tau)(e'\tau)(e\tau) = (ee'e)\tau = e\tau. \end{aligned}$$

It follows that

$$e\tau = (e\tau)(e\tau) = (e\tau)(e'\tau) = (ee')\tau,$$

and similarly that $f\tau = (ff')\tau$. Hence

$$(e\tau)(f\tau) = (ee'ff')\tau = (ff'ee')\tau = (f\tau)(e\tau),$$

exactly as required. Hence $\mathfrak{I}^*S_i$ is an inverse semigroup.

The result now stated is analogous to Proposition 1.2, and establishes the coproduct property mentioned above.

Proposition 4.5. *Let $\{S_i: i \in I\}$ be a family of disjoint inverse semigroups, and let $Q = \mathfrak{I}^*S_i$ be F/τ, where $F = \Pi^*S_i$ is the free product of the semigroups S_i and τ is the congruence*

$$[\{(ww'w, w): w \in F\} \cup \{(ww'zz', zz'ww'): w, z \in F\}]^\#$$

on F. Then

(i) *Q is an inverse semigroup;*
(ii) *there exists for each i in I a monomorphism $\theta_i: S_i \to Q$;*
(iii) *if T is an inverse semigroup for which a homomorphism $\psi_i: S_i \to T$ exists for each i in I then there is a unique homomorphism $\gamma: Q \to T$ with the property that the diagram*

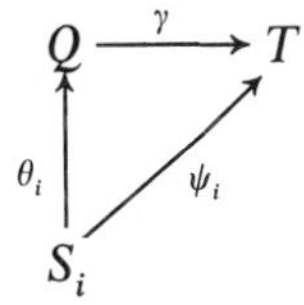

(4.6)

commutes for every i in I.

Proof. Part (i) is already established. To prove part (ii) we define

$$s_i\theta_i = s_i\tau \qquad (s_i \in S_i).$$

(Notice that we are identifying S_i with the isomorphic copy of it in F.) It is immediate that $\theta_i: S_i \to Q$ is a well-defined homomorphism. To see that it is a monomorphism notice that if $s_i\theta_i = t_i\theta_i$ for s_i and t_i in S_i, then there is a sequence

$$s_i \to \dots \to t_i$$

of elementary **T**-transitions connecting s_i to t_i. Now an elementary **T**-transition, whether based on a pair $(ww'w, w)$ or on a pair $(ww'zz', zz'ww')$, cannot change an element of S_i to anything other than an element of S_i. But since S_i is an inverse semigroup and since $w' = w^{-1}$ for any w in S_i, a sequence of elementary **T**-transitions leaves an element of S_i unaltered. Hence $s_i = t_i$ and so θ_i is a monomorphism as required.

To prove part (iii), suppose that T and $\{\psi_i : i \in I\}$ are given. We define $\gamma: Q \to T$ by

$$[(a_1 a_2 \ldots a_m)\tau]\gamma = (a_1\psi_{\sigma(a_1)})(a_2\psi_{\sigma(a_2)})\ldots(a_m\psi_{\sigma(a_m)}).$$

The mapping γ is well-defined, since any sequence of elementary **T**-transitions connecting two elements $a_1 a_2 \ldots a_m$ and $b_1 b_2 \ldots b_n$ of F can be mirrored by a sequence of equalities connecting the two elements

$$(a_1\psi_{\sigma(a_1)})(a_2\psi_{\sigma(a_2)})\ldots(a_m\psi_{\sigma(a_m)}) \quad \text{and} \quad (b_1\psi_{\sigma(b_1)})(b_2\psi_{\sigma(b_2)})\ldots(b_n\psi_{\sigma(b_n)})$$

of the inverse semigroup T. That γ is homomorphic, that the diagram (4.6) commutes, and that γ is unique with respect to the stated properties all follow as in the proof of Proposition 1.2. The details may safely be left to the reader. ■

If now we have an inverse semigroup amalgam

$$\mathfrak{A} = [\{S_i : i \in I\}; U; \{\phi_i : i \in I\}],$$

we define the *free inverse product* $\mathfrak{J}^*_U S_i$ of the amalgam $\mathfrak{A}$ to be the quotient F/ξ of $F = \Pi^* S_i$ by the congruence ξ generated by $\mathbf{R} \cup \mathbf{T}$, where

$$\mathbf{R} = \{(u\phi_i, u\phi_j) : u \in U, i, j \in I\}$$

and **T** is as in (4.4). Since $\xi \supseteq \tau$ it follows that F/ξ is a homomorphic image of $Q = F/\tau$ and so is an inverse semigroup. In fact $F/\xi \simeq (F/\tau)/(\xi/\tau)$, and by Proposition I.5.11 the congruence ξ/τ on $Q = F/\tau$ is generated by the subset $[\mathbf{R} \cup \mathbf{T}]/\tau$ of $Q \times Q$. That is, since $\mathbf{T}/\tau$ reduces simply to the identical relation, the congruence ξ/τ on Q is generated by the subset

$$\mathbf{R}/\tau = \{((u\phi_i)\tau, (u\phi_j)\tau) : u \in U, i, j \in I\}$$

of $Q \times Q$.

By analogy with Proposition 1.10 we have

Proposition 4.7. *If*

$$\mathfrak{A} = [\{S_i : i \in I\}; U; \{\phi_i : i \in I\}]$$

is an inverse semigroup amalgam, then the free inverse product $J = \mathfrak{J}^*_U S_i$ *is the pushout in the category of inverse semigroups of the diagram* $\{U \to S_i\}_{i \in I}$. *That is:*

(a) *there exists for each* i *in* I *a homomorphism* $\mu_i : S_i \to J$ *such that the diagram* $\{U \xrightarrow{\phi_i} S_i \xrightarrow{\mu_i} J\}_{i \in I}$ *commutes, i.e. such that* $\phi_i\mu_i = \phi_j\mu_j$ *for all* i, j *in* I;

(b) *if* K *is an inverse semigroup for which homomorphisms* $\nu_i : S_i \to K$ $(i \in I)$ *exist such that* $\phi_i\nu_i = \phi_j\nu_j$ *for all* i, j *in* I, *then there exists a unique homomorphism* $\delta: J \to K$ *such that the diagram*

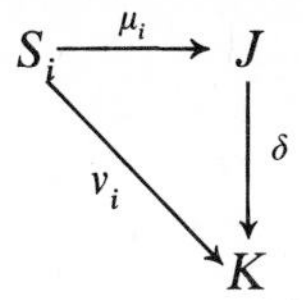

commutes for each i in I.

Proof. We define μ_i: $S_i \to J$ by

$$s_i\mu_i = s_i\xi \qquad (s_i \in S_i);$$

then certainly $u\phi_i\mu_i = u\phi_i\xi = u\phi_j\xi = u\phi_j\mu_j$ for all u in U and all i, j in I.

To show that (b) also holds, note that by Proposition 4.5 (iii) we have a homomorphism γ: $Q \to K$ (where $Q = \mathfrak{J}^*S_i = F/\tau$) such that the diagram

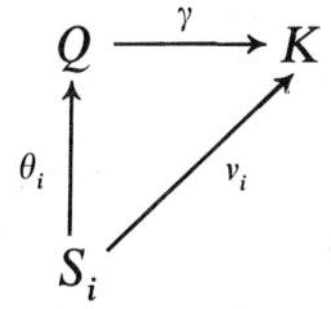

(4.8)

commutes for each i in I. Now for all u in U and for all i, j in I,

$$u\phi_i\theta_i\gamma = u\phi_i\nu_i = u\phi_j\nu_j \quad \text{(by assumption)}$$
$$= u\phi_j\theta_j\gamma;$$

hence $(u\phi_i\theta_i, u\phi_j\theta_j) \in \gamma \circ \gamma^{-1}$. That is,

$$((u\phi_i)\tau, (u\phi_j)\tau) \in \gamma \circ \gamma^{-1}$$

for all u in U and all i, j in I. Thus $\mathbf{R}/\tau \subseteq \gamma \circ \gamma^{-1}$ and so $\xi/\tau = (\mathbf{R}/\tau)^{\#} \subseteq \gamma \circ \gamma^{-1}$. It follows that the homomorphism γ: $Q \to K$ factors through $Q/(\xi/\tau)$, i.e. effectively through J. That is, if as in Theorem I.5.4 we denote the natural isomorphism from $(F/\tau)/(\xi/\tau)$ onto F/ξ by α—so that

$$w\tau^{\natural}(\xi/\tau)^{\natural}\alpha = w\xi^{\natural} \quad (w \in F)$$

—there is a homomorphism δ: $J \to K$ such that the diagram

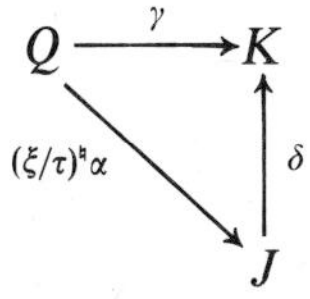

(4.9)

commutes. It now follows from the definition of μ_i and from the commutativity of the diagrams (4.8) and (4.9) that, for all s_i in S_i,

$$s_i\mu_i\delta = s_i\xi^{\natural}\delta = s_i\tau^{\natural}(\xi/\tau)^{\natural}\alpha\delta = s_i\tau^{\natural}\gamma = s_i\theta_i\gamma = s_i\nu_i,$$

i.e. that the diagram

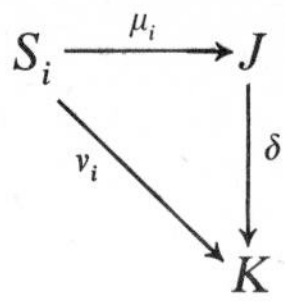

commutes for each i.

The uniqueness of δ follows from the fact that J is generated by the images of the elements of S_i ($i \in I$) within it. If $\delta' : J \to K$ is to have the desired commuting property then it must map each $s_i\mu_i$ to $s_i\nu_i$ for all s_i in S_i and for all i in I, i.e. it must coincide with δ over all the sets $S_i\mu_i$. Now each element of J is expressible as a product of elements from the various $S_i\mu_i$ and so it follows that δ' coincides with δ over the whole of J.∎

We have now succeeded in reproducing for inverse semigroups precisely the same coproduct and pushout apparatus that we constructed for semigroups in Section 1. The proofs of Proposition 2.1 and Theorem 2.3 involve only this apparatus, and so we can now simply write down a combined restatement for inverse semigroups:

Theorem 4.10. *Let S be an inverse semigroup and let U be an inverse subsemigroup of S. Let S' be an inverse semigroup disjoint from S and let $\alpha : S \to S'$ be an isomorphism. Let $J = S \mathfrak{I}_U S'$ be the free inverse product of the amalgam*

$$\mathfrak{A} = [\{S, S'\}; U; \{\iota, \alpha | U\}],$$

where ι is the inclusion mapping from U into S. Then the natural mappings $\mu : S \to J$ and $\mu' : S' \to J$ defined by

$$s\mu = s\xi \ (s \in S), \quad s'\mu' = s'\xi \ (s' \in S')$$

are both monomorphisms. Moreover,

$$(S\mu \cap S'\mu')\mu^{-1} = \mathrm{Dom}_S(U).∎$$

Notes. (1) The slight difference in form between the definition $s\mu = (s\theta)\rho$ and the present definition $s\mu = s\xi$ arises from the fact that we are now identifying the isomorphic copy $S\theta$ of S in the free product $S * S'$ with S itself.

(2) The dominion $\mathrm{Dom}_S(U)$ means exactly the same as it did before. Dominions are defined in terms of homomorphisms, and by formula V.1.7 a semigroup homomorphism between inverse semigroups is automatically an inverse semigroup homomorphism—i.e. it preserves inverses as well as products.

We now combine the information from this theorem with the information in Theorem 2.14 that an inverse subsemigroup of an (inverse) semigroup is always its own dominion, and obtain

Theorem 4.11. *Let U be an inverse subsemigroup of a semigroup S. Let S' be an isomorphic copy of S, disjoint from S, and let $\alpha: S \to S'$ be an isomorphism. Then the amalgam*

$$[\{S, S'\}; U; \{\iota, \alpha | U\}]$$

(where ι is the inclusion mapping of U into S) is embeddable in the inverse semigroup $J = S \mathfrak{I}_U S'$.■

In the terminology of T. E. Hall (1975) and Jónsson (1965), we have shown that inverse semigroups have the *special amalgamation property.*

Let us now consider a general inverse semigroup amalgam $[S, T; U]$, and let us adopt the point of view that each of S and T is an inverse semigroup containing U as an inverse subsemigroup. Let $\phi_1: S \to \mathscr{I}(X)$ be a faithful representation of S—the most natural choice for ϕ_1 is the Vagner–Preston representation with $X = S$, but it is not necessary to choose ϕ_1 in this way. We now apply Theorem V.8.13 to the restriction $\phi_1 | U: U \to \mathscr{I}(X)$ and obtain a representation $\psi'_1: T \to \mathscr{I}(X \cup Y'_1)$ with the property that

$$(\psi'_1 | U)_X = \phi_1 | U.$$

We define $\psi_1 = \psi'_1 \oplus \chi$, where $\chi: T \to \mathscr{I}(Z)$ is any faithful representation of T with $Z \cap (X \cup Y'_1) = \phi$. (Again, one way is to take χ as equivalent to a Vagner–Preston representation of T.) Then if we write $Y_1 = Y'_1 \cup Z$ we have a faithful representation $\psi_1: T \to \mathscr{I}(X \cup Y_1)$ with the property that

$$(\psi_1 | U)_X = \phi_1 | U, \tag{4.12}$$

and such that

$$\psi_1 | U = (\psi_1 | U)_X \oplus (\psi_1 | U)_{Y_1}.$$

Next, we apply Theorem V.8.13 to the representation

$$(\psi_1 | U)_{Y_1}: U \to \mathscr{I}(Y_1)$$

and obtain a representation $\phi_2: S \to \mathscr{I}(Y_1 \cup Z_1)$ with the properties that

$$\phi_2 | U = (\phi_2 | U)_{Y_1} \oplus (\phi_2 | U)_{Z_1} \quad \text{and} \quad (\phi_2 | U)_{Y_1} = (\psi_1 | U)_{Y_1}.$$

Then apply Theorem V.8.13 to $(\phi_2 | U)_{Z_1}$ and obtain a representation $\psi_2: T \to \mathscr{I}(Z_1 \cup Y_2)$ with the properties that

$$\psi_2 | U = (\psi_2 | U)_{Z_1} \oplus (\psi_2 | U)_{Y_2} \quad \text{and} \quad (\psi_2 | U)_{Z_1} = (\phi_2 | U)_{Z_1}.$$

We continue this process and obtain representations

$$\phi = \phi_1 \oplus \phi_2 \oplus \phi_3 \oplus \ldots : S \to \mathscr{I}(X \cup Y_1 \cup Z_1 \cup Y_2 \cup Z_2 \cup \ldots),$$
$$\psi = \psi_1 \oplus \psi_2 \oplus \psi_3 \oplus \ldots : T \to \mathscr{I}(X \cup Y_1 \cup Z_1 \cup Y_2 \cup Z_2 \cup \ldots).$$

In general we have (for $i = 1, 2, 3, \ldots$)

$$\phi_{i+1}: S \to \mathscr{I}(Y_i \cup Z_i), \qquad \psi_{i+1}: T \to \mathscr{I}(Z_i \cup Y_{i+1}),$$
$$(\phi_{i+1} | U)_{Y_i} = (\psi_i | U)_{Y_i}, \qquad (\psi_{i+1} | U)_{Z_i} = (\phi_{i+1} | U)_{Z_i}. \tag{4.13}$$

Both ϕ and ψ are faithful, since ϕ_1 and ψ_1 are faithful. Hence the semigroup $P = \mathscr{I}(X \cup Y_1 \cup Z_1 \cup Y_2 \cup Z_2 \cup \ldots)$ contains an isomorphic copy $S\phi$ of S and an isomorphic copy $T\psi$ of T. We consider the intersection $S\phi \cap T\psi$.

We shall show that $\phi | U = \psi | U$, i.e. that $u\phi = u\psi$ for every u in U. This will imply that $S\phi \cap T\psi$ contains at the very least the appropriate isomorphic copy of U.

Notice that, by (4.12)

$$(\phi | U)_X = (\phi_1 | U)_X = \phi_1 | U = (\psi_1 | U)_X = (\psi | U)_X.$$

Also, for $i = 1, 2, 3, \ldots$, it follows from (4.13) that

$$(\phi | U)_{Y_i} = (\phi_{i+1} | U)_{Y_i} = (\psi_i | U)_{Y_i} = (\psi | U)_{Y_i},$$
$$(\phi | U)_{Z_i} = (\phi_{i+1} | U)_{Z_i} = (\psi_{i+1} | U)_{Z_i} = (\psi | U)_{Z_i}.$$

Hence $\phi | U = \psi | U$ as required. We have shown

Theorem 4.14. *Let $[S, T; U]$ be an inverse semigroup amalgam. Then there is an inverse semigroup P with the property that there are monomorphisms $\phi: S \to P$, $\psi: T \to P$ for which $U\phi \subseteq S\phi \cap T\psi$.* ■

In the terminology of Jónsson (1965), we have shown that inverse semigroups have the *weak amalgamation property*. We can now fairly quickly complete the proof of our main result (Theorem 4.2), showing that inverse semigroups have the *strong amalgamation property*.

If $[S, T; U]$ is an inverse semigroup amalgam we begin by embedding S and T simultaneously in P in accordance with Theorem 4.14. We may suppose therefore that P is an inverse semigroup containing S and T as inverse subsemigroups and with $U \subseteq S \cap T$. Let P' be an isomorphic copy of P, disjoint from P, and let $\alpha: P \to P'$ be an isomorphism. Then by Theorem 4.11 (the *special amalgamation property*) the amalgam

$$[\{P, P'\}; U; \{\iota, \alpha | U\}]$$

is embeddable in its free inverse product $J = P \mathfrak{I}_U P'$. Now the inverse semigroup J may be thought of as containing inverse subsemigroups P and P' intersecting in U. Thus the copy of S in P and the copy of T in P' intersect precisely in U and so the amalgam $[S, T; U]$ is embedded in J. This completes the proof of Theorem 4.2.

Diagrammatically we can represent the proof as follows:

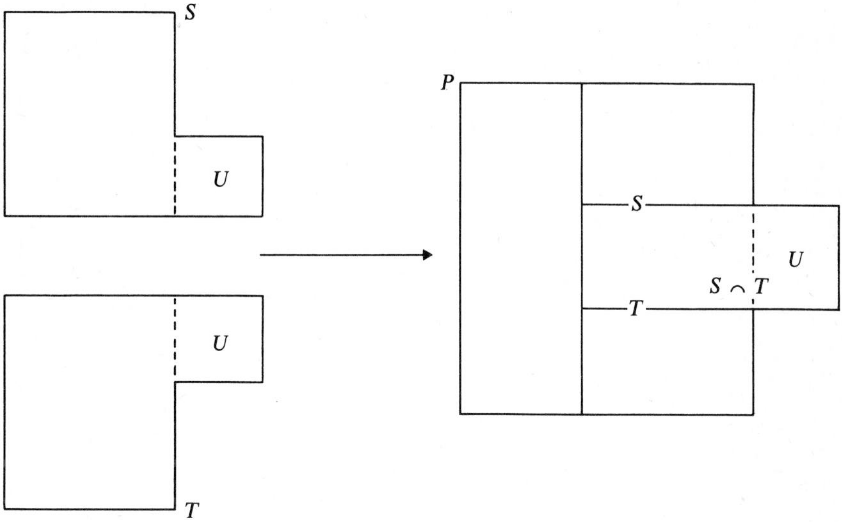

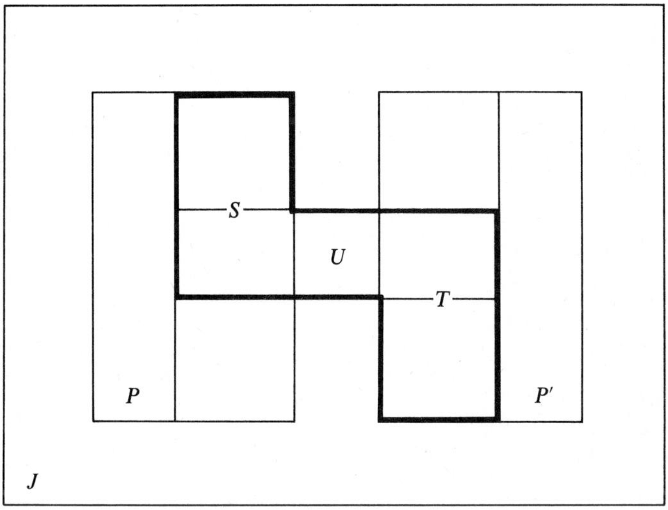

We remark finally that it is of the essence of the approach that the inverse semigroup P in Theorem 4.14 is infinite, even if S and T are finite. In group theory a finite amalgam is embeddable in a finite semigroup (B. H. Neumann 1954). The method above does not give the corresponding information for inverse semigroups, and indeed the corresponding result is not true. The following example from (T. E. Hall 1975) is attributed by Hall to C. J. Ash.

Example 4.15. Let $S = \{0, e, f, a, b\}$, $T = \{0, e, f, g, x, y\}$ be the semigroups whose multiplication tables are:

	0	e	f	a	b
0	0	0	0	0	0
e	0	e	0	a	0
f	0	0	f	0	b
a	0	0	a	0	e
b	0	b	0	f	0

	0	e	f	g	x	y
0	0	0	0	0	0	0
e	0	e	0	0	x	0
f	0	0	f	g	0	y
g	0	0	g	g	0	y
x	0	0	x	x	0	e
y	0	y	0	0	g	0

It is easy to see that these are semigroups, either by a straightforward (though tedious) verification of associativity, or by observing that each may be represented by matrices as follows: for S,

$$0 = \begin{bmatrix} 0 & 0 \\ 0 & 0 \end{bmatrix}, \quad e = \begin{bmatrix} 1 & 0 \\ 0 & 0 \end{bmatrix}, \quad f = \begin{bmatrix} 0 & 0 \\ 0 & 1 \end{bmatrix},$$

$$a = \begin{bmatrix} 0 & 1 \\ 0 & 0 \end{bmatrix}, \quad b = \begin{bmatrix} 0 & 0 \\ 1 & 0 \end{bmatrix};$$

for T,

$$0 = \begin{bmatrix} 0 & 0 & 0 \\ 0 & 0 & 0 \\ 0 & 0 & 0 \end{bmatrix}, \quad e = \begin{bmatrix} 1 & 0 & 0 \\ 0 & 0 & 0 \\ 0 & 0 & 0 \end{bmatrix}, \quad f = \begin{bmatrix} 0 & 0 & 0 \\ 0 & 1 & 0 \\ 0 & 0 & 1 \end{bmatrix},$$

$$g = \begin{bmatrix} 0 & 0 & 0 \\ 0 & 1 & 0 \\ 0 & 0 & 0 \end{bmatrix}, \quad x = \begin{bmatrix} 0 & 1 & 0 \\ 0 & 0 & 0 \\ 0 & 0 & 0 \end{bmatrix}, \quad y = \begin{bmatrix} 0 & 0 & 0 \\ 1 & 0 & 0 \\ 0 & 0 & 0 \end{bmatrix}.$$

In fact both S and T are easily seen to be inverse semigroups, and $U = \{0, e, f\}$ is a common inverse subsemigroup.

Suppose now that the inverse semigroup amalgam $[S, T; U]$ is embedded in a semigroup Q, i.e. that Q is a semigroup containing S and T, with $S \cap T = U$. Then $g < f$ in Q. But in Q we also have

$$g = gfg, \qquad f = ba = bea = bxya = bxgya,$$

and so $(g, f) \in \mathscr{J}$ in Q. The principal factor $J_f \cup \{0\}$ of Q is certainly not null, since it contains the idempotent f. Hence, by proposition III.1.6, it is 0-simple. If it were finite it would be completely 0-simple and all its non-zero idempotents would be primitive (Section III.3). Since $g < f$ in $J_f \cup \{0\}$ this is not so; hence $J_f \cup \{0\}$, and so *a fortiori* Q, is infinite.

This example of a finite semigroup amalgam that is embeddable but not finitely embeddable incidentally answers negatively a question in (Howie 1964a).

EXERCISES

1. (Isbell 1965). Show that $U \mapsto \mathrm{Dom}_S(U)$ is a closure operation on the set of subsemigroups of a semigroup S. That is, show that

$$U \subseteq \mathrm{Dom}_S(U),$$

$$U \subseteq V \Rightarrow \mathrm{Dom}_S(U) \subseteq \mathrm{Dom}_S(V),$$

$$\mathrm{Dom}_S(\mathrm{Dom}_S(U)) = \mathrm{Dom}_S(U).$$

2. Let $U = \langle a \rangle$, an infinite monogenic semigroup.
   (i) If S is the infinite cyclic group generated by a, show that
   $$U \subset \mathrm{Dom}_S(U) = S.$$
   (ii) If S is the direct product $\langle a \rangle \times \langle b \rangle$ of $U = \langle a \rangle$ with a second infinite monogenic semigroup $\langle b \rangle$ show that
   $$U = \mathrm{Dom}_S(U) \subset S.$$
   (iii) If P and Q are infinite cyclic groups generated by a, b, respectively, and if S is the 0-direct union of P^0 and Q^0, show that
   $$U \subset \mathrm{Dom}_S(U) \subset S.$$

3. (Howie and Isbell 1967). Show that left simple semigroups (i.e. semigroups S in which $\mathscr{L} = S \times S$) are absolutely closed. [Hint: show that any zigzag in any T over S can be replaced by a left-inner zigzag.]

4. (Howie and Isbell 1967). Let S be a semigroup in which there exist elements a_1, a_2, a_3 for which

$$a_1 S \cap a_2 S = S a_2 \cap S a_3 = \varnothing.$$

(This is a property possessed in particular by the rectangular band $\{1, 2\} \times \{1, 2\}$, with $a_1 = (1, 1)$, $a_2 = (2, 1)$, $a_3 = (1, 2)$.) Let F be the free semigroup on four generators x_1, x_2, y_1, y_2, and let $T = (S * F)/\mathbf{R}$, where

$$\mathbf{R} = \{(a_1, x_1 a_1), (a_1 y_1, a_2 y_2), (x_1 a_2, x_2 a_3), (a_3 y_2, a_3)\}.$$

   (i) Show by induction that any element of $S * F$ obtained from an element s of S by elementary $\mathbf{R}$-transitions is of the form $w_1 z_1 w_2 z_2 \cdots z_{n-1} w_n$, where:
   (a) $w_1, \ldots, w_n \in S^1$;

(b) each of $z_1, \ldots, z_{n-1}$ is either x_1 or y_2;
(c) if $z_i = x_1$ and $w_{i+1} \neq 1$ then $w_{i+1} \in a_1S$;
(d) if $z_i = y_2$ and $w_i \neq 1$ then $w_i \in Sa_3$;
(e) $s = w_1 w_2 \ldots w_n$.

(ii) Deduce that $\mathbf{R}^\# \cap (S \times S) = 1_S$, and hence that T contains S up to isomorphism.
(iii) Show that the element $d = (a_1y_1)\mathbf{R}^\#$ of T is such that $d \notin S$, $d \in \mathrm{Dom}_T S$, and deduce that S is not absolutely closed.

This example shows that Theorem 2.14 cannot generalize to regular, or even to orthodox semigroups.

5. (Kimura 1957). Let $U = \{u, v, w, 0\}$ be a four-element null semigroup. Let $S = U \cup \{a\}$, with $au = ua = v$ and all other products equal to 0. Let $T = U \cup \{b\}$, with $bv = vb = w$ and all other products equal to 0. Then S and T are semigroups. If P is a semigroup containing the amalgam $[S, T; U]$, show that $w = z$ in P, and deduce that $[S, T; U]$ is not embeddable.

6. (Howie 1964a). Let $[S, T; U]$ be an amalgam. Let

$$N = U \cup (S \backslash U) \cup (T \backslash U) \cup \{0\},$$

and define a binary operation $\circ$ in N by

$$x \circ y = \begin{cases} xy \text{ if } x, y \text{ are both in } S \text{ or both in } T, \\ 0 \text{ otherwise.} \end{cases}$$

Show that $(N, \circ)$ is a semigroup if and only if U is *consistent* in S and T, i.e. if and only if in both S and T

$$xy \in U \Rightarrow x \in U \quad \text{and} \quad y \in U.$$

Deduce that a (finite) amalgam $[S, T; U]$ in which U is consistent both in S and in T is embeddable in a (finite) semigroup.

7. Show that a subsemigroup of a group is a subgroup if and only if it is a (left) unitary subsemigroup.

8. Let $G = \{e, a\}$, $H = \{f, b\}$ and $K = \{g, c\}$ be groups, and let S be the semilattice of groups $G \cup H \cup K$ with multiplication table

	e	a	f	b	g	c
e	e	a	f	b	g	c
a	a	e	b	f	c	g
f	f	b	f	b	g	c
b	b	f	b	f	c	g
g	g	c	g	c	g	c
c	c	g	c	g	c	g

i.e. in which $e > f > g$ and the structure homomorphisms are all isomorphisms.

Show that the subgroup H is not unitary in S but is unitary in fSf.

9. Let U, V be disjoint semigroups with identities e, f respectively, and let $S = U \cup V \cup \{0\}$, the 0-direct union of U and V. (See Section III.3.) Show that U is a unitary subsemigroup of S. Thus U is almost unitary in S, with associated mappings $\lambda = \rho = 1_S$. By Proposition 3.9, U is unitary in eSe, but $\lambda S\rho = S$, while $eSe = U$.

10. It is known (B. H. Neumann 1954, Section 15) that if

$$\mathfrak{A} = [\{S_1, S_2\}; U; \{\phi_1, \phi_2\}]$$

is a *group* amalgam in which $U\phi_1 \subset S_1$, $U\phi_2 \subset S_2$ (properly), then the free product of $\mathfrak{A}$ is infinite. Show that this is not the case for semigroups. More precisely, let S_1 be a finite semigroup with identity element 1, let S_2 be a finite semigroup with zero element 0, and let $U = \{e\}$ be a one element semigroup. Let $e\phi_1 = 1$, $e\phi_2 = 0$.

Show that the free product P of the amalgam $\mathfrak{A}$ is a semigroup of order $|S_1| + |S_2| - 1$. (Notice that since the core of the amalgam is a group, $\mathfrak{A}$ is naturally embedded in P.)

11. (N. R. Reilly, T. E. Hall, private communication). Show that any inverse semigroup S with semilattice of idempotents E can be embedded in an inverse semigroup P with semilattice of idempotents E and such that

$$\mathscr{D}^P \cap (E \times E) = \mathscr{U}$$

(i.e. such that $Ee \simeq Ef \Rightarrow (e, f) \in \mathscr{D}^P$). [Hint: Let $P = S \,\mathfrak{Z}_E T_E$.]

References

ALLEN, D., JR.
1971 A generalization of the Rees Theorem to a class of regular semigroups, *Semigroup Forum* **2** (1971), 321–331.

ANDERSEN, O.
1952 "Eine Bericht über die Struktur abstrakter Halbgruppen", Thesis (Staatsexamensarbeit), Hamburg, 1952.

BAER, R. and LEVI, F.
1932 Vollständige irreduzibele Systeme von Gruppenaxiomen (Beitrage zur Algebra, No. 18), *Sitzber. Heidelberger Akad. Wiss.* Abh. 2 (1932), 1–12.

BAIRD, G. R.
1972a On a sublattice of the lattice of congruences on a simple regular ω-semigroup, *J. Australian Math. Soc.* **13** (1972), 461–471.
1972b Congruences on simple regular ω-semigroups, *J. Australian Math. Soc.* **14** (1972), 155–167.

BIRJUKOV, A. P. (=Бирюков, А. П.)
1970 Varieties of idempotent semigroups, *Algebra i Logika* **9** (1970) 255–273 (Russian).

BIRKHOFF, G.
1935 On the structure of abstract algebras, *Proc. Cambridge Phil. Soc.* **31** (1935), 433–454.
1948 "Lattice Theory", American Math. Soc. Colloquium Publications, Vol. 25, 2nd Edition, New York, 1948.

BRUCK, R. H.
1958 "A Survey of Binary Systems", Ergebnisse der Math. Heft 20, Springer, Berlin, 1958.

CLIFFORD, A. H.
1941 Semigroups admitting relative inverses, *Ann. of Math.* **42** (1941), 1037–1049.
1953 A class of d-simple semigroups, *American J. Math.* **75** (1953), 547–556.
1972a The structure of bisimple orthodox semigroups as ordered pairs, Tulane University, New Orleans, 1972.
1972b The structure of orthodox unions of groups, *Semigroup Forum* **3** (1972), 282–337.
1974 The fundamental representation of a regular semigroup, Tulane University, New Orleans, 1974.
1975 The partial groupoid of idempotents of a regular semigroup, *Semigroup Forum* **10** (1975) 262–268.

CLIFFORD, A. H. and PRESTON, G. B.
1961 "The Algebraic Theory of Semigroups", vol. I, Math. Surveys of the American Math. Soc. 7, Providence, R. I., 1961.

1967 "The Algebraic Theory of Semigroups", vol. II, Math. Surveys of the American Math. Soc. 7, Providence, R. I., 1967.

COHN, P. M.

1965 "Universal Algebra", Harper and Row, New York, 1965.

CROISOT, R.

1953 Demi-groupes inversifs et demi-groupes réunions de demi-groupes simples, *Ann. Sci. Ecole Norm. Sup.* (3) **70** (1953), 361–379.

DUBREIL, P.

1941 Contribution à la theorie des demi-groupes, *Mém. Acad. Sci. Inst. France* (2) **63** No. 3 (1941), 1–52.

DUBREIL–JACOTIN, M.-L., LESIEUR, L. and CROISOT, R.

1953 "Lecons sur la Théorie des Treillis", Paris, Gauthier-Villars, 1953

EBERHART, C. and SELDEN, J.

1972 One parameter inverse semigroups, *Trans. American Math. Soc.* **168** (1972), 53–66.

EVANS, T.

1968 The number of semigroup varieties, *Quart. J. Math. Oxford* (2) **19** (1968).

1971 The lattice of semigroup varieties, *Semigroup Forum* **2** (1971), 1–43.

FANTHAM, P. H. H.

1960 On the classification of a certain type of semigroup, *Proc. London Math. Soc.* (3) **10** (1960), 409–427.

FENNEMORE, C. F.

1971a All varieties of bands, I, *Math. Nachr.* **48** (1971), 237–252.

1971*b* All varieties of bands, II, *Math. Nachr.* **48** (1971), 253–262.

FITZ-GERALD, D. G.

1972 On inverses of products of idempotents in regular semigroups, *J. Australian Math. Soc.* **13** (1972), 335–337.

GERHARDT, J. A.

1970 The lattice of equational classes of idempotent semigroups, *J. Algebra,* **15** (1970), 195–224.

GLUSKIN, L. M. (=Глускин, Л. М.)

1956 Completely simple semigroups, *Uč. Zap. Kharkov. Ped. Inst.* **18** (1956), 41–55 (Russian).

1957a Normal series in completely simple semigroups, *Uč. Zap. Kharkov.Ped. Inst.* **21** (1957), 99–106 (Russian).

1957b Elementary generalised groups, *Mat. Sbornik* **41** (83) (1957), 23–36 (Russian).

GREEN, J. A.

1951 On the structure of semigroups, *Ann. of Math.* **54** (1951), 163–172.

GREEN, J. A. and REES, D.

1952 On semigroups in which $x^r = x$, *Proc. Cambridge Phil. Soc.* **48** (1952), 35–40.

GRILLET, P.-A.

1974a The structure of regular semigroups. I. A representation. *Semigroup Forum* **8** (1974) 177–183.

1974*b* The structure of regular semigroups. II. Cross-connections. *Semigroup Forum* **8** (1974) 254–259.

1974c The structure of regular semigroups. III. The reduced case. *Semigroup Forum* **8** (1974) 260–265.

1974d The structure of regular semigroups. IV. The general case, *Semigroup Forum* **8** (1974) 368–373.

GRILLET, P.-A. and PETRICH, M.
1970 Free products of semigroups amalgamating an ideal, *J. London Math. Soc.* (2), **2** (1970), 389–392.

HALL, M.
1959 "The Theory of Groups", Macmillan, 1959.

HALL, T. E.
1969 On regular semigroups whose idempotents form a subsemigroup, *Bull. Australian Math. Soc.* **1** (1969), 195–208.
1970a On orthodox semigroups and uniform and antiuniform bands, *J. Algebra,* **16** (1970), 204–217.
1970b On the natural ordering of $\mathscr{J}$-classes and of idempotents in a regular semigroup, *Glasgow Math. J.* **11** (1970), 167–168.
1971 Orthodox semigroups, *Pacific J. Math.* **39** (1971), 677–686.
1972 Congruences and Green's relations on regular semigroups, *Glasgow Math. J.* **13** (1972), 167–175.
1973 On regular semigroups, *J. Algebra,* **24** (1973), 1–24.
1975 Free products with amalgamation of inverse semigroups, *J. Algebra* **34** (1975) 375–385.

HALMOS, P. R.
1960 "Naive Set Theory", Van Nostrand, New York, 1960.

HARDY, G. H. and WRIGHT, E. M.
1938 "An Introduction to the Theory of Numbers", Oxford University Press, London, 1938.

HICKEY, J. B.
1970 "Bisimple Inverse Semigroups and Uniform Semilattices", Ph.D. Thesis, University of Glasgow, 1970.

HOWIE, J. M.
1962 Embedding theorems with amalgamation for semigroups, *Proc. London Math. Soc.* (3), **12** (1962), 511–534.
1963a An embedding theorem with amalgamation for cancellative semigroups, *Proc. Glasgow Math. Assoc.* **6** (1963), 19–26.
1963b Subsemigroups of amalgamated free products of semigroups, *Proc. London Math. Soc.* (3), **13** (1963), 672–686.
1963c Embedding theorems for semigroups, *Quart. J. Math. Oxford* (2), **14** (1963), 254–258.
1964a The embedding of semigroup amalgams, *Quart. J. Math. Oxford* (2), **15** (1964), 55–68.
1964b Subsemigroups of amalgamated free products of semigroups. II, *Proc. London Math. Soc.* (3), **14** (1964), 537–544.
1964c The maximum idempotent-separating congruence on an inverse semigroup, *Proc. Edinburgh Math. Soc.* (2), **14** (1964), 71–79.
1968 Commutative semigroup amalgams, *J. Australian Math. Soc.* **8** (1968), 609–630.
1975 Semigroup amalgams whose cores are inverse semigroups, *Quart. J. Math. Oxford* (2), **26** (1975), 23–45.

HOWIE, J. M. and ISBELL, J. R.
1967 Epimorphisms and dominions. II, *J. Algebra* **6** (1967), 7–21.

HOWIE, J. M. and LALLEMENT, G.
1966 Certain fundamental congruences on a regular semigroup, *Proc. Glasgow Math. Assoc.* **7** (1966), 145–159.

HOWIE, J. M. and SCHEIN, B. M.
1969 Anti-uniform semilattices, *Bull. Australian Math. Soc.* **1** (1969), 263–268.

ISBELL, J. R.
1965 Epimorphisms and dominions, "Proc. Conference on Categorical Algebra, La Jolla, 1965" (Springer-Verlag, 1966).
1970 Two examples in varieties of monoids, *Proc. Cambridge Phil. Soc.* **68** (1970), 265–266.

JÓNSSON, B.
1965 Extensions of relational structures. "The Theory of Models", (Proceedings of the 1963 International Symposium at Berkeley), North Holland, Amsterdam (1965), 146–157.

KAPLANSKY, I.
1954 "Infinite Abelian Groups", University of Michigan, 1954.

KAPP, K. M. and SCHNEIDER, H.
1969 "Completely 0-simple Semigroups", W. A. Benjamin, New York, 1969.

KIMURA, N.
1957a "On Semigroups", Doctoral Thesis, the Tulane University of Louisiana, 1957.
1957b Note on idempotent semigroups. I, *Proc. Japan Acad.* **33** (1957), 642–645.
1958a Note on idempotent semigroups. III, *Proc. Japan Acad.* **34** (1958), 113–114.
1958b Note on idempotent semigroups. IV, Identities of three variables, *Proc. Japan Acad.* **34** (1958), 121–123.
1958c The structure of idempotent semigroups. (I), *Pacific J. Math.* **8** (1958), 257–275.

KOČIN, B. P. (=Кочин, Б. П.)
1968 The structure of inverse ideal-simple ω-semigroups. *Vestnik Leningrad. Univ.* **23**, No. 7, (1968), 41–50 (Russian).

KUROSH, A. G.
1956 "The Theory of Groups", volumes I and II, (translated from the Russian by K. A. Hirsch) Chelsea, 1956.

LAJOS, S.
1969 On semilattices of groups, *Proc. Japan Acad.* **45** (1969), 383–384.
1970a Characterization of completely regular inverse semigroups, *Acta Sci. Math. (Szeged)* **31** (1970), 229–231.
1970b Notes on regular semigroups, *Proc. Japan Acad.* **46** (1970), 253–254.
1970c Notes on regular semigroups. II, *Proc. Japan Acad.* **46** (1970), 929–931.
1971a On semilattices of groups. II, *Proc. Japan Acad.* **47** (1971), 36–37.
1971b Notes on regular semigroups. III, *Proc. Japan Acad.* **47** (1971), 185–186.
1972a A remark on semilattices of groups, *Ann. Univ. Sci. Budapest* **15** (1972), (1972), 121–122.
1972b A note on semilattices of groups, *Acta Sci. Math.* (Szeged) **33** (1972), 315–317.

LALLEMENT, G.
1966 Congruences et équivalences de Green sur un demi-groupe régulier, *C.R. Acad. Sci. Paris, Sér. A,* **262** (1966), 613–616.
1967 Demi-groupes réguliers, *Ann. Mat. pura ed appl,* **77** (1967), 47–129.
1972 Structure theorems for regular semigroups, *Semigroup Forum* **4** (1972), 95–123.

1974 A note on congruences on Rees matrix semigroups, *Semigroup Forum* **8** (1974), 89–92.

LJAPIN, E. S. (=Ляпин, Е. С.)

1969 The independence of subsemigroups of a semigroup, *Doklady Akad. Nauk SSSR* **185** (1969), 1229–1231 (Russian).

1970 Intersections of independent subsemigroups of a semigroup, *Izv. Vysš. Učebn. Zaved. Matematika,* No. 4 (95) (1970), 67–73 (Russian).

1974 Independent semigroup extensions of partial groupoids, *Modern Algebra*, 2nd issue (ed. Ljapin), Leningrad, 1974 (Russian).

MCALISTER, D. B.

1973 Groups, semilattices and inverse semigroups, *Trans. American Math. Soc.* **192** (1973), 1–18.

1974a Groups, semilattices and inverse semigroups. II, *Trans. American Math. Soc.* **196** (1974), 351–370.

1974b 0-bisimple inverse semigroups, *Proc. London Math. Soc.* (3) **28** (1974), 193–221.

MCLEAN, D.

1954 Idempotent semigroups, *American Math. Monthly* **61** (1954), 110–113.

MCLEAN, P.

1973 "Contributions to the Theory of 0-Simple Inverse Semigroups", Ph.D. Thesis, University of Stirling, 1973.

MEAKIN, J. C.

1971 Congruences on orthodox semigroups, *J. Australian Math. Soc.* **12** (1971), 323–341.

1972 The maximum idempotent-separating congruence on a regular semigroup, *Proc. Edinburgh Math. Soc.* (2), **18** (1972/3), 159–163.

MILLER, D. D. and CLIFFORD, A. H.

1956 Regular $\mathscr{D}$-classes in semigroups, *Trans. American Math. Soc.* **82** (1956), 270–280.

MITCHELL, B.

1965 "Theory of Categories", Academic Press, London and New York, 1965.

MUNN, W. D.

1955 On semigroup algebras, *Proc. Cambridge Phil. Soc.* **51** (1955), 1–15.

1964 A certain sublattice of the lattice of congruences on a regular semigroup, *Proc. Cambridge Phil. Soc.* **60** (1964), 385–391.

1966a The lattice of congruences on a bisimple ω-semigroup, *Proc. Royal Soc. Edinburgh* (*A*) **67** (1966), 175–184.

1966b Uniform semilattices and bisimple inverse semigroups, *Quart. J. Math. Oxford* (2), **17** (1966), 151–159,

1967 The idempotent-separating congruences on a regular 0-bisimple semigroup, *Proc. Edinburgh Math. Soc.* (2), **15** (1967), 233–240.

1968 Regular ω-semigroups, *Glasgow Math. J.* **9** (1968), 46–66.

1970a Fundamental inverse semigroups, *Quart. J. Math, Oxford* (2), **21** (1970), 157–170.

1970b 0-Bisimple inverse semigroups, *J. Algebra,* **15** (1970), 570–588.

1970c On simple inverse semigroups, *Semigroup Forum* **1** (1970), 63–74.

1972 Embedding semigroups in congruence-free semigroups, *Semigroup Forum* **4** (1972), 46–60.

1974a Free inverse semigroups, *Proc. London Math. Soc.* (3), **29** (1974), 385–404.

1974b Congruence-free inverse semigroups, *Quart. J. Math. Oxford* (2), **25** (1974), 463–84.

1976 A note on *E*-unitary inverse semigroups, *Bull. London Math. Soc.* **8** (1976), 71–76.

MUNN, W. D. and REILLY, N. R.

1966 Congruences on a bisimple ω-semigroup, *Proc. Glasgow Math. Assoc.* **7** (1966), 184–192.

NAMBOORIPAD, K. S. S.

1973 "Structure of Regular Semigroups", Dissertation, University of Kerala (Kariavattom, Trivandrum, India), 1973.

1975a Structure of regular semigroups. I. Fundamental regular semigroups, *Semigroup Forum*, **9**, (1975) 354–363,

1975b Structure of regular semigroups. II. The general case, *Semigroup Forum* **9** (1975) 364–371.

NEUMANN, B. H.

1954 An essay on free products of groups with amalgamations, *Phil. Trans. Roy. Soc. A,* **246** (1954), 503–554.

1970 Some remarks on cancellative semigroups, *Math. Zeit.* **117** (1970), 97–111.

VON NEUMANN, J.

1936 On regular rings, *Proc. Nat. Acad. Sci. U.S.A.* **22** (1936) 707–713.

O'CARROLL, L.

1974a A note on free inverse semigroups, *Proc. Edinburgh Math. Soc.* (2), **19** (1974/75), 17–23.

1974b Quasi-reduced inverse semigroups, *J. London Math. Soc.* **9** (1974) 142–150.

PETRICH, M.

1965 The structure of a class of semigroups which are unions of groups, *Notices American Math. Soc.* **12**, No. 1, Part 1 (1965), p. 102.

1967 "Topics in Semigroups", Pennsylvania State University, 1967.

1968 The translational hull of a completely 0-simple semigroup, *Glasgow Math. J.* **9** (1968), 1–11.

1971 A construction and a classification of bands, *Math. Nach.* **48** (1971), 263–274.

1973 Regular semigroups which are subdirect products of a band and a semilattice of groups, *Glasgow Math. J.* **14** (1973), 27–49.

1974 The structure of completely regular semigroups, *Trans. American Math. Soc.* **189** (1974), 211–236.

PHILIP, J. M.

1974 A proof of Isbell's Zigzag Theorem, *J. Algebra* **32** (1974), 328–331.

PRESTON, G. B.

1954a Inverse semi-groups, *J. London Math. Soc.* **29** (1954), 396–403.

1954b Inverse semi-groups with minimal right ideals, *J. London Math. Soc.* **29** (1954), 404–411.

1954c Representations of inverse semi-groups, *J. London Math. Soc.* **29** (1954), 411–419.

1959 Embedding any semigroup in a $\mathscr{D}$-simple semigroup, *Trans. American Math. Soc.* **93** (1959), 351–355.

1961 Congruences on completely 0-simple semigroups, *Proc. London Math. Soc.* (3), **11** (1961), 557–576.

1973a Inverse semigroups: some open questions, "Proc. Symposium on Inverse

Semigroups and Their Generalisations", Northern Illinois University, 1973, pp. 122–139.
1973b Free inverse semigroups, (Collection of articles dedicated to the memory of Hanna Neumann, IV), *J. Australian Math. Soc.* **16** (1973), 443–453.

REES, D.
1940 On semi-groups, *Proc. Cambridge Phil. Soc.* **36** (1940), 387–400.

REILLY, N. R.
1965 Embedding inverse semigroups in bisimple inverse semigroups, *Quart. J. Math. Oxford* (2), **16** (1965), 183–187.
1966 Bisimple ω-semigroups, *Proc. Glasgow Math. Assoc.* **7** (1966), 160–167.
1968 Bisimple inverse semigroups, *Trans. American Math. Soc.* **132** (1968), 101–114.
1971 Congruences on a bisimple inverse semigroup in terms of *RP*-systems, *Proc. London Math. Soc.* (3), **23** (1971), 99–127.
1972 Free generators in free inverse semigroups, *Bull. Australian Math. Soc.* **7** (1972), 407–424.
1973 Corrigenda: Free generators in free inverse semigroups, *Bull. Australian Math. Soc.* **9** (1973), 479–480.

REILLY, N. R. and CLIFFORD, A. H.
1968 Bisimple inverse semigroups as semigroups of ordered triples, *Canadian J. Math.* **20** (1968), 25–39.

REILLY, N. R. and SCHEIBLICH, H. E.
1967 Congruences on regular semigroups, *Pacific J. Math.* **23** (1967), 349–360.

SAITÔ, T.
1965 Proper ordered inverse semigroups, *Pacific J. Math.* **15** (1965), 649–666.

SCHEIBLICH, H. E.
1972 Free inverse semigroups, *Semigroup Forum* **4** (1972), 352–359.
1973 Free inverse semigroups, *Proc. American Math. Soc.* **38** (1973), 1–7.

SCHEIBLICH, H. E. and MOORE, K. C.
1974 $\mathscr{T}_X$ is absolutely closed, *Semigroup Forum* **6** (1973), 216–226.

SCHEIN, B. M. (=Šaĭn, B. M. =Шайн, Б. М.)
1962 Representations of generalised groups, *Izv. Vysš. Učebn. Zaved Matematika* No. 3 (28) (1962), 164–176 (Russian).
1963 On the theory of generalized groups, *Doklady Akad. Nauk SSSR* **153** (1963), 296–299 (Russian).
1964 Generalized groups with the well-ordered set of idempotents, *Mat. Fyz. Casopis Sloven. Akad. Vied.* **14** (1964), 259–262.
1975 Free inverse semigroups are not finitely presentable, *Studia Sci. Math. Hungar.*, **26** (1975), 179–195.

SCHUBERT, H.
1972 "Categories", Springer-Verlag, New York, 1972.

SILVER, L. Non-commutative localization and applications, *J. Algebra,* **7** (1967), 44–76.

STENSTRÖM, B.
1971 Flatness and localization over monoids, *Math. Nach.* **48** (1971), 315–333.
A. K. (Suškevič = Сушкевич, А. К.)
1928 Über die endlichen Gruppen ohne das Gesetz der eindeutigen Umkehrbarkeit, *Math. Ann.* **99** (1928), 30–50.

TAMURA, T.

1960 Decompositions of a completely simple semigroup, *Osaka Math. J.* **12** (1960), 269–275.

TAMURA, T. and KIMURA, N.

1954 On decompositions of a commutative semigroup, *Kodai Math. Sem. Rep.* (1954), 109–112.

VAGNER, V. V. (=Вагнер, В. В.)

1952 Generalised groups, *Doklady Akad. Nauk SSSR* **84** (1952), 1119–1122 (Russian).

1953 Theory of generalised heaps and generalised groups, *Mat. Sbornik* (*N.S.*) **32** (1953), 545–632 (Russian).

WARNE, R. J.

1966a A class of bisimple inverse semigroups, *Pacific J. Math.* **18** (1966), 563–577.

1966b On certain bisimple inverse semigroups, *Bull. American Math. Soc.* **72** (1966), 679–682.

1966c The idempotent-separating congruences of a bisimple inverse semigroup with identity, *Publ. Math. Debrecen,* **13** (1966), 203–206.

1967a Bisimple inverse semigroups mod groups, *Duke Math. J.* **34** (1967), 787–811.

1967b Errata: "On certain bisimple inverse semigroups", *Bull. American Math. Soc.* **73** (1967), 496.

1968 *I*-bisimple semigroups, *Trans. American Math. Soc.* **130** (1968), 367–386.

1969 Congruences on ω^n-bisimple semigroups, *J. Australian Math. Soc.* **9** (1969), 257–274.

1970 $\omega^n I$-bisimple semigroups, *Acta Math. Acad. Sci. Hungar.* **21** (1970), 121–150.

1971a *E*-bisimple semigroups, *J. Natur. Sci. and Math.* **11** (1971), 51–81.

1971b On the structure of regular bisimple semigroups, *Tamkang J. Math.* **2** (1971), 187–190.

YAMADA, M.

1958 Note on idempotent semigroups. V, *Proc. Japan Acad.* **34** (1958), 668–671.

1963a Inversive semigroups. I, *Proc. Japan Acad.* **39** (1963), 100–103.

1963b Inversive semigroups. II, *Proc. Japan Acad.* **39** (1963), 104–106.

1965 Inversive semigroups. III, *Proc. Japan Acad.* **41** (1965), 221–224.

1966 Note on the structure of regular semigroups, *Proc. Japan Acad.* **42** (1966), 136–140.

1967 Regular semigroups whose idempotents satisfy permutation identities, *Pacific J. Math.* **21** (1967), 371–392.

1970 On a regular semigroup in which the idempotents form a band, *Pacific J. Math.* **33** (1970), 261–272.

1971 Construction of inversive semigroups, *Mem. Fac. Lit. & Sci., Shimane Univ., Nat. Sci.* **4** (1971), 1–9.

1973a Orthodox semigroups whose idempotents satisfy a certain identity, *Semigroup Forum* **6** (1973), 113–128.

1973b Note on a certain class of orthodox semigroups, *Semigroup Forum* **6** (1973), 180–188.

YAMADA, M. and KIMURA, N.

1958 Note on idempotent semigroups. II, *Proc. Japan Acad.* **34** (1958), 110–112.

YAMURA

1956 Indecomposable completely simple semigroups except groups, *Osaka Math. J.* **8** (1956) 35–42.

List of Special Symbols

B_A	the free band on a set A	104
$BR(T, \theta)$	the Bruck–Reilly extension of T determined by θ	153
$\mathscr{B}(X)$	the semigroup of binary relations on a set X	14
C_ω	the semilattice $\{e_0, e_1, e_2, \ldots\}$, with $e_0 > e_1 > e_2 > \ldots$	144
$\mathscr{C}(S)$	the lattice of congruences on a semigroup S	27
$C(w)$	the content of a word w	104
$\mathscr{D}$	Green's relation	39
D_a	the $\mathscr{D}$-class containing a	40
$\mathscr{D}^S$	Green's relation on the semigroup S	50
D_a^S	the $\mathscr{D}^S$-class containing a	50
dom (ρ)	the domain of a binary relation ρ	15
$\mathrm{Dom}_S(U)$	the dominion of U in S	220
$E(e)$	the $\mathscr{J}^B$-class containing an element e of a band B	188
$\mathbf{E}^\flat$	the largest congruence contained in an equivalence $\mathbf{E}$	27
$E(S)$	the set of idempotents of a semigroup S	131
$\mathscr{E}(X)$	the lattice of equivalences on a set X	27
F_A	the free semigroup on a set A	29
$\mathscr{G}(X)$	the symmetric group on a set X	6
γ	the minimum inverse semigroup congruence on an orthodox semigroup	190
$\mathscr{H}$	Green's relation	39
H_a	the $\mathscr{H}$-class containing a	40
$\mathscr{H}^S$	Green's relation on the semigroup S	50
H_a^S	the $\mathscr{H}^S$-class containing a	50
$\mathscr{H}(B, T, \psi)$	the Hall–Yamada semigroup	207
$I(a)$	$J(a)\backslash J_a$	60
$\mathbf{I}(S)$	the set of all identical relations satisfied in a semigroup S	110
$\mathfrak{I}^*S_i$	the free inverse product of inverse semigroups S_i	244
$\mathfrak{I}^*_U S_i$	the free inverse product of an inverse semigroup amalgam	246
$\mathscr{I}(X)$	the symmetric inverse semigroup on a set X	133
$\mathscr{J}$	Green's relation	39
J_a	the $\mathscr{J}$-class containing a	40
$\mathscr{J}^S$	Green's relation on the semigroup S	50
J_a^S	the $\mathscr{J}^S$-class containing a	50
$J(a)$	the principal ideal generated by a	59
K_a	the kernel of a monogenic semigroup $\langle a\rangle$	9
ker ϕ	the kernel of a mapping ϕ	19

$K(S)$	the kernel of a semigroup S	59
$\mathscr{L}$	Green's relation	38
L_a	the $\mathscr{L}$-class containing a	40
$\mathscr{L}^S$	Green's relation on the semigroup S	50
L_a^S	the $\mathscr{L}^S$-class containing a	50
$\mathscr{L}\mathscr{N}$	the variety of left normal bands	116
$\mathscr{L}\mathscr{Z}$	the variety of left zero semigroups	114
λ_a	the left translation $x \mapsto ax$	43
$\min_J$	the minimal condition on principal ideals	41
$\min_L$	the minimal conditions on principal left ideals	41
$\min_R$	the minimal condition on principal right ideals	41
$M(m, r)$	the monogenic semigroup with index m and period r	11
$\mathscr{M}[G; I, \Lambda; P]$	a completely simple Rees matrix semigroup	68
$\mathscr{M}^\circ[G; I, \Lambda; P]$	a completely 0-simple Rees matrix semigroup	62
μ	the maximum idempotent-separating congruence	141
$\mathbf{N}$	the set $\{1, 2, 3, \ldots\}$ of natural numbers	1
$\mathscr{N}$	the variety of normal bands	119
$\Omega(S)$	the translational hull of a semigroup S	98
$\mathscr{P}\mathscr{T}(X)$	the semigroup of partial mappings of a set X	16
$\mathscr{P}\mathscr{T}^*(X)$	the dual semigroup of $\mathscr{P}\mathscr{T}(X)$	194
$\Pi^* S_i$	the free product of semigroups S_i	212
$\Pi_U^* S_i$	the free product of a semigroup amalgam	217
$\mathbf{Q}$	the set of rational numbers	149
$\mathscr{R}$	Green's relation	38
R_a	the $\mathscr{R}$-class containing a	40
$\mathscr{R}^S$	Green's relation on the semigroup S	50
R_a^S	the $\mathscr{R}^S$-class containing a	50
$[\mathbf{R}]$	the variety determined by a set $\mathbf{R}$ of identical relations	109
$\mathbf{R}^e$	the equivalence generated by a relation $\mathbf{R}$	19
$\mathbf{R}^\infty$	the transitive closure of a relation $\mathbf{R}$	19
$\mathbf{R}^\#$	the congruence generated by a relation $\mathbf{R}$	24
$\mathbf{R}^v$	the fully invariant congruence generated by a relation $\mathbf{R}$	110
ran (ρ)	the range of a binary relation ρ	15
$\mathscr{R}\mathscr{B}$	the variety of rectangular bands	119
$\mathscr{R}\mathscr{N}$	the variety of right normal bands	119
$\mathscr{R}\mathscr{Z}$	the variety of right zero semigroups	114
ρ_a	the right translation $x \mapsto xa$	7
$\rho^\natural$	the natural mapping associated with an equivalence ρ	19
S^1	the semigroup obtained from S by adjoining an identity	2
S^0	the semigroup obtained from S by adjoining a zero	3
S/I	the Rees quotient of S by an ideal I	31
$\mathscr{S}$	the variety of semilattices	114
$\mathscr{S}(Y; S_\alpha; \phi_{\alpha,\beta})$	a strong semilattice of semigroups	90
σ	the minimum group congruence on an inverse semigroup	139
$\sigma(a_r)$	the index of a_r	212
T_E	the Munn semigroup of a semilattice E	143

$T_{e,f}$	the set of isomorphisms from Ee onto Ef	142
$\mathscr{T}(X)$	the full transformation semigroup on a set X	6
$\mathscr{T}^*(X)$	the dual semigroup of $\mathscr{T}(X)$	101
$\mathscr{U}$	$\{(e,f) \in E \times E : Ee \simeq Ef\}$	142
$V(a)$	the set of inverses of an element a	49
W_B	the Hall semigroup of a band B	199
$\mathbf{Z}$	the set of integers	225

Subject Index

A

absolutely closed semigroup, 228
admissible triple, 75
almost unitary subsemigroup, 232
amalgam, 212, 217
  free inverse product of, 246
  free product of, 217
  inverse semigroup, 243
amalgamation property
  special, 249
  strong, 250
  weak, 220, 250
anti-representation, 194
anti-uniform band, 204
anti-uniform semilattice, 144, 146
associated mappings, 232, 255
associative, 1
atom, 114
automorphism, 5

B

band, 90, 95
  anti-uniform, 204
  free, 104
  (left, right) normal, 116, 119
  rectangular, 3, 96, 119
  trivial, 119
  uniform, 203
  (left, right) zero, 3, 119
bicyclic semigroup, 144
bijection, 6
binary relation(s), 14
  semigroup of, 14
(0-) bisimple semigroup, 63
Bruck–Reilly extension, 153

C

centralizer, 183
chain of groups, 163
Clifford semigroup, 93, 163
closed subsemigroup, 220
closed subset, 139
closure, 138, 220
codomain, 5
comparable, 148
completely (0-) simple semigroup, 57, 60, 68
congruence(s), 21
  fully invariant, 110
  generated by a relation, 24
  group, 139, 185
  idempotent-separating, 52, 140, 141
  identical, 14
  intersection of, 24
  inverse semigroup, 190
  join of, 27, 28
  left (right), 21
  proper, 73
  Rees, 31
  semilattice, 93
  universal, 14
congruence-free semigroup, 83
consistent subsemigroup, 254
content, 104, 105
converse (of a relation), 15
coproduct, 217, 244
core, 217
(right) coset (of an inverse subsemigroup), 172
cyclic group, 9

D

$\mathscr{D}$ (Green's equivalence), 39
decreasing step, 231
defining relations, 30
dense subsemigroup, 220
diagonal, 116
direct product, 5, 6, 116

domain (of a mapping), 5
  of a binary relation, 15
dominate, 220
dominion, 220

E

elementary **R**-transition, 26
endomorphism, 5
equivalence(s), 18
  classes, 19
  commuting, 29, 31
  compatible, 21
  generated by a relation, 19
  idempotent-separating, 52
  identical, 14
  intersection of, 19
  join of, 27, 28
  left (right) compatible, 21
  universal, 14
eggbox diagram, 42
equality relation, 14
E-unitary
  regular semigroup, 55, 125
  inverse semigroup, 181, 182
extension, 16
extract, 74

F

finitely presented semigroup, 30
formal inverse, 244
free,
  band, 104
  inverse product, 244
inverse semigroup, 129
  product of amalgam, 217
  product of semigroups, 212
  semigroup, 29
full transformation semigroup, 6

G

generators, 7, 30
Green's equivalences, 38, 39
  in an inverse semigroup, 131
  in a regular semigroup, 49, 50
Green's Lemmas, 43
Green's Theorem, 44
greatest lower bound, 12
group, 4
  congruence, 139
  left (right), 54
  minimum congruence, 139
  symmetric, 6
  0-, 4
groupoid, 1

H

$\mathscr{H}$ (Green's equivalence), 39
Hall semigroup, 199
Hall–Yamada semigroup, 207
Hasse diagram, 14
homorphism, 5
  Rees, 31

I

ideal (left, right, two-sided), 5
  (0-) minimal, 58
  minimum, 59
  principal, 38
  proper, 5
idempotent, 5
  central, 93
  primitive, 68
idempotent-separating congruence, 52, 140, 141
  maximum, 52, 141, 192
identical relation, 109
identity element, 1
  left (right), 35
incomparable, 148
increasing step, 230
index (of an element in a free product), 212
index (of an element of finite order), 8
initial, 106
  mark, 106
internalized sequence, 235
  fully, 235
  left (right), 235
inverse of an element, 45
inverse semigroup, 129
  bisimple, 150
  free, 129
  fundamental, 141

inverse semigroup—*contd.*
  minimum congruence, 190
  simple, 159
  symmetric, 134
inverse subsemigroup, 139
  closed, 139
  full, 145
  normal, 181
  subtransitive, 160
  transitive, 151
irregular $\mathscr{D}$-class, 45
Isbell's Zigzag Theorem, 228
isomorphic, 5
isomorphism, 5

J

$\mathscr{J}$(Green's equivalence), 39
join, 12

K

kernel
  of a homomorphism, 22, 31
  of a mapping, 19
  of a semigroup, 9, 59

L

$\mathscr{L}$(Green's equivalence), 38
Lallement's Lemma, 52
lattice(s), 12
  complete, 12
  direct product of, 34
  modular, 32
  semimodular, 33
least upper bound, 12
lexicographic order, 149

M

maximal, 12
maximum, 12
meet, 12
minimal, 11
minimal conditions $\min_L$, $\min_R$, $\min_J$, 12, 41
minimum, 11
modular lattice, 32
monogenic semigroup, 8
monoid, 2
monomorphism, 5
morphism (of S-systems), 224
multiplication, 1
Munn semigroup, 143

N

(left, right) normal band, 116, 119
null semigroup, 3

O

(right) ω-coset, 172
ω-semigroup, 152
order
  of an element, 8
  of a semigroup, 1
order relation, 11
  partial, 11
  total, 11
ordered set, 11
  partially, 11
  totally, 11
  well-, 12, 147
ordinal product, 184
orthodox semigroup, 186

P

P-semigroup, 182
partition, 18
partial mapping(s), 16
  semigroup of, 16
period, 8
periodic semigroup, 11, 40
primitive idempotent, 68
principal factor, 60
principal (left, right, two-sided) ideal, 38
principal series, 86
proper inverse semigroup—see under *E-unitary*
pushout, 218, 246

Q

quotient semigroup, 21, 22
quotient set, 19

R

$\mathscr{R}$ (Green's equivalence), 38
rectangular band, 3, 96, 119
range (of a mapping), 5
  of a binary relation, 15
regular
  $\mathscr{D}$-class, 45
  element, 44
  sandwich matrix, 60, 62
  semigroup, 44
Rees congruence, 31
Rees homomorphism, 31
Rees matrix semigroup, 62, 68
Rees's Theorem, 63
relation, 14
  antisymmetric, 11
  binary, 14
  compatible, 21
  congruence generated by, 24
  congruence, 21
  converse of, 15
  equality, 14
  equivalence generated by, 19
  equivalence, 18
  identical, 109
  left (right) compatible, 21
  order, 11
  reflexive, 11, 18
  symmetric, 18
  transitive closure of, 19
  transitive, 11, 18
  universal, 14
representation(s) (by mappings), 6, 135, 168
  decomposable, 176
  effective, 169
  equivalent, 170
  extended right regular, 7
  faithful, 6, 168
  sum of, 170
  transitive, 169
  Vagner–Preston, 135, 168
restriction, 16

S

sandwich matrix, 62
semigroup, 1
  absolutely closed, 228
  archimedean, 125
  bicyclic, 144
  (0-) bisimple, 63
  (right, left) cancellative, 53, 54
  Clifford, 93, 163
  commutative, 1
  completely (0-) simple, 57, 60, 68
  congruence-free, 83
  finite, 11
  finitely presented, 30
  free, 29
  inverse, 129
  left (right) simple, 54
  left (right) zero, 3
  monogenic, 8
  null, 3
  orthodox, 186
  periodic, 11, 40
  relatively free, 112
  with identity, 1
  with zero, 2
semisimple, 86
(0-) simple, 57
trivial, 2
semilattice, 14
  anti-uniform, 144, 146
  complete lower, 12
  complete upper, 12
  congruence, 93
  lower, 12
  minimum congruence, 93
  subuniform, 160
  uniform, 144, 146
  upper, 12
semilattice of semigroups, 89
semimodular lattice, 33
(0-) simple semigroup, 57
spined product, 118
(right, left) S-system, 223
stable step, 230, 231
(S, T)-bisystem, 223
standard embedding, 29
strong semilattice of semigroups, 90
subdirect product, 116
subgroup, 5
sublattice, 12
subsemigroup, 5
  inverse, 139
subuniform semilattice, 160
subword, 230

syllable, 230
symmetric group, 6
symmetric inverse semigroup, 134

T

tensor product, 224
terminal, 106
  mark, 106
transitive
  closure, 19
  relation, 11, 18
  representation, 169
transitivity classes, 169
transitivity relation, 168
(left, right) translation, 98, 232
  inner, 98
  linked left and right, 98, 232
translational hull, 98

U

uniform
  band, 203
  semilattice, 144, 146
union of groups, 92
(left, right) unitary subset, 55
universal relation, 14

V

Vagner–Preston Representation Theorem, 135
variety, 109

W

weak direct product, 116
word, 29, 213
  bi-internal, 234
  (left-, right-) internal, 234
  internal, 234

Z

zero-divisor, 58
0-direct union, 71
zero element, 2
  left (right), 35
zigzag, 223, 228
  left-inner, 228